BROADBAND WIRELESS MULTIMEDIA NETWORKS

WILEY SERIES ON INFORMATION AND COMMUNICATION TECHNOLOGY

Series Editors: T. Russell Hsing and Vincent K. N. Lau

The Information and Communication Technology (ICT) book series focuses on creating useful connections between advanced communication theories, practical designs, and end-user applications in various next generation networks and broadband access systems, including fiber, cable, satellite, and wireless. The ICT book series examines the difficulties of applying various advanced communication technologies to practical systems such as WiFi, WiMax, B3G, etc., and considers how technologies are designed in conjunction with standards, theories, and applications.

The ICT book series also addresses application-oriented topics such as service management and creation and end-user devices, as well as the coupling between end devices and infrastructure.

T. Russell Hsing, PhD, is the Executive Director of Emerging Technologies and Services Research at Telcordia Technologies. He manages and leads the applied research and development of information and wireless sensor networking solutions for numerous applications and systems. Email: thsing@telcordia.com

Vincent K.N. Lau, PhD, is Associate Professor in the Department of Electrical Engineering at the Hong Kong University of Science and Technology. His current research interest is on delay-sensitive cross-layer optimization with imperfect system state information. Email: eeknlau@ee.ust.hk

BROADBAND WIRELESS MULTIMEDIA NETWORKS

BENNY BING

Georgia Institute of Technology, Atlanta, GA

A JOHN WILEY & SONS, INC., PUBLICATION

Library of Congress Cataloging-in-Publication Data is available.

ISBN 978-0-470-92354-2

10 9 8 7 6 5 4 3 2 1

CONTENTS

PREFACE

Wireless access has evolved rapidly over the last two decades and has become the dominant medium for network connectivity due to the recent proliferation of sleek personal devices, such as smartphones and tablets. These devices primarily target high-resolution video consumption but have been employed in diverse applications. Ultra-thin touchscreen tablets may well replace flip-to-open laptops in the future, and this will eradicate the need for bulky wired network interfaces. It started with the cellular revolution (1990), followed by Wi-Fi (2000), and many personal devices are now demanding ever-increasing wireless bandwidth to enable high-quality video service. When the cellphone became the first wireless innovation to be available for widespread public use, it helped consumers in many ways, ranging from the convenience of casual conversations to mobile entertainment to emergency calls to preventing burglars from "cutting" the wire phone line before entering a house. Since then, impressive improvements in wireless data rates and coverage have directly benefited enterprise, home, and public access networks. This book covers a broad range of key wireless technologies with a comparative assessment of the strengths and weaknesses. It also discusses the role of high-speed indoor networks and broadband 4G systems in driving the utility of new personal devices with mobile video support. A specific emphasis will be placed on challenging deployments and implementation, such as dealing with the antenna constraints for handheld devices and overcoming the rate mismatch and variation in multiuser video transmission.

Despite recent progress, fundamental challenges in wireless deployment remain. For example, the communication link remains unpredictable, connections can be intermittent, and the bandwidth can be varying due to mobility and possible interference from on-site and non-network entities. A fixed network topology based on the star topology (with a wireless router or base station for relaying user traffic) is now the norm. Self-organizing ad hoc wireless networks are rare but direct point-to-point wireless connections between two user devices are occasionally used. Because there are many competing standards, selecting the appropriate wireless technology can be a difficult task. While the attributes and application of the technology play important roles, affordable pricing and value to consumers are the common denominators that determine success. Thus, a significant percentage of wireless users still prefers to use Wi-Fi even though smartphones and tablets have become 4G-ready. Besides the cost consideration, the connectivity of these personal devices does not work well without Wi-Fi. Wi-Fi has become the bedrock of many consumer electronics devices and the dominant technology for high-speed in-building and in-home networking. Some wireless carriers are even bundling Wi-Fi services to their cellular and Internet service offerings, allowing subscribers free Wi-Fi access in stores and coffeeshops all over the United States. Currently,

LTE, WiMAX, and HSPA+ are forming a cellular ecosystem, providing broadband wireless Internet access in urban cities as well as underserved and unserved remote areas, and enabling new applications for health care, education, and public safety communications.

Spectrum availability is a key consideration to improving wireless data rates. LTE is unlikely to provide 100 Mbit/s on the downlink and 50 Mbit/s on the uplink anytime soon until more spectrum becomes available. According to the Federal Communications Commission (FCC), by 2014 mobile data traffic is expected to be 35 times greater than 2009 levels. The FCC's national broadband plan calls for releasing 300 MHz of spectrum within the next 5 years and 500 MHz within the next 10 years to meet rising mobile data demands. In January 2010, the FCC opened a proceeding to explore expanded terrestrial mobile use in the mobile satellite service (MSS) bands (up to 90 MHz more). In May 2010, the commission approved an order that changes the rules governing 25 MHz of the Wireless Communications Services (WCS) spectrum in the 2.3 GHz band, which is now available for mobile broadband use. The rules are put in place to avoid interference issues. In September 2010, it approved the use of unlicensed white space spectrum, clearing the way for new classes of devices that take advantage of "super Wi-Fi." In November 2010, the FCC approved a notice of proposed rulemaking that lays the groundwork for reallocating 120 MHz of broadcast TV spectrum for wireless, via incentive auctions with broadcasters. The rules will create a licensing framework for spectrum in the UHF/VHF bands and will allow for voluntary channel sharing. The TV broadcast spectrum rulemaking allows all spectrum in the TV bands to be opened to fixed, mobile, and broadcast services. The rulemaking also proposes rules that permit two or more TV broadcasters to share a single 6 MHz channel, although stations will retain rights to mandatory broadcast carriage. In April 2012, the FCC released an order that details how TV stations can share the same channel. All stations utilizing a shared channel retain spectrum usage rights sufficient to ensure enough capacity to operate one standard-definition TV program stream at all times. The order may motivate TV stations that agree to share channels to turn over broadcast TV spectrum to the FCC, which will in turn auction off the spectrum to support the bandwidth demand from smartphones, tablets, and other mobile data products. TV stations that contribute spectrum to such an auction may share the proceeds. The move will allow FCC to recover up to 120 MHz or 40% of all TV spectrum.

Although the data rate is an important metric to gauge the effectiveness of any new wireless technology, the availability of connected devices and applications are key drivers for adopting the technology. Mobile entertainment has become a key application due to the widespread adoption of video-capable smartphones and tablets. These devices have become ubiquitous with greatly expanded computing power and memory, improved displays, and wireless connectivity. The iPhone was the first mobile device to combine voice and Web applications successfully. Fifty million iPhones are expected to be sold in the final quarter of 2012. The iPad tablet shipped over 28 million units in 2011. These figures can be compared with iPod (released in October 2001 with 220 million units sold by September 2009) and Blackberry (160,000 units in 2001 with 36 million in use by 2009). 4G smartphones (e.g. Sprint's Evo and Apple's iPhone 5) and tablets (e.g. Motorola's Xoom, Apple's new iPad, and Samsung's Galaxy Tab 10.1) have now become mainstream. The latest iPad tablet not only supports LTE, but also works with a number of carriers, such as Verizon, AT&T, Rogers, Bell, and Telus. It provides 10 hours of battery life on active usage and 9 hours on LTE usage. It also supports Wi-Fi personal hotspots and can share a network connection with up to 5 devices.

In the third quarter of 2010, the United States accounted for nearly 50% of smartphone sales worldwide. During the same quarter, Android emerged as the dominant mobile operating system in the United States with a market share of just over 40%. The utility of Android lies in the large community of developers writing apps that extend the functionality of the devices. There are currently over 100,000 Android apps, which control the device via Google-developed Java libraries. Apple iPhone apps are equally popular and are proliferating as well (over 10 billion downloads to date with 300,000 apps). They range from multiuser games and social apps to creative ones that allow the user to watch four camera views of the World Series on the iPhone. TV remains the single most important source of information and entertainment. In 2010, U.S. teenagers spent more than three times the amount of their spare time watching TV than they spent on social media. According to online video, which includes online TV, represents one of the key traffic types to be carried over personal smartphones and tablets with high-resolution displays. The year 2012 could be the first year that U.S. consumers watch more movies online than on DVDs, Blu-ray discs, and other physical video formats. However, supporting high-quality video presents a significant challenge to wireless carriers due to the higher bandwidth demands compared with data and voice traffic. Thus, many carriers switch from unlimited data plans to usage-based plans, enforcing caps in bandwidth usage. For example, AT&T currently charges 3G subscribers $15 for 200 MB of data and $25 for 2 GB. Verizon Wireless currently charges LTE subscribers $50 for 5 GB and $80 for 10 GB, with $10 per GB for over the limit usage. These figures should be compared with the monthly data usage caps currently imposed by AT&T for wireline access networks: 150 GB for DSL and 250 GB for U-Verse. Subscribers will be charged an additional $10 for every 50 GB beyond those limits. Clearly, there is a big gap between wireline and wireless caps, but note that there is no major difference between the DSL and LTE data rates. Streaming just two 2-hour movies can max out a 5 GB/month data plan. If music is streamed at a bit rate of 160 Kbit/s for an hour a day, this equates to about 2.2 GB of data per month. Adaptive streaming is a key technology to achieve reliable video delivery over heterogeneous wireless networks and multiscreen consumer devices. It provides autonomous bandwidth management and maintains quality of service even as wireless link conditions fluctuate and network congestion on the Internet vary.

So what's next for wireless? Although there are some encouraging breakthroughs, improving the capacity and range of wireless networks remain paramount, and this must be properly balanced with implementation, deployment, and power consumption considerations. I have the privilege of reviewing many interesting research articles focusing on cognitive radio, vehicular, millimeter radio, sensor, nano, and bio networks, as well as the current buzz—network science involving identifying synergies in different network types. Unfortunately, many real-world issues are often brushed aside in these research articles. Allow me to provide some illustrations. Wireless sensor networks have received a lot of research attention. These networks aim to remove the wiring for both data connectivity and power supply. As an example, ultra-low power Zigbee switches are able to help turn on lights without wire cabling to connect the switch to the light. The use of Wi-Fi or Bluetooth switches will require replacement of hundreds or thousands of sensor batteries in a large building on a regular basis, which makes these sensors challenging to install and maintain. There are enormous benefits in sensors collecting time-critical information. Environmental sensor applications (e.g., factory/home/office automation and oil leak/earthquake detection), biosensors tracking human conditions and collecting vital signs of patients, and networked sensors (i.e., sensors that collect data from various locations and then independently upload this information to a network for human analysis) are just a few examples. However, I am somewhat skeptical about the practicality of self-autonomous sensor networks, which is the major thrust of many research articles. Here, I am referring to sensors directly managing other sensors via an ad hoc multihop network. Sensors are simple devices that are designed with limited functionality to reduce power consumption and cost requirements. For example, a typical temperature sensor sends only 200 bytes of data every 5 minutes. These sensors cannot be relied upon to control other sensors. For instance, there is a light sensor in my office that tends to put me to sleep whenever there are dark clouds outside. I am aware of other light sensors that automatically turn off the lights when the room is well lit with sunlight, but you can figure out how such sensors will fail in other ways. As another example, the temperature sensor in my home defies logic when it decides to turn on the air conditioner less frequently for a long duration of time, and at other times, turn it on more frequently for a shorter duration of time. In addition to the reliability issue, self-configuring and self-healing wireless sensor networks may trigger frequent network topology reconfiguration when channel conditions change and sensor nodes randomly run out of battery. This will in turn require more rerouting messages to be exchanged, draining battery power and causing more node failures. Thus, setting up a fail-safe, self-organizing wireless sensor network involving concurrent link and route establishments of hundreds of sensors (as in some military deployments and first responder operations) may be unrealistic. More practical validation work has to be conducted on such large-scale sensor networks using cheap and disposable life-size sensors.

The usefulness of cognitive radio is another area that deserves further validation. Like multiple antenna systems, cognitive radio addresses the issue of limited radio spectrum. The technology has been employed for wireless systems operating in the TV bands recently. However, one-way broadcast operations may be more suited for these bands due to the signal propagation delay. Sensing ongoing transmissions complicates two-way system deployment, as many spurious signals have to be filtered; and not all interfering signals can be detected, since the sensing range (usually based on detecting signal power levels) can be longer than the RF reception range (usually based on decoding the desired signals). Simply setting a predefined signal threshold may potentially lead to many false alarms, especially in wide-area deployments. Wi-Fi technology can be considered a form of cognitive radio since the medium must be sensed idle before a packet is transmitted. Because Wi-Fi is designed for use in short-range networks (typically spanning between 100 and 200 ft), faster signal propagation allows carrier sensing to work well at these distances. Its performance degrades dramatically in an exhibit hall when many transmitters try to coexist in the same frequency band. Cognitive radio networks suffer from a similar problem, which must be solved before efficient management of limited spectrum resources can be achieved. There are other practical issues that need to be overcome: cost-effective wideband antennas to deal with collocated interference, multiband interactions, insertion loss; low-noise wideband power amplifiers to process RF and IF processing; high-performance analog-to-digital and digital-to-analog conversion; and high-performance DSPs based on FPGA and ASIC designs. Finally, there is a nontrivial task of dynamically switching and adapting the modulation and coding schemes according to time-varying interference and fading conditions.

It is common for wireless communications researchers to associate technical depth with mathematical dexterity. More often than not, this clouds the motivation and usefulness of the research. In addition, many simplifying (and often unrealistic) assumptions arise when solving complex mathematical problems. Technical depth should be related to real-world engineering impact. Technically profound papers, such as the pioneering TCP/IP paper authored by Vince Cerf and Bob Kahn, would probably be rejected today as lack of technical depth (no equations). Fortunately, this IEEE transactions paper was published in 1974 and continues to drive the most important engineering invention for the last four decades or so—the Internet. This is the only paper that has become an interoperable data networking standard, and I am amazed how it has become even more relevant today, forming the basis to support high-quality video streaming on the Internet as well as mobile wireless networks, whereas many newer protocols that were designed for real-time multimedia streaming have failed when it comes to practical implementation. Good engineering research can be compared with good movies—you need a good storyline (motivation) with a supporting cast of good actors/actresses (researchers) and audio/visual effects (practical tools). It is not possible to break new grounds in nascent engineering research without real-world considerations. For example, many prominent scientists have been predicting pervasive wireless personal communications for over two decades but it was only the recent availability of touchscreen devices that really drives the demand for high-speed wireless networks. Similarly, the improved spectral efficiency promised by high-order multiple input multiple output (MIMO) systems has been severely challenged. It is like expecting voice recognition technology to work well when several people near the source are talking loudly at the same time. In addition, battery-operated user devices pose space and power constraints when building multiantenna transmitters. LTE handsets currently support single-antenna transmission. Although 802.11n access points and home gateways have just started using three spatial antenna streams, they will switch to a lower number of streams if end-user devices that support one or two streams are detected. Since the vast majority of smartphones and tablets only support a single stream, the higher capacity of three streams (or even two streams) cannot be attained for these devices. Judging from the 2012 CES Show, it appears many vendors are already moving toward single-antenna 802.11ac user devices and adapters with broader channel bandwidths and higher-order modulation instead of developing the complex option of four transmit spatial streams that is available in 802.11n. Thus, future wireless communications may well be dictated by single-transmitter user devices operating on wideband channels.

Transformative engineering research should lead to physical discoveries. Consider the following quote from Michael Faraday in 1827: "I could not imagine much progress by reading only, without experimental facts and trials . . . I was never able to make a fact my own without seeing it." Even more remarkable is the quote from Hermann von Helmholtz in 1881: "Faraday performed in his brain the work of a great mathematician without using a single mathematical formula." James Clerk Maxwell's unifying electromagnetic equations may not hold much significance if they were formulated prior to the physical discoveries made by Faraday. To this end, I am glad that many wireless standard committees have taken the lead in commercializing many breakthrough wireless technologies. Although OFDM was tested in the labs in the 1980s, the first real-world demonstration of the technology actually began with 802.11a in 1999. The 802.11 and 802.16 Working Groups were the first to commercialize multiantenna MIMO technology. The 802.11 Working Group is now leading pioneering efforts to improve network throughput to gigabits per second using multiuser MIMO and 60 GHz technologies. The 802.16 and LTE standard groups should be credited for demonstrating the practicality of orthogonal frequency division multiple access (OFDMA) and frequency domain equalization, respectively. These wireless standards undergo numerous revisions from various vendors (many with working prototypes) before the standard is ratified. This is the benchmark for successful technology innovation because it helps validate new concepts, drive cost-effective implementation, ensure interoperability, and improve performance. An important wireless frontier for the future could be the removal of the power cord and batteries. Wireless electric power will solve many RF design constraints associated with many mobile devices. It will eliminate chargers and eventually, batteries. Since the human body is affected by electric fields and not magnetic fields, the magnetic field can be used to transport electric power wirelessly in a safe and efficient manner. Several promising experiments have been conducted to demonstrate the feasibility of wireless electricity. For example, Intel's magnetic energy resonant wireless electric power prototype is able to light up a 60 W bulb that uses more power than a typical laptop.

I would like to thank Dr. Simone Taylor of John Wiley & Sons for her encouragement and patience in overseeing this book project. The excellent layout of the book is due to the professional efforts of Janet Hronek of Toppan Best-set Premedia Ltd. and Diana Gialo. I would also like to acknowledge my past collaborators and students who have been unselfish in sharing many useful comments. I wish to thank Eric Levine of the IEEE Communications Society for his tireless efforts in securing industry sponsors for eight of my online tutorials. The valuable feedback received from the tutorial participants helped to shape the current contents of the book. The advantage of writing a book as opposed to a research paper is that you can obtain critical but constructive reviews from leading experts whom you know. To this end, I am indebted to the following reviewers. They are Professors Tim Brown, Andrew Paplinski, Jiang Xie, Admela Jukan, and Michael Fang and Dr. Zhensheng Zhang. I am also grateful to Dr. Richard Van Nee, Dr. Christopher Hansen, Dr. Bjorn Bjerke,

Dr. Chirag Patel, Naftali Chayat, Frank Gonzalez, and John Civiletto for their generous insights on various wireless topics. Finally, I wish to thank Dr. Frank Caimi for his contribution on 4G antenna design and Dr. Bob Heile for his chapter on ZigBee and green communications.

I hope this book will serve as a useful resource to many practicing engineers, as well as researchers and students. I have attempted to cover only the more popular wireless technologies and standards, which should equip the reader with a strong foundation to understand emerging technologies. Unsolved real-world issues in wireless networking are emphasized. I have included over 180 homework problems to help the reader master the concepts described in each chapter and discover new insights and utility of wireless technologies. The majority of these problems do not require math but are designed to stimulate critical thinking. Hopefully, this will inspire new breakthroughs. The solutions to these homework problems are available to instructors. I hope the concise writing style resonates with the reader, and I have attempted to spice that up with many interesting technology snippets. I wish I could include a collection of cartoons related to wireless that would make the book lighthearted, engaging, and hard to put down. I guess I will have to reserve them for the courses that I teach. I have an uneasy intuition that certain cartoon creators have greater technical vision and sharper minds than any of us. As a substitute, I have included 200 illustrative figures to aid comprehension and reinforce concepts. These figures are again available to instructors. Please feel free to send your comments, corrections, and questions to bennybing@yahoo.com. They will be gratefully received.

Benny Bing

CHAPTER 1

OVERVIEW OF BROADBAND WIRELESS NETWORKS

Mobility and flexibility make wireless networks effective extensions and attractive alternatives to wired networks. Wireless networks provide all the functionality of wired networks, but without the physical constraints of the wire itself. However, the wireless link possesses some unique obstacles that need to be solved. For example, the medium is a scarce resource that must be shared among network users. It can be noisy and unreliable where transmissions from mobile users interfere with each other to varying degrees. The transmitted signal power dissipates in space rapidly and becomes attenuated. Physical obstructions may block or generate multiple copies of the transmitted signal. The received signal strength normally changes slowly with time because of path loss, more quickly with shadow fading and very quickly because of multipath fading. The most distinguishing issues in wireless network design are the constraints placed on bandwidth and power efficiency.

The broadcast nature of wireless transmission offers ubiquity and immediate access for both fixed and mobile users, clearly a vital element of quad-play (voice, video, data, and mobile) services. Moving from one location to another does not lead to disruptive reconnections at the new site. Wireless technology overcomes the need to lay cable, which is difficult, expensive, and time consuming to install, maintain, and especially, modify. Providing wireline connectivity in rural or remote areas runs the risk of someone pulling the cable (and accessories such as amplifiers) out of the ground to sell! A wireless network avoids underutilizing the access infrastructure. Unlike wired access (copper, coax, and fiber), a large portion of wireless deployment costs is incurred only when a customer signs up for service. The Fiber-to-the-Home (FTTH) Council reported that in September 23, 2008, there were 13.8 million FTTH networks in North America but the adoption rate is only 3.76 million (about 27%) even though many of these homes are located in strategic neighborhoods. The take up rate improved marginally to 34% (7.1 million connected homes) with 20.9 million homes passed on March 30, 2011. The cable industry's capital expenditure over the last 15 years is estimated at $172 billion. Broadband usage for cable services fared better but still fall below 50%. According to the National Cable and Telecommunications Association (NCTA), there were 129.3 million homes passed by cable video service in June 2011 (which translates to over 96% of U.S. households passed), but the take up rate is 45.5%. These numbers are unlikely to

Broadband Wireless Multimedia Networks, First Edition. Benny Bing.
© 2013 John Wiley & Sons, Inc. Published 2013 by John Wiley & Sons, Inc.

increase significantly in future with high-speed wireless and free broadcast services becoming widely available.

Terrestrial wireless access may offer portable and mobile service without the need for a proprietary customer premise equipment (CPE), such as a set-top box. This facilitates voice, TV, and Internet connectivity inside and outside the residential home. For instance, such connectivity can be made available on virtually any open space (e.g., on a fishing boat!), on fast moving vehicles and trains, and even when the subscriber moves to a foreign location. The ability to connect disparate end-user devices quickly and inexpensively remains one of the key strengths of wireless. New smartphones and tablets all come with two or more wireless network interfaces but no wired interfaces, thus making wireless connectivity indispensable. These devices demand higher wireless rates to support multimedia applications, including high-quality video streaming, which is in contrast to low bit rate voice applications supported by legacy cellular systems.

Because cellular systems cover long distances, they involve costly infrastructures, such as base stations (BSs) and require users to pay for bandwidth on a time or usage basis. Each BS may potentially serve a large number of mobile handsets. Coordination between BSs as users move across wireless coverage boundaries is achieved using a mobile backhaul, which also carries a variety of user traffic. The BS may prioritize near and far handsets. For example, the BS can reduce interference by transmitting at a lower power to closer handsets. In contrast, on-premise and geographically limited wireless local area networks (wireless LANs) require no usage fees, employ lower transmit power, and provide higher data rates than cellular systems. Wireless LANs are built around cheaper access points (APs) that connect a smaller number of stationary user devices, such as laptops or tablets to a wired network. However, achieving reliable high-speed wireless transmission is a challenging task. Besides the need to overcome traditional issues, such as multipath fading and interference from known and unknown sources, broadband wireless transmission also demands new methods to support highly efficient use of limited radio spectrum and handset battery power. This chapter discusses several fundamental topics related to broadband wireless networks. These include environmental factors, frequency bands, multicarrier operation, multiple antenna systems, medium access control, duplexing, and deployment considerations.

1.1 INTRODUCTION

Mobile broadband represents a multibillion dollar market. Service providers, including incumbent cable/telephone wireline providers, can increase the number of subscribers significantly by leveraging on broadband wireless solutions (e.g., in areas not currently served or served by competitors). The performance of a broadband wireless network is heavily dependent on the characteristics of the wireless channel, such as signal fading, multipath distortion, limited bandwidth, high error rates, rapidly changing propagation conditions, mutual interference of signals, and the vulnerability to eavesdrop and unauthorized access. Moreover, the performance observed by each individual user in the network is different and is a function of its location as well as the location of other interacting users. In order to improve spectral efficiency and hence, the overall network capacity, wireless access techniques need to be closely integrated with various interference mitigation techniques including the use of smart antennas, multi-user detection, power control, channel state tracking, and coding. Broadband wireless networks must also adequately address the combined requirements of wireless and multimedia communications. On one hand, the network must allow users to share the limited bandwidth resource efficiently to achieve higher rates. This implies two criteria: maximizing the utilization of the radio frequency spectrum and minimizing the delay experienced by the users. On the other hand, because the network supports multimedia traffic, it is expected to handle a wide range of bit rates together with various types of real-time and non–real-time traffic attributes and quality of service (QoS) guarantees.

More than 60% of Americans are using a wireless device to talk, send email, take pictures, watch video, listen to music, and play online games. Compressed video is a key traffic type that needs to be accommodated due to the emergence of many personal smartphones and tablet computers. Despite the smaller displays, many of these devices can support high-definition (HD) video with 720p (1280 × 720 pixels) picture resolutions. Highly efficient video coding standards, such as H.264/MPEG-4 Advanced Video Coding (AVC), are normally used to compress these videos for wireless delivery. This enables efficient use of radio spectrum, but the higher compression efficiency may also result in higher bit rate variability. In addition, compressed video is very sensitive to packet loss, with very limited time for packet retransmission, and wireless channels tend to be more error-prone than wired networks. Although wireless rates are typically lower than its wired counterparts, serving bandwidth-intensive applications, such as HD videos, may not always be an issue since such videos can be downloaded in the background. Users tend to watch videos in their own time, rather than according to broadcast schedules. However, real-time video applications (e.g., Skype video chat) may pose a problem depending on bandwidth availability.

Figure 1.1 shows the evolution of wireless access standards. Since the late 1990s, there were numerous digital cellular standards supporting second-generation (2G), 2.5 generation (2.5G), and third-generation (3G) services. These standards can be broadly categorized under code division multiple access (CDMA) or time division

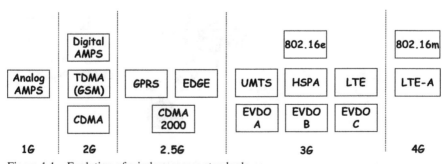

Figure 1.1   Evolution of wireless access standards.

multiple access (TDMA), and there was much debate on the individual merits and capacity of these systems. However, high-speed fourth-generation (4G) wireless standards are converging towards multicarrier transmission as the defacto method and currently there are only two 4G standards. Unlike legacy digital cellular services that primarily support voice and low rate data, the demand for 4G wireless is driven by millions of personal devices that require high-speed Internet connectivity. These sleek devices exude mass appeal due to their usability, and many devices employ an open software platform for users to program their own applications or download other applications. 4G wireless is the missing link that allows multimedia applications running on these devices to become portable, thus enabling on-the-go entertainment.

1.2   RADIO SPECTRUM

To encourage pervasive use of a wireless technology, the operating radio frequency (RF) band should be widely available. Locating a harmonized band is a difficult task because spectrum allocation is strictly controlled by multiple regulatory bodies in different countries. These include the Federal Communications Commission (FCC) in the United States, the European Committee of Post and Telecommunications Administrations (CEPT), Ofcom in the United Kingdom, the Radio Equipment Inspection and Certification Institute (MKK) in Japan, the Australian Communications and Media Authority (ACMA), and others.

1.2.1   Unlicensed Frequency Bands

Many wireless networks operate on unlicensed frequency bands, as illustrated in Figure 1.2. Many of these bands are available worldwide. Since the allocated spectrum is not licensed, large-scale frequency planning is avoided and ad hoc

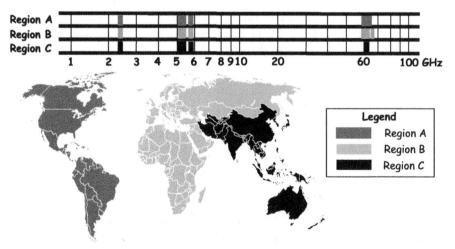

Figure 1.2   Popular unlicensed frequency bands at 2.4 GHz, 5 GHz, and 60 GHz.

deployments are possible. Perhaps the most popular band is the 2.4 GHz industrial, scientific and medical (ISM) band, which has been adopted by the IEEE 802.11 wireless LAN standard. The rules for operating in this band were first released by the FCC in 1985, which also includes the 900 MHz and the 5.8 GHz bands. Another band that has become popular is the 5 GHz Unlicensed National Information Infrastructure (U-NII) frequency band. The large amount of radio spectrum in this band enables the provision of high-speed Internet and multimedia services. The rules for operating in the 5 GHz U-NII band were released by the FCC in 1997. The band is subdivided into three blocks of 100 MHz each, corresponding to the lower, middle, and upper U-NII bands. The FCC subsequently expanded the middle U-NII band when 255 MHz of bandwidth was added in 2003. Thus, the 5 GHz U-NII band offers substantially more bandwidth than the 2.4 GHz band (580 MHz vs. 83.5 MHz). Recently, the 60 GHz unlicensed band has emerged, providing about 7–8 GHz of bandwidth, which is significantly higher than the 2.4 and 5 GHz bands.

1.2.2   The 2.4 GHz Unlicensed Band

The 2.4 GHz channel sets and center frequencies are shown in Table 1.1. The operating power for the 2.4 GHz band is limited to 100 mW (U.S.), 100 mW (Europe), and 10 mW/MHz (Japan). If transmit power control is employed, up to 1 W operation is permissible in the United States. A radiated power of 30 mW is normally used in 802.11 wireless LANs. In France, the output power for outdoor operation between 2454 and 2483.5 MHz is restricted to 10 mW. The total available bandwidth is 72 MHz in the United States, but generally higher in Europe and other parts of the world (83.5 MHz total bandwidth). The 2.4 GHz channels are defined on a

TABLE 1.1   2.4 GHz Channel Sets and Center Frequencies

Channel ID	North America (GHz)	Europe, Japan, and Australia (GHz)
1	2.412	2.412
2	2.417	2.417
3	2.422	2.422
4	2.427	2.427
5	2.432	2.432
6	2.437	2.437
7	2.442	2.442
8	2.447	2.447
9	2.452	2.452
10	2.457	2.457
11	2.462	2.462
12		2.467
13		2.472

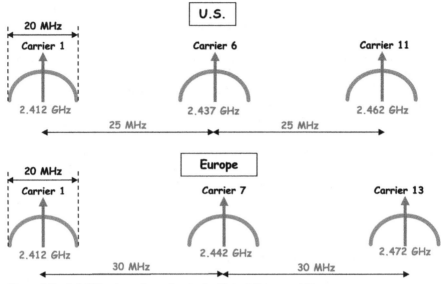

Figure 1.3   2.4 GHz channel spacing in the United States and Europe.

5 MHz channel grid (i.e., each channel has a bandwidth of 5 MHz). This results in 11-13 nonoverlapping channels. However, some wireless systems such as 802.11 wireless LANs require a higher channel bandwidth of 20 MHz. This gives rise to 11-13 overlapped 20 MHz channels or only 3 nonoverlapping 20 MHz channels in the 2.4 GHz band, as illustrated in Figure 1.3. These nonoverlapping channels need not be assigned sequentially and not all channels need to be assigned.

Interference must be carefully evaluated in the 2.4 GHz band especially when deploying large-scale 802.11 networks, such as in conference or exhibit halls and airport terminals. Besides 802.11 devices, many other devices operate in the same GHz band. They include 802.15.1 (Bluetooth) and 802.15.4 (ZigBee) devices, as well as digital cordless phones and microwave ovens that do not have in-built interference avoidance mechanisms. Microwave ovens remain one of the detrimental sources of interference in the 2.4 GHz band due to the high operating power (typically 750 to 1000 W). These ovens are present in almost every home and office. Due to the difficulty in controlling interference, devices must observe etiquette during transmission so that incompatible systems may co-exist. Basic elements of etiquette are to listen before transmit (to detect ongoing transmissions), and to limit transmit time and transmit power. The carrier (activity)-sensing mechanism used in 802.11 provides natural etiquette support.

1.2.3   The 5 GHz Unlicensed Band

Unlike the 2.4 GHz band, 5 GHz channels are defined on a 20 MHz grid. The channel sets, center frequencies, and operating power for the 5 GHz band are shown

TABLE 1.2   5 GHz U-NII Channel Sets, Center Frequencies, and Operating Power

Frequency band	Channel numbers	Center frequency (GHz)	Maximum output power in the United States	Maximum output power in Europe (EIRP)
5.15 to 5.25 GHz	36	5.18	50 mW[a]	200 mW[c]
(100 MHz)	40	5.20	(2.5 mW/MHz)	
(U-NII lower	44	5.22	(Indoor)	
band or	48	5.24		
U-NII-1)				
5.25–5.35 GHz	52	5.26	250 mW[a]	200 mW[c]
(100 MHz)	56	5.28	(12.5 mW/MHz)	
(U-NII middle	60	5.30	(Indoor/outdoor)	
band or	64	5.32		
U-NII-2)				
5.470–5.725 GHz	100	5.50	250 mW[a]	1 W[c]
(255 MHz)	104	5.52	(12.5 mW/MHz)	
(U-NII-2	108	5.54	(Indoor/outdoor)	
extended)	112	5.56		
(worldwide)	116	5.58		
	120	5.60		
	124	5.62		
	128	5.64		
	132	5.66		
	136	5.68		
	140	5.70		
5.725–5.825 GHz	149	5.745	1 W[b]	4 W[c]
(100 MHz)	153	5.765	(50 mW/MHz)	
(U-NII upper	157	5.785	(Outdoor	
band or	161	5.805	point-to-point)	
U-NII-3)				

[a]Maximum antenna gain is 6 dBi.

[b]Maximum antenna gain is 23 dBi.

[c]Requires DFS.

in Table 1.2. The 5 GHz channel assignment and bandwidth are shown in Figure 1.4. The U-NII lower and upper bands (four channels each) are normally employed in the United States, whereas the U-NII middle band is normally supported in Europe. Note that the center frequencies of the 20 MHz channels start (and end) at different points in each band. The highest number of contiguous channels of 11 is available in the 5.470–5.725 GHz middle band. A significant part of the 5.8 GHz ISM band (ranging from 5.725 to 5.850 GHz) has been absorbed in the upper U-NII band. However, ISM channel 165 with a center frequency of 5.825 GHz falls outside of the U-NII band. Thus, the total available bandwidth at 5 GHz (ISM + U-NII) is 580 MHz, giving 24 nonoverlapping channels. The upper U-NII band holds the most

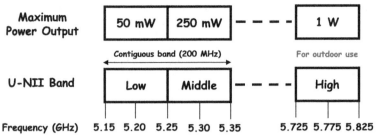

Figure 1.4 5 GHz channel assignment and bandwidth.

promise, as it allows the possibility of longer operational range without the need for a range extender.

Some 5 GHz wireless channels overlap with radar frequencies. Unlike the 2.4 GHz band, however, most interference sources at 5 GHz are located outdoors and may be attenuated sufficiently if penetrated indoors. Nevertheless, dynamic frequency selection (DFS) and transmit power control (TPC) are interference mitigation techniques recommended for unlicensed 5 GHz operation. With DFS, a 5 GHz device can automatically detect radar transmissions and change to a different channel. With TPC, the transmit power can be reduced by several dB below the maximum permitted level.

1.2.4 The 60 GHz Unlicensed Band

A higher frequency band generally implies a higher amount of available bandwidth. At high frequencies, oxygen absorption in the atmosphere leads to rapid fall off in signal strength. Although the operating range becomes limited, this facilitates frequency reuse (i.e., bandwidth reclamation) in high-capacity, picocell (very small cell) wireless systems. There are two oxygen absorption bands ranging from 51.4 to 66 GHz (band A) and 105 to 134 GHz (band B). There are also peaks in water vapor absorption at 22 and 200 GHz. The oxygen absorption is lower in band B than band A, while the water vapor attenuation is higher. These observations suggest that band A is more suitable for communications than band B. The Millimeter Wave Communications Working Group first recommended the use of the 60 GHz band in a report released by the FCC in 1996 [1]. The 60 GHz frequency band is also available worldwide (Figure 1.5). It is uniquely suited for carrying extremely high data rates (multi-Gbit/s) over short distances. Transmit power is typically limited to 10 mW, and the band is subdivided into four nonoverlapping channels of 2.16 GHz each. Nearly 7 GHz of unlicensed spectrum is available in the United States and Japan, whereas 8 GHz of bandwidth is available in Europe. Currently, there are no coexistence issues.

1.2.5 Licensed Frequency Bands

The popular 3GPP Evolved Universal Terrestrial Radio Access (E-UTRA) licensed frequency bands are listed in Table 1.3. The downlink (DL) band is employed in the

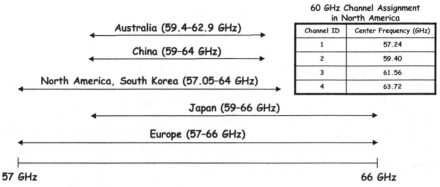

Figure 1.5   60 GHz frequency allocation.

TABLE 1.3   Major E-UTRA Frequency Bands

E-UTRA band	Band name	UL band (MHz)		DL band (MHz)		Duplex mode
1	2.1 GHz	1920	1980	2110	2170	FDD
2	PCS 1900	1850	1910	1930	1990	FDD
3	1800 MHz	1710	1785	1805	1880	FDD
4	AWS	1710	1755	2110	2155	FDD
5	850 MHz	824	849	869	894	FDD
7	2.6 GHz	2500	2570	2620	2690	FDD
8	900 MHz	880	915	925	960	FDD
9	1700 MHz (Japan)	1749.9	1784.9	1844.9	1879.9	FDD
10	1.7/2.1 GHz	1710	1770	2110	2170	FDD
12	Lower 700 MHz	699	716	729	746	FDD
13	Upper C 700 MHz	777	787	746	756	FDD
14	Upper D 700 MHz	788	798	758	768	FDD
17	Lower B, C 700 MHz	704	716	734	746	FDD
18	850 MHz (Japan)	815	830	860	875	FDD
19	850 MHz (Japan)	830	845	875	890	FDD
20	CEPT800	832	862	791	821	FDD
21	1500 MHz (Japan)	1447.9	1462.9	1495.9	1510.9	FDD
24	U.S. L band	1626.5	1660.5	1525	1559	FDD
33	TDD 2000 lower	1900	1920	1900	1920	TDD
34	TDD 2000 upper	2010	2025	2010	2025	TDD
35	TDD 1900 lower	1850	1910	1850	1910	TDD
36	TDD 1900 upper	1930	1990	1930	1990	TDD
37	PCS center gap	1910	1930	1910	1930	TDD
38	IMT Ext	2570	2620	2570	2620	TDD
40	2300 MHz	2300	2400	2300	2400	TDD
41	U.S. 2600	2496	2690	2496	2690	TDD
42	3500 MHz	3400	3600	3400	3600	TDD
43	3700 MHz	3600	3800	3600	3800	TDD

Source: 3GPP TS.104 V10.2.0.

transmission from the BS to the handset. Conversely, the uplink (UL) band is employed in the transmission from the handset to the BS. High-speed packet access (HSPA) systems are deployed in the major E-UTRA cellular bands. For example, there are over 400 tri-band 850/1900/2100 MHz HSPA devices that support global roaming. Many HSPA devices also support legacy global system for mobile communications (GSM), general packet radio service (GPRS), and Enhanced Data rates for GSM Evolution (EDGE), giving rise to quad-band 850/900/1800/1900 MHz devices. A combination of higher spectrum (e.g., 1800/1900/2100 MHz) for improved capacity and sub-1 GHz spectrum (e.g., 700/850/900 MHz) for improved coverage in rural areas and urban in-buildings, is highly desirable. However, with 4G wireless standards employing larger bandwidths (such as 40 MHz), it is important to be able to use bands that offer wider bandwidths. Thus, the International Telecommunication Union (ITU) identified the 2.6 GHz band for supporting mobile broadband services. This extension band is large enough to allow operators to deploy wideband channels to achieve faster data speeds. In addition, some 700 MHz spectrum (also known as digital dividend spectrum) is released for 4G wireless services as analog TV broadcasters migrate to more efficient digital TV platforms. In the 2007 ITU World Radio Conference, the allocation of 700 MHz spectrum for mobile service has been harmonized in the following regions:

- 698–806 MHz for the Americas
- 790–862 MHz for Europe, Middle East, and Africa
- 698–862 MHz or 790–862 MHz for Asia.

The United States is currently the only country in the world that can build ubiquitous wireless Internet access and communications using the 700 MHz and 2.5 GHz (2.496 to 2.69 GHz) Educational Broadband Service (EBS) spectrum.

1.3   SIGNAL COVERAGE

Wireless networks employ either radio or infrared electromagnetic waves to transfer information from one point to another. The use of a wireless link introduces new restrictions not found in conventional wired networks. The quality of the wireless link varies over space and time. Objects in a building (e.g., structures, equipment, and people) can block, reflect, and scatter transmitted signals. In addition, problems of noise and interference from both intended and unintended users must also be solved. While wired networks are implicitly distinct, there is no easy way to physically separate different wireless networks. Well-defined network boundaries or coverage areas do not exist since users may move and transmissions can occur in various locations of the network. Wireless networks lack full connectivity and are significantly less reliable than the wired physical layer (PHY). Thus, one of the most important aspects of wireless system design is to ensure that sufficient signal levels are accessible from most of the intended service areas. To support mobility, separate wireless coverage areas or cells must be properly overlapped to ensure service continuity. Estimating signal coverage requires a good understanding of the communication channel, which comprises the antennas and the propagation medium. Usually, additional signal power is needed to maintain the desired channel quality and to offset the amount of received signal power variation about its average level. These power variations can be broadly classified under small-scale or large-scale fading effects. Small-scale fades are dominated by multipath propagation (caused by RF signal reflections), Doppler spread (caused by relative motion between transmitter and receiver), and movement of surrounding objects. Large-scale fades are characterized by attenuation in the propagation medium and shadowing caused by obstructing objects. These effects are explained in the following sections.

1.3.1 Propagation Mechanisms

Signal propagation patterns are unpredictable and changes rapidly with time. Consequently, signal coverage is not uniform, even at equal distances from the transmitter. A transmitted RF signal diffuses as it travels across the wireless medium. As a result, a portion of the transmitted signal power arrives directly at the receiver, while other portions arrive via reflection, diffraction, and scattering. Reflection occurs when the propagating signal impinges on an object that is large compared to the wavelength of the signal (e.g., buildings, walls, and surface of the earth). When the path between the transmitter and receiver is obstructed by sharp, irregular objects, the propagating wave diffracts and bends around the obstacle even when a direct line-of-sight (LOS) path does not exist. Finally, scattering takes place when obstructing objects are smaller than the wavelength of the propagating signal (e.g., people, foliage).

1.3.2 Multipath

Among the various forms of radio signal degradations, multipath fading assumes a high degree of importance. Multipath is a form of self-interference that occurs when the transmitted signal is reflected by objects in the environment such as walls, trees, buildings, people, and moving vehicles. When a signal takes multiple paths to reach the receiver, the received signal becomes a superposition of different components (Figure 1.6), each with a different delay, amplitude, and phase. These components form different clusters, and, depending on the phase of each component, interfere constructively and destructively at the receiving antenna, thereby producing a phenomenon called multipath fading. Such fading produces a variable bit error rate that can lead to intermittent network connectivity and significant delay variation (jitter).

Multipath fading represents the quick fluctuations in received power and is therefore commonly known as fast (or Rayleigh) fading. In addition, it is often classified as small-scale fading because the rapid changes in signal strength only occur over a small area or time interval. Multipath fading is affected by the location of the transmitter and receiver, as well as the movement around them. Such fading tends to be frequency selective or frequency dependent. Of considerable importance to

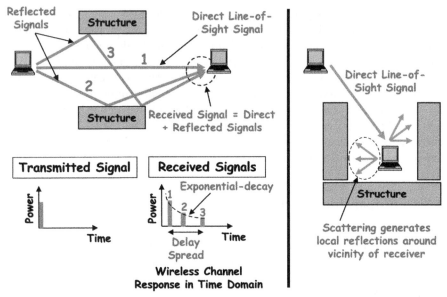

Figure 1.6 Multipath propagation and signal scattering.

wireless network designers is not only the depth but also the duration of the fades. Fortunately, it has been observed that the deeper the fade, the less frequently it occurs and the shorter the duration when it occurs. The severity of the fades tends to increase as the distance between the transmitter and receiver, and the number of reflective surfaces in the environment, increase. Multipath fading can be countered effectively using diversity techniques, in which two or more independent channels are somehow combined. The motivation here is that only one of the channels is likely to suffer a fade at any instant of time.

Since multipath propagation results in varying travel times, signal pulses are broadened as they travel through the wireless medium. This limits the speed at which adjacent data pulses can be sent without overlap, and hence, the maximum information rate a wireless system can operate. Thus, in addition to frequency-selective fading, a multipath channel also exhibits time dispersion. Time dispersion leads to intersymbol interference (ISI) while fading induces periods of low signal-to-noise ratio (SNR), both effects causing burst errors in wireless digital transmission. Figure 1.7 shows the impact of ISI. In this case, the same delay spread is assumed. Thus, while multipath propagation causes fast fading at low data rates, at high data rates (i.e., when the delay spread becomes comparable with the symbol interval), the received signals become indistinguishable, giving rise to ISI. Lowering the data symbol rate and/or introducing a guard time interval (also known as a cyclic prefix [CP]) between symbols can help mitigate the impact of time dispersion.

The performance metric for a wireless system operating over a multipath channel is either the average probability of error or the probability of outage. The average probability of error is the average error rate for all possible locations in the

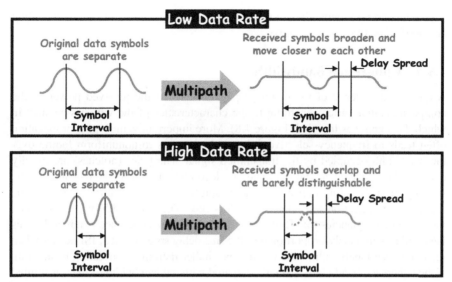

Figure 1.7   Intersymbol interference.

cell. The probability of outage represents the error probability below a predefined signal threshold for all possible locations in the cell.

1.3.3   Delay Spread and Time Dispersion

Delay spread is caused by differences in the arrival time of a signal from the various paths when it propagates through a time-dispersive multipath channel. The net effect of the arrival time difference is to spread the signal in time. The delay spread is proportional to the length of the path, which is in turn affected by the span of the propagating environment, as well as the location of the objects around the transmitter and receiver. The delay spread decreases at higher frequencies due to greater signal attenuation and absorption. A negative effect of delay spread is that it results in ISI. This causes data symbols (each representing one or more bits) to overlap in varying degrees at the receiver. Such overlap results in bit errors that increase as the symbol period approaches the delay spread. The effect becomes worse at higher data rates and cannot be solved simply by increasing the power of the transmitted signal. To avoid ISI, the duration of the delay spread should not exceed the duration of a data symbol, which carries a set of information bits.

The root mean square (rms) delay spread is often used as a convenient measure to estimate the amount of ISI caused by a multipath wireless channel. The maximum achievable data rate depends primarily on the rms delay spread and not the shape of the delay spread function. The rms delay spread in an indoor environment can vary significantly from 30 ns in a small room to 250 ns in a large hall. In outdoor environments, a delay spread of 10 μs or less is common (for a range of 1000 ft or less), although for non-LOS cases, a delay spread in the order of 100 μs is possible. If the product of the rms delay spread and the signal bandwidth is much less than

1, then the fading is called flat fading. If the product is greater than 1, then the fading is classified as frequency selective.

1.3.4   Coherence Bandwidth

A direct consequence of multipath propagation is that the received power of the composite signal varies according to the characteristics of the wireless channel in which the signal has traveled (Figure 1.8). More importantly, multipath propagation often leads to frequency-selective fading, which refers to nonuniform fading over the bandwidth occupied by the transmitted signal. The fades (notches) are usually correlated at adjacent frequencies and are decorrelated after a few megahertz. The severity of such fading depends on how rapidly the fading occurs relative to the round-trip propagation time on the wireless link. The bandwidth of the fade (i.e., the range of frequencies that fade together) is called the coherence bandwidth. This bandwidth is inversely proportional to the rms delay spread. Thus, ISI occurs when the coherence bandwidth of the channel is smaller than the modulation bandwidth. If the coherence bandwidth is small compared with the bandwidth of the transmitted signal, then the wireless link is frequency selective, and different frequency components are subject to different amplitude gains and phase shifts. Conversely, a wireless link is nonfrequency selective if all frequency components are subject to the same attenuation and phase shift. Frequency-selective fading is a more serious problem since matched filters that are structured to match the undistorted part of the spectrum will suffer a loss in detection performance when the attenuated portion of the spectrum is encountered. Either the data rate must be restricted so that the signal bandwidth falls within the coherence bandwidth of the link or other techniques such as spread spectrum must be used to suppress the distortion. The delay spread caused by multipath is typically greater outdoors than indoors due to the wider coverage area. This gives rise to a higher coherence bandwidth in indoor environments. For example, an indoor channel with a delay spread of 250 ns corresponds to a coherence

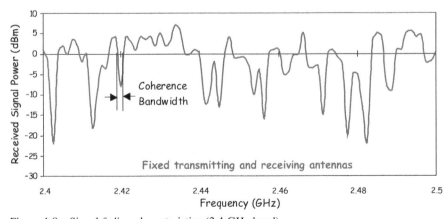

Figure 1.8   Signal fading characteristics (2.4 GHz band).

bandwidth of 4 MHz. An outdoor channel with a larger delay spread of 1 μs implies a smaller coherence bandwidth of 1 MHz.

Signals with bandwidth larger than the coherence bandwidth of the channel may make effective use of multipath by resolving (isolating) many independent signal propagation paths to provide better SNR at the receiver. This is exploited by some multiple antenna systems. On the other hand, multipath interference can be avoided by keeping the symbol rate low, thereby reducing the signal bandwidth below the coherence bandwidth. Although a wideband receiver can resolve more paths than another receiver with a narrower bandwidth, this may be done at the expense of receiving less energy and more noise per resolvable path.

1.3.5   Doppler Spread

Doppler spread is primarily caused by the relative motion between the transmitter and receiver. It introduces random frequency or phase shifts at the receiver that can result in loss of synchronization but affects LOS and reflected signals independently. Reflected signals affected by Doppler shifts are perceived as noise contributing to intercarrier interference (ICI) in multicarrier transmission. The Doppler effect may also be due to the movement of reflecting objects (e.g., vehicles, humans) that causes multipath fading. In an indoor environment for instance, the movement of people is the main cause of Doppler spread. A person moving at 10 km/h can induce a Doppler spread of ±22 Hz at 2.4 GHz. The Doppler spread for indoor channels is highly dependent on the local environment, providing different shapes for different physical layouts. On the other hand, outdoor Doppler spreads consistently exhibit peaks at the limits of the maximum Doppler frequency. Typical values for Doppler spread are 10–250 Hz (suburban areas), 10–20 Hz (urban areas), and 10–100 Hz (office areas).

Just as coherence bandwidth is inversely related to the delay spread, coherence time is defined as the inverse of the Doppler spread. The coherence time determines the rate at which fading occurs. Fast fading occurs when the fading rate is higher than the data symbol rate. The coherence time is a key parameter that affects channel feedback mechanisms in high-speed mobile systems. For example, the delay in sending channel feedback information from the handset to the BS may exceed the coherence time. This renders the feedback outdated by the time the BS processes the information. The relationship between coherence time and coherence bandwidth is shown in Figure 1.9. This relationship forms the basis for fading channel classification.

1.3.6   Shadow Fading

Besides multipath fading, large physical obstructions (e.g., walls in indoor environments, buildings in outdoor environments) can cause large-scale shadow fading. In this case, the transmitted signal power is blocked and hence severely attenuated by the obstruction. The severity of shadow fading is dependent on the relative positions of the transmitter and receiver with respect to the large obstacles in the propagation environment, as well as the number of obstructing objects and the dielectric

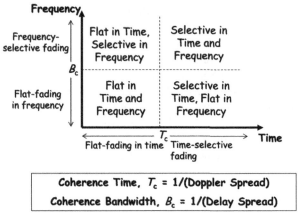

Figure 1.9   Coherence time and coherence bandwidth.

properties of the objects. Unlike multipath fading (which is usually represented by a Rician or Rayleigh distribution), shadow fading is generally characterized by the probability density function of a log-normal or Gaussian distribution. Increasing the transmit power can help to mitigate the effects of shadow fading although this places additional burden on the handset battery and can cause interference for other users.

1.3.7   Radio Propagation Modeling

The transmitted signal power normally radiate (spread out) in all directions and hence attenuates quickly with distance. Thus, very little signal energy reaches the receiver, giving rise an inverse relationship between distance and path loss. Depending on the severity, the decay in signal strength can make the signal become unintelligible at the receiver. Radio propagation analysis allows the appropriate power or link budget to be determined between the RF transmitter and receiver. It can be very complex when the shortest direct path between the transmitter and receiver is blocked by fixed or moving objects, and the received signal arrives by several reflected paths. The degree of attenuation depends largely on the frequency of transmission. For example, lower frequencies tend to penetrate objects better, while high frequency signals encounter greater attenuation.

For clear LOS paths in the vicinity of the receiving antenna, signal attenuation is close to free space. This is the simplest signal loss model where the received signal power decreases with the square of the distance between the transmitter and receiver. For instance, the signal strength at 2 m is a quarter of that at 1 m. At longer distances away from the receiving antenna, an increase in the attenuation exponent is common (Figure 1.10). In this case, the signal attenuation is dependent not only on distance and transmit power but also on reflecting objects, physical obstructions, and the amount of mutual interference from other transmitting users. Small changes in position or direction of the antenna, shadowing caused by blocked signals and moving obstacles (e.g., people and doors) in the environment may also lead to drastic fluctuations in signal strength. Similar effects occur regardless of whether a user is

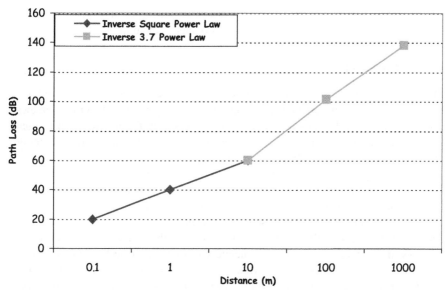

Figure 1.10   Signal attenuation for omnidirectional antenna with spherical (isotropic) radiation pattern.

stationary or mobile. Hence, while the free-space exponent may be relevant for short distance transmission (e.g., up to 10 m), the path loss is usually modeled with a higher-valued exponent of 3 to 5 for longer distances. For indoor environments, where objects move very slowly, fading is primarily due to the receiver. Thus, the path loss (a distance-related phenomenon) is independent of fast fading (a time-related phenomenon).

An accurate characterization of the propagation mechanism is difficult since this is greatly influenced by a number of factors, such as antenna height, terrain, and topology. Radiowave propagation modeling is usually based on the statistics of the measured channel profiles (time and frequency domain modeling) or on the direct solution of electromagnetic propagation equations based on Maxwell's equations. The most popular models for indoor radio propagation are the time domain statistical models. In this case, the statistics of the channel parameters are collected from measurements in the propagation environment of interest at various locations between the transmitter and receiver. Another popular method involves ray tracing, which assumes that all objects of interest within the propagation environment are large compared with the wavelength of propagation, thus removing the need to solve Maxwell's equations. Its usefulness is ultimately dependent on the accuracy of the site-specific representation of the propagation environment.

Modeling the channel characteristics of narrowband and wideband signals is different. For narrowband signals, the emphasis is on the received power whereas for wideband communications, both the received signal power and multipath characteristics are equally important. A further distinction exists between models that describe signal strength as a function of distance as opposed to a function of time.

The former is used to determine coverage areas and intercell interference while the latter is used to determine bit error rates and outage probabilities.

1.3.8   Channel Characteristics

Many wireless systems, including multiantenna systems, show large performance improvements when the channel characteristics are known. The extent to which these gains are achieved depends on the accuracy with which the receiver can estimate the channel parameters. In practical implementations, the channel characteristics are often affected by fading, which can be time arying. In some cases, the channel is assumed to be reciprocal. This implies that the channel model in the forward direction (directed at the receiver) is identical to that in the reverse direction (directed at the transmitter). To learn the behavior of the wireless channel and build a model, a channel sounding method can be used. The transmitter provides a periodic multitone test sequence to excite the channel. At the receiver, the arriving test signal is correlated with a local copy of the test sequence. Due to the impulse-like autocorrelation function of the test sequence, the correlator output at the receiver provides the measured channel impulse response, which must be acquired real time for correct estimation of the signal path statistics. In general, the channel impulse response is stationary at short distances, but can be highly variable over longer distances. It can also be time varying due to the presence of nonstationary objects (e.g., people and vehicles).

1.3.9   Gaussian Channel

The Gaussian channel provides an upper bound on the wireless system performance and accurately describes many physical channels, including time-varying channels. Often referred to as the additive white Gaussian noise (AWGN) channel, it is typically used to model the noise generated in the receiver when the transmission path is ideal (i.e., LOS). The noise is modeled with a constant power spectral density over the channel bandwidth and a Gaussian probability density function. If the user is stationary, the model is valid even when multipath propagation is present. In this case, the channel is approximated as Gaussian with the effects of multipath fading represented as a path loss.

The Gaussian channel model has received considerable attention because of its importance for single-transmitter and single-receiver channels. However, the Gaussian model may not be suitable when applied to multiple access channels where users transmit intermittently. In this case, the desired model is one where the number of active users on the network is a random variable, and the Gaussian signals are conditioned on the fact that some user is transmitting. In addition, Doppler spread, multipath fading, shadowing, and mutual interference from multiple transmitting users make the channel far from Gaussian.

1.3.10   Rayleigh Channel

There are two kinds of channel fading, namely, long-term (log-normal) and short-term (Rayleigh) fading. Long-term or slow fading is characterized by the envelope

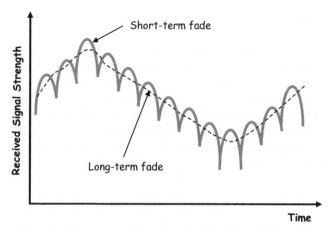

Figure 1.11   Short- and long-term fading characteristics.

of the fading signal, which is related to the distance and the received power. Short-term or fast fading is primarily caused by reflections of a transmitted signal and refers to the rapid fluctuations of the received signal amplitude. Short-term fading occurs when there is no direct LOS signal component. Conversely, if there is a LOS component, long-term fading occurs. In addition, second-order statistics on the fades exist and are usually functions of time (e.g., level crossing rates, average duration of fades, and distribution of fade duration). The received signal is a product of the long-term fading and short-term fading characteristics. The short-term fading signal is superimposed on an average value that varies slowly as the receiver moves (Figure 1.11).

1.3.11   Rician Channel

In some wireless channels, a dominant path (normally the LOS path) exists between the transmitter and receiver, in addition to many scattered (nondirect) paths. This dominant path reduces the delay spread and may significantly decrease the fading depth, thus requiring a much smaller fading margin in system design. The probability density function of the received signal envelope is said to be Rician. Rayleigh fading can be considered as a special case of Rician fading. This is because Rician fading is essentially a superposition of a Rayleigh distributed signal and a LOS signal. The strong LOS signal reaches the receiver at about the same time as the reflected signals generated by local scatterers near the receiver. Rician channels are highly desirable because they require a much smaller fade margin and supports significantly larger bandwidths than Rayleigh channels. The Rician fading function is very similar to the Rayleigh fading function with the exception of two additional variables: the arrival angle of the LOS signal and the K factor. The LOS angle specifies the relative location of the direct path with respect to the faded spectrum by changing the static frequency shift. The K factor represents the energy ratio of the LOS signal to the scattered signal. Thus, a higher value for K shows a greater proportion of signal energy in the dominant path.

1.4   MODULATION

The maximum data rate is not only bounded by the multipath characteristics of the channel but also the modulation technique used. With a bandwidth-efficient modulation technique, a greater number of bits can be transmitted in each symbol interval, resulting in higher data rates without increasing the symbol rate. As shown in Figure 1.12, in multilevel modulation, each symbol is mapped into multiple data bits. A 2 Mbit/s bit stream using 64-QAM modulation requires a symbol rate of 33 Ksymbol/s. Unfortunately, a higher-order modulation, such as 64-QAM, demands a high SNR for good performance and may not be appropriate for noisy or power limited wireless systems. The modulation is normally chosen together with a coding scheme that provides some level reliability in the transmission of the bit stream. As we shall see later, the modulation and coding scheme (MCS), hence the data rate, is often adjusted dynamically to cope with changes in the SNR. This process is called link or rate adaptation, which is not required in wired networks. Other considerations for selecting a MCS include bandwidth (spectral) and power efficiency, resistance against multipath, cost/complexity of implementation, envelope variation, and out-of-band radiation.

1.4.1   Linear versus Constant Envelope

Constant envelope (e.g., binary phase-shift keying [BPSK] and quadrature phase-shift keying [QPSK]) and linear modulation (e.g., quadrature amplitude modulation [QAM]) are common digital modulation methods employed in wireless transmission. Constant envelope modulation involves only the phase, whereas linear modulation requires both the phase and the amplitude to be modulated (thus resulting in a

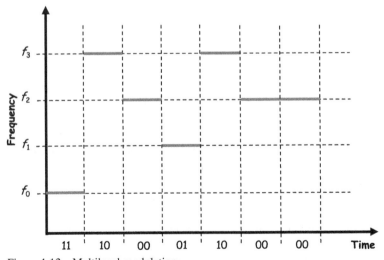

Figure 1.12   Multilevel modulation.

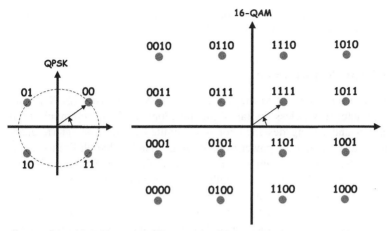

Figure 1.13   16-QAM and QPSK constellations.

non-constant envelope). For QAM, a constellation diagram can be plotted where each constellation point specifies an amplitude and a phase of the modulation, and is mapped to a modulated data symbol. For example, 16-QAM codes 4 bits per symbol, and there are 16 points on the constellation diagram, whereas QPSK comprises 4 points, each representing 2 bits per symbol (Figure 1.13). Thus, the symbol rate is equal or less than the actual bit rate. A lower symbol rate reduces instances of ISI.

Constant envelope modulation provides gains in power efficiency since it allows more efficient power amplifiers to be used. On the other hand, linear modulation is spectrally more efficient than constant envelope modulation but requires the use of linear power amplifiers. The concern for linearity is primarily due to the possibility of higher intermodulation and out-of-band power emissions. There is considerable debate as to whether multilevel modulation, such as QAM, provides any real gain in frequency reuse situations, since it requires an increase in SNR. This results in an increase in frequency reuse distance, which more than offsets the modulation improvement.

1.4.2   Coherent versus Noncoherent Detection

There are two basic techniques of detecting data symbols from the demodulated signals, namely, coherent (synchronous) and noncoherent (envelope) detection. In coherent detection, the receiver first processes the signal with a local carrier of the same frequency and phase, and then cross-correlate with other replicated signals received before performing a match (within a predefined threshold) to make a decision. In a multipath environment, phase synchronization is not easy, and hence, noncoherent detection is preferred. This can be achieved by comparing the phase of a received symbol with a previous symbol to obtain the relative phase difference. Since noncoherent detection does not depend on a phase reference, it is less complex but exhibits worse performance when compared with coherent methods. Although no absolute phase reference is needed for the demodulation of differential phase modulation, the receiver is sensitive to carrier frequency offset.

1.4.3 Bit Error Performance

The ratio of the energy per bit to the noise power spectral density (E_b/N_0) is an important parameter in wireless digital communications. E_b/N_0 is commonly used with modulation and coding designed for noise-limited rather than interference-limited communication, and for power-limited rather than bandwidth-limited communications (e.g., spread spectrum systems). It normalizes the SNR and is also known as the "SNR per bit." It is especially useful when comparing the bit error rate (BER) performance of different digital modulation schemes without taking bandwidth into account. E_b/N_0 is equivalent to the SNR divided by the link spectral efficiency in bit/s/Hz, where the transmitted data bits are inclusive of overheads due to error coding and network protocols, such as medium access control (MAC) protocols. Thus, E_b/N_0 is generally used to relate actual transmitted power to noise. The noise spectral density N_0, is usually expressed in W/Hz. Therefore, E_b/N_0 is a non-dimensional ratio.

1.5 MULTIPATH MITIGATION METHODS

If symbol rates greater than the inverse of the delay spread are required, then several techniques can be used to compensate for ISI caused by a time-dispersive channel and regenerate the data symbols correctly. They include equalization, multicarrier transmission, and wideband transmission, error control, and antenna diversity.

1.5.1 Equalization

Equalizers essentially correct the phase differences between the direct and reflected signals (Figure 1.14) by subtracting the delayed and attenuated images of the direct signal from the received signals. To do this, equalizers employ a training sequence at the start of each packet transmission to derive the impulse response (transfer characteristic) of the channel. This response must be frequently remeasured as the wireless channel can change rapidly in both time and space. When applied to slowly varying indoor wireless links, equalization typically requires a minimum overhead of 14–30 symbols for every 150–200 data symbols. For rapid variations in the wireless link, equalization can be difficult because these problems have to be alleviated using a long equalizer training sequence. Nevertheless, by using joint data and link estimation algorithms, data symbols can still be reliably recovered even after the link undergoes fading.

The simplest equalization technique (also known as linear equalization) amplifies the attenuated part and attenuates the amplified part of the spectrum. This technique attempts to invert and neutralize the effects of the medium. This can be achieved by passing the received signal through a filter with a frequency response

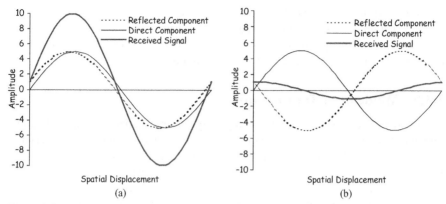

Figure 1.14   Phase correction needed for multipath signals. (a) In-phase received signal. (b) Out-of-phase received signal.

that is the inverse of the frequency response of the medium. For time-varying wireless channels, the linear equalizer can be made adaptive by employing an adaptive filter at the receiver whose frequency response adapts to the inverse of the channel response, further suppressing the multipath components. Although linear equalizers are the simplest to implement, their performance is limited because the inverse filter of the wireless channel may not exist in all cases (e.g., channels with nulls) or may cause the signal to be restored to its original form at the expense of boosting the received noise to a very high level. For example, when additional interference or frequency-selective fades are present at the input to the equalizer, the equalizer must minimize distortion without adversely enhancing noise. In this case, more sophisticated equalizers are needed.

The decision feedback equalizer (DFE) is a nonlinear equalizer that does not enhance the noise because it estimates the frequency response of the link rather than inverting it. It can also handle channels with nulls. Like spread spectrum transmission, a DFE can isolate the arriving paths and take advantage of them as a source of implicit diversity to improve performance. A DFE feeds past decisions through a feedback filter and subtracts the results from the input to the decision device. This nonlinear process results in less noise enhancement than linear equalization. A DFE also exploits decisions about the received data to cancel ISI from the received signal. It is easier to implement than a linear equalizer since the input to its feedback filter is a data symbol. Hence, convolution can be implemented based solely on additions, and no multiplications are required. However, DFEs are prone to error propagation and are useful only when the medium exhibits reasonably low bit error rate without coding.

The maximum likelihood sequence estimator (MLSE), also known as the Viterbi equalizer, operates by testing hypotheses of the transmitted data sequence, combined with knowledge about the channel response, against the signal that was received. The signal that gives the closest match is chosen. Sequence detection offers performance improvements over traditional equalization when the channel response is known and time-invariant. MLSE equalizers are well suited for spread spectrum systems that employ different code sequences to distinguish the transmission from multiple users. Although these equalizers achieve the lowest error probability, they are complex and may not be feasible at very high data rates.

1.5.2   Multicarrier Transmission

At high data rates, the computation complexity for an equalizer increases. In addition, the overhead for channel estimation increases when the channel is time-varying. Multicarrier or multitone systems rely on sending several data streams in parallel at low symbol (signaling) rates that combine to the desired high rate transmission. For example, a 100 Msymbol/s data stream (corresponding to 100 Mbit/s using BPSK modulation) can be converted to 100 streams of 1 Msymbol/s using 100 subcarriers. In doing so, multipath delay spread is compensated without the need for complex equalization. The principle behind multicarrier schemes is that since ISI caused by multipath is only a problem for very high symbol rates (typically above 1 Msymbol/s), the data rate can be reduced until there is no degradation in performance due to ISI. The overall frequency channel is divided into a number of subchannels (which can range from 64 in indoor networks to 2048 in outdoor networks), each modulated on a separate subcarrier at a lower symbol rate than a single carrier scheme using the entire channel. Hence, the delay spread now impacts a smaller proportion of the lengthened symbol time. Since each subchannel is narrow enough to cause only flat (nonfrequency-selective) fading, this makes a multicarrier system less susceptible to ISI. By adding a guard interval (GI) to each symbol, such interference can be completely eradicated.

Multicarrier transmission is spectrally efficient because the subcarriers can be packed close together. It also allows considerable flexibility in the choice of different modulation methods. On the negative side, multicarrier transmission is more sensitive to frequency offset and timing mismatch than single-carrier systems. Fast synchronization of multiple subcarriers at the receiver is crucial in reducing processing delays and overheads. This is normally achieved by allocating a specific set of pilot subcarriers for synchronization purposes. The location of the pilots may be changed from symbol to symbol and the power of these subcarriers is boosted to minimize errors from recovering the training signal. In general, either the overall data rate is increased or a larger delay spread can be tolerated if a larger number of subcarriers is used.

Like multilevel modulation, a major drawback of multicarrier systems is the high peak-to-average power ratio (PAPR). The power amplifier in a multicarrier system is more costly since it must back off more than a single-carrier system, especially for user devices operating near the edge of a cell due to the higher transmit power. The large amplifier backoff requires the use of a highly linear (and inefficient) amplifier with precise gain control, which leads to high power consumption. Such an amplifier is also needed to suppress ICI when subcarriers become misaligned due to multipath propagation. Since a lower number of subcarriers leads to reduced PAPR, the number of subcarriers and the subcarrier spacing have to be chosen appropriately. In addition to linear amplifiers, signal distortion techniques, such as peak clipping or filtering, spectrum shaping or windowing (which reduces output spectrum), may have to be implemented to counter the PAPR effect. Low-phase noise oscillators and high-resolution analog-to-digital converters are also needed.

1.5.3 Orthogonal Frequency Division Multiplexing

Orthogonal frequency division multiplexing (OFDM) is a special form of multicarrier transmission involving orthogonal subcarriers. Although its adoption has become widespread in many wireless network standards, OFDM also forms the basis of some wireline standards, including discrete multitone digital subscriber line (DSL) and HomePlug powerline communication systems. OFDM employs equally spaced subchannels, each containing a subcarrier that carries a portion of the user information. An OFDM system uses less bandwidth than an equivalent frequency division multiple access (FDMA) system because OFDM subchannels are overlapped. In this way, OFDM offers some degree of frequency diversity since if the subchannel becomes degraded by interference, noise, or fading, the data rate is reduced but the connection is maintained. Error control further enhances the reliability of each subcarrier. OFDM can be combined with frequency hopping, allowing nonsequential transmission of subcarriers, thereby achieving a greater degree of frequency diversity and interference averaging.

In OFDM, because data symbols are transmitted in parallel rather than serially, the duration of each OFDM symbol is longer than the corresponding symbol on a single carrier system of an equivalent data rate. This reduces the effect of ISI. In addition, the orthogonality between the subcarriers mitigates ICI. The baseband frequency of each subcarrier is then carrier-modulated using the same RF carrier. By switching subcarriers after each symbol time, frequency selective fading is minimized. If the symbol interval is T and the number of subcarriers is N, then the subcarriers are spaced at intervals of $1/T, 2/T, \ldots, N/T$. Thus, the baseband frequencies of the OFDM subcarriers become an integral multiple of the base frequency $(1/T)$. This leads to orthogonal subcarriers where the peak of one modulated subcarrier can overlap on the null of every other subcarrier without causing unnecessary interference within the subchannel (Figure 1.15). When the receiver samples at the center frequency of each subcarrier, the desired signal is obtained. Only sinusoidal subcarriers can ensure orthogonality, allowing the receiver to distinguish between the waveforms.

The use of overlapped subchannels results in higher spectral efficiency. As can be seen in Figure 1.16, three modulated subcarriers can occupy the same bandwidth as two modulated subcarriers. Unlike FDMA, OFDM allows separation of subcarriers without a guard band. OFDM reduces ISI compared with high-speed single carrier transmission because the symbol rate can be reduced over multiple subcarriers, leading to flat fading, which is easier to deal than frequency-selective fading. A larger number of subcarriers allow the symbol rate for each subcarrier to be reduced further. This can, in turn, mitigate a larger multipath delay spread in outdoor environments. For example, the 802.16 wireless access standard employs 256 OFDM subcarriers compared with 64 subcarriers in the 802.11 standard.

At the transmitter, if there are N subcarriers (N samples), the time-domain subcarrier sequence $f(n)$ can be generated from the frequency-domain sequence $F(k)$ using the inverse discrete Fourier transform (DFT):

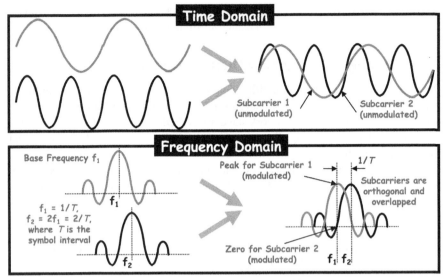

Figure 1.15   OFDM subcarriers in time and frequency domains.

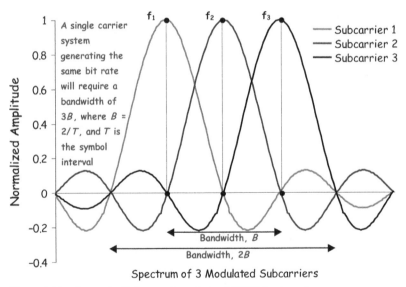

Figure 1.16   Spectral efficiency of overlapped OFDM subcarriers.

$$f(n) = \sum_{k=0}^{N-1} F(k) e^{+\frac{j2\pi nk}{N}}, \quad n = 0, 1, 2, \ldots, N-1. \tag{1.1}$$

At the receiver, the DFT can be used to generate $F(k)$.

$$F(k) = \sum_{n=0}^{N-1} f(n) e^{-\frac{j2\pi nk}{N}}, \quad k = 0, 1, 2, \ldots, N-1. \tag{1.2}$$

The resulting Fourier spectrum has discrete frequencies at k/NT, where T is the sampling interval in the time domain, and N is the number of samples. The sampling frequency in the frequency domain is $1/T$. The DFT operations in OFDM can be implemented using fast Fourier transform (FFT), which can be adapted to different data rates and link conditions. The FFT size corresponds to N. FFT allows efficient digital signal processing implementation and significantly reduces the amount of required hardware compared with FDMA systems. For example, OFDM achieves reduced processing complexity compared with single carrier systems, achieving $\log N$ per sample complexity using FFT. As long as the channel is linear, the subcarriers can be demodulated without interference and without the need for analog filtering to separate the received subcarriers.

Due to the simplicity of equalizer process, OFDM reduces the need to maintain a LOS path. A CP absorbs late-arriving symbols in the presence of delay spread caused by multipath. The CP is essentially a GI that is longer than the delay spread, thereby preventing one symbol from interfering with the next. A short GI reduces the operating range of the OFDM system. It is an important component of practical OFDM implementations because it leads to simple equalization. A copy of the last portion of the data symbol is usually appended to the front of the next symbol to form the CP. Figure 1.17 shows the typical OFDM transmitter and receiver implementation.

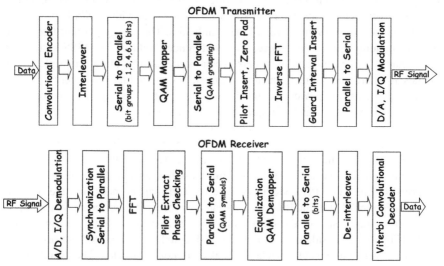

Figure 1.17   OFDM transmitter and receiver.

1.5.4   Wideband Systems

In wideband systems, such as spread spectrum or ultra-wideband (UWB) systems, the transmission bandwidth is much greater compared with the signal bandwidth. Spread spectrum is a powerful combination of bandwidth expansion (beyond the coherence bandwidth) and coding, whereby the former combats ISI and the latter allows individual symbols smeared by multipath to be received correctly. The general concept of spread spectrum has been in existence for more than 50 years and was originally developed by the military for reliable and secure communications. Spread spectrum refers to signaling schemes that are based on some form of coding (that is independent of the transmitted information) and which use a much wider bandwidth beyond what is required to transmit the information. The wider bandwidth means that interference and multipath fading typically affect only a small portion of a spread spectrum transmission. The received signal energy is therefore relatively constant over time, and frequency diversity is achieved. This in turn produces a signal that is easier to detect provided the receiver is synchronized to the parameters of the spread spectrum signal. Spread spectrum signals are able to resist intentional jamming since they are hard to detect and intercept, thus ensuring some level of PHY security. Instead of an equalizer, spread spectrum systems often use a simpler device, called a correlator, to counter the effects of multipath.

There are two spread spectrum techniques: direct-sequence spread spectrum (DSSS) and frequency-hopping spread spectrum (FHSS). In DSSS, the data stream is combined with a higher speed digital code. Each data bit is mapped into a common pattern of bits known only to the transmitter and intended receiver. The bit pattern is called a pseudonoise (PN) code, and each bit of the code is called a chip (the term chip emphasizes that one bit in the PN code forms part of the actual data bit). The sequence of chips within each bit period is random, but the same sequence is repeated in every bit period, thus making it pseudorandom or partially random. The chipping rate of an n-chip PN code is n times higher than the data rate. Thus, a high rate for the chipping sequence results in a very wide bandwidth. DSSS spreads the energy (power) of the signal over a large bandwidth. The energy per unit frequency is correspondingly reduced. Hence, the interference produced by DSSS systems is significantly lower compared with narrowband systems. This allows multiple DSSS signals to share the same frequency channel. To an unintended receiver, DSSS signals appear as low-power wideband noise and are rejected by most narrowband receivers. Conversely, this technique diminishes the effect of narrowband interference sources by spreading them at the receiver.

As an example, 802.11b devices employ DSSS using the 11-chip Barker PN code (Figure 1.18). Information bits are spread out 11 times by the PN code before being modulated. The chip rate is 11 times faster than the data rate. Chips are despread by the same PN code at the receiver and mapped back into the original data bit (Figure 1.19). If the PN code generated by receiver is perfectly synchronized to the transmitted signal, the despreading process produces high autocorrelation peaks (Figure 1.20). Sidelobes of the received signal will either add constructively or destructively, but the autocorrelation peaks will not be affected. The overall result

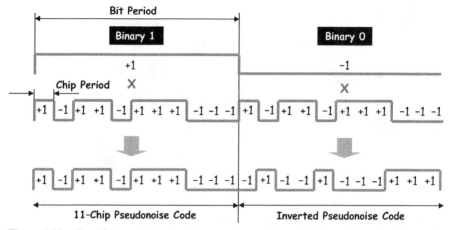

Figure 1.18   Combining a higher-speed PN code with lower-speed digital data.

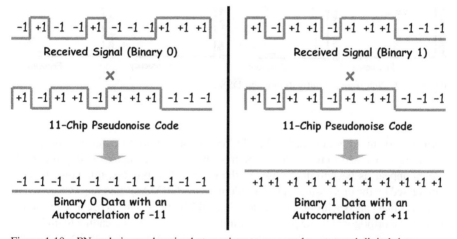

Figure 1.19   PN code is synchronized at receiver to recover lower-speed digital data.

is a single peak within each bit interval. The energy of noise and narrowband interference that may have been added during the transmission are spread and hence suppressed by the PN code (Figure 1.21).

Clearly, DSSS systems take up more bandwidth due to the spreading. To achieve higher aggregate efficiency, different PN codes can be used, thereby allowing concurrent transmissions to take place. This gives rise to CDMA. Code division refers to the fact that transmission with orthogonal PN codes may overlap in time

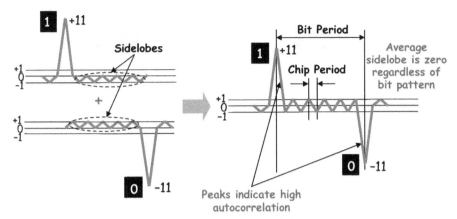

Figure 1.20 Autocorrelation and sidelobes in despread digital data.

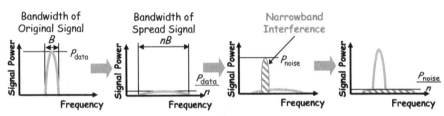

Figure 1.21 Interference mitigation using DSSS.

with little or no effect on each other. Different users transmit using unique codes. For a receiver tuned to the code of one transmission, other signals (using other PN codes) appear as background noise. In the despreading process, this noise will be suppressed by the processing gain. A random access CDMA system requires complex receivers that can synchronize or demodulate all PN codes. This is because the codes may not be orthogonal in the presence of multipath delay spread, leading to multiuser interference that ultimately limits the capacity of the system. A long PN code supports more concurrent transmissions. However, for a fixed channel bandwidth, the rate of each transmission is reduced. CDMA systems are often used to estimate the distance between the transmitter and receiver in a process called ranging. DSSS wireless LANs utilize the same PN code and therefore do not have a set of codes available as is required for CDMA operation. A single code enables information to be broadcast easily. In addition, the code can be made shorter, thereby increasing data bandwidth.

Like narrowband systems, multipath, shadow fading, and interference impose limits on the coverage region and on the number of users in spread spectrum systems. For example, if a 100-Mchip/s chip rate is selected, then multipath components must be separated by at least 10 ns in order for these components to be resolved. Hence, the appropriate area of coverage should have a propagation delay greater than

10 ns, or, equivalently, a minimum span of 3×10^8 m/s $\times$ 10 ns or 3 m. In general, a bigger coverage area is preferred for spread spectrum communications since this results in more resolvable multipaths. However, the opposite is true for narrowband systems. Furthermore, polarization antenna diversity becomes more effective since there is significant coupling between the vertical and horizontal polarization directions when the coverage area is large.

1.5.5 Error Control

There are two general types of error control in wireless communications. Forward error correction (FEC) techniques employ error control codes (e.g., convolutional codes and block codes) that can detect with high probability the error bit's location and correct the error. Due to the additional bit overheads, FEC tends to degrade data throughput (i.e., the usable bit rate available to applications). FEC may also require long bit interleaving intervals to randomize and spread the error bits so that they can be corrected easily. This increases the transmission delay. Code rates of 1/2, 2/3, 3/4, 5/6, are common. A code rate of 1/2 implies that for every 1 data bit, an additional FEC bit is added and so 2 bits are transmitted. Thus, a code rate of 1/2 results in a 50% reduction in data throughput. A high code rate of 5/6 is more efficient than a low code rate of 1/2 but demands a higher SNR and hence, a shorter operating range or higher transmit power. Low code rates are usually used with QPSK because high-order modulations are not effective with low rate codes. Thus, a more efficient code rate is normally chosen with a high-order modulation. For example, a code rate of 5/6 can be chosen with 64-QAM to send a group of 6 bits (representing one data symbol). The coded bits may be repeated (which further reduces coding efficiency) or punctured to remove some of the parity bits.

A convolutional code is an example of FEC that can be implemented with a constraint length of x. In this case, x memory registers are needed, each holding 1 input bit. The encoder contains n modulo-2 adders and n polynomial generators, one for each adder. Using the generators and the existing values in the remaining registers, the encoder outputs n bits. Convolutional codes can be decoded using several algorithms. For small values of x, the Viterbi algorithm provides maximum likelihood performance and is highly parallelizable. Puncturing is often used with the Viterbi algorithm to improve coding efficiency. Larger values for x are more practically decoded with sequential decoding algorithms (e.g., Fano algorithm). Unlike Viterbi decoding, sequential decoding is not maximum likelihood. However, since its complexity increases only slightly with x, this allows the use of stronger long constraint-length codes. Turbo codes are a class of high-performance FEC codes. While normal FEC codes typically achieve a channel capacity of 3 dB below the theoretical limit (i.e., Shannon bound), Turbo codes may potentially achieve a better performance of 0.5 dB below that limit. Two encoders with an interleaver are employed (Figure 1.22).

Low-density parity-check (LDPC) coding is another example of FEC. It is a class of linear block codes whose parity-check matrix contains only a few 1's in comparison with the number of 0's. This requires the number of 1's in each row or column of the matrix to be much less than the dimensions of the matrix. Like Turbo

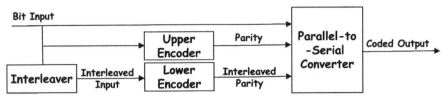

Figure 1.22 Turbo coding.

codes, LDPC codes offer near-capacity performance for many channels. However, constructing good-performing LDPC codes can be challenging, especially in implementation, which requires a high degree of parallelism to shorten the processing time. Theoretically, iterative decoding algorithms of sparse LDPC codes perform close to the optimal maximum likelihood decoder and better than most Turbo codes, but are complex to encode and decode. As such, LDPC has been offered as an optional feature in many wireless standards.

In automatic repeat request (ARQ), the receiver employs simpler error detection codes (which incur less bit overheads than FEC) to detect bit errors in the received packets and then request the transmitter to resend any error packet. Thus, ARQ may achieve a packet error rate of zero. The simplest ARQ method is the Stop-and-Wait ARQ. In this case, each packet transmitted by the sender is accompanied by an acknowledgment (ACK). If the ACK is not received after a predetermined interval (i.e., a timeout), the packet is retransmitted by the sender. Instead of a single packet, other more complex ARQ methods (e.g., selective repeat ARQ) employ ACKs that deal with a group of packets. In general, ARQ schemes are simpler and more flexible to implement than FEC but may suffer from variable delay. Many wireless standards employ a combination of FEC and ARQ, but these methods operate independently, and hence may not give the best performance. To this end, hybrid ARQ (HARQ) has been developed to balance and optimize the tradeoff in FEC overheads and ARQ retransmissions.

Since FEC, ARQ, or HARQ may not correct packet losses caused by buffer overflow at the receiver, packet retransmission by higher network layers (e.g., TCP) or error concealment by the application (e.g., video frame copy, image pixel interpolation) may still be needed.

1.6 MULTIPLE ANTENNA SYSTEMS

Wireless communication is characterized by random propagation conditions commonly referred to as fading channels. A fading channel can be modeled as a linear time-varying channel whose impulse response is a random process. Reliable communication over a fading channel can be achieved through the use of diversity techniques. Since the channel response at any given time and frequency (i.e., channel dimension or "degree of freedom") is a random variable, it is essential that each bit of information is spread over several dimensions. Diversity is traditionally achieved by coding in the time or frequency domains. Because time duration and frequency bandwidth are expensive resources, the number of channel degrees of freedom for diversity can be increased in the space domain, by adding spatially separated antennas at the receiver and the transmitter.

Under poor propagating conditions, the channel can be changed using antenna diversity. The basic principle for using multiple antennas is that if one antenna receives a deeply faded signal, the other antenna(s) receive only a slightly faded version of the signal. Since the angle of departure of the signal from the transmit antenna is typically narrower than the angle of arrival at the receiver, this gives the receiver the possibility of recovering different signals even if the antennas are closely spaced. Macroscopic (spatial) diversity is important when the fading is slow (long-term). It is often the best approach because it allows receivers to choose independent channels. Among the various forms of spatial diversity, selection combining is the simplest to implement since the receiver chooses to process only one channel—the channel with the highest SNR. Maximal ratio combining provides the largest SNR improvement because the signals from all available channels are weighted and then summed. This requires coherent modulation since the phase of each channel must be taken into account before the combining can occur. Microscopic diversity using two frequencies at the same (collocated) antenna site can help reduce short-term fading. Polarization diversity can also be used to select the best channel at a particular location. For example, circular polarized directional antennas, when used in LOS channels, can provide much lower delay spread than linear polarized antennas with similar directionality.

1.6.1 Receive Diversity versus Transmit Diversity

In multiantenna receive diversity, signal combining techniques are employed at the BS or AP to improve the UL performance (i.e., transmission from handset to BS or AP). The simplest method involves choosing the best signal among two or more antennas. Two antennas separated by an odd multiple of a quarter of a wavelength is sufficient to cause almost independent fades at the receiving antennas. Alternatively, different antenna polarization can be employed to keep the antenna dimensions small. Such diversity techniques are effective in overcoming multipath fading without any bandwidth penalty, and no additional transmit power is required. The overall impact is to reduce the fade margin and enhance interference rejection. It is difficult to implement receive diversity on the DL (i.e., transmission from BS or AP to handset). This is due to the size and battery power limitations of the handset. Mobility causes further problems. Thus, new multiantenna systems focus on transmit diversity where multiple antennas at the BS or AP transmit simultaneous data streams on the DL to handset.

1.6.2 Switched Antenna Receive Diversity

Switched antenna receive diversity is the simplest spatial diversity technique involving the use of two or more receive antennas at the user device. To avoid concurrent long-term fading at the collocated receive antennas, the antennas are separated by a

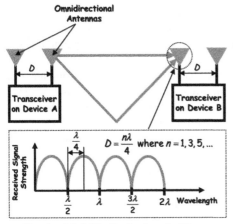

Figure 1.23 Switched antenna diversity.

distance that is a multiple of a quarter of the operating wavelength (Figure 1.23). At any time, only one of the two internal antennas is actively transmitting or receiving. If the receive SNR for the active antenna becomes poor, the antenna is deactivated and the other internal antenna becomes active. A 5 dBi external antenna can also be connected to improve the range and provide more coverage. In this case, the internal antennas are turned off, which removes receive diversity. The different antenna spacings at 2.4 and 5 GHz imply that two different radios may be needed to operate in these bands.

1.6.3 Multiple Input Multiple Output Systems

Unlike single antenna systems, which are also known as single input single output (SISO) systems, multiple input multiple output (MIMO) systems employ multiple antennas to transmit and receive simultaneous data streams on the same frequency channel. Unlike CDMA, MIMO is spectrally efficient even when one user is transmitting since it allows concurrent transmission of data streams without spreading the radio spectrum. When combined with OFDM, MIMO systems can realize impressive spectral efficiencies. For example, up to 15 bits/s/Hz can be realized for the 802.11n wireless LAN standard. This is in contrast to prior standards, such as 802.11b (0.5 bit/s/Hz) and 802.11a/g (2.7 bit/s/Hz), which are all based on single-antenna transmission. Because no switching of transmit antennas is required in MIMO, this may lead to faster response times. Some MIMO configurations are highly recommended for indoor environments due to the rich multipath signal reflections from walls and structures. Such reflections may not always be present in outdoor environments.

Reducing the MIMO antenna size is a key challenge for small personal handsets, such as smartphones. These devices may restrict the achievable benefits of MIMO due to limited space to separate the antennas in the microwave frequency bands. On the other hand, MIMO can be beneficial to handsets because it provides robustness or capacity without the need for a second radio transceiver (using a different frequency band) or additional frequency channels (using the same frequency band). This, in turn, reduces battery power consumption and handset cost. In addition, MIMO's greater reliability or higher rate minimizes the time to transmit or retransmit packets. Thus, the radio is turned on for less time when transmitting a given amount of data, conserving additional battery power. This benefit can also be achieved using broader channel bandwidths and higher-order modulation schemes.

Spatial diversity is available if the individual paths from the transmit to receive antennas fade more or less independently. This implies the probability of all signal paths fading at the same time at the receiver is significantly smaller than the probability of a single path experiencing fading (Figure 1.24). Spatial demultiplexing at the receive antennas separate the interfering streams. This improves signal processing capabilities at the same time because the overall MIMO processed signal exhibits less instances of deep fading than a single received signal. However, joint processing of multiple spatial streams may be required, and this increases receiver complexity and power consumption. The coupling relationship between signals received at the antennas (i.e., path correlation) determines the difficulty in recovering the individual data streams.

Using MIMO, the overall signal power is split equally among the transmit antennas. Accurate channel knowledge may be needed. The receiver decoding complexity is a critical factor in determining the actual performance. By selecting a

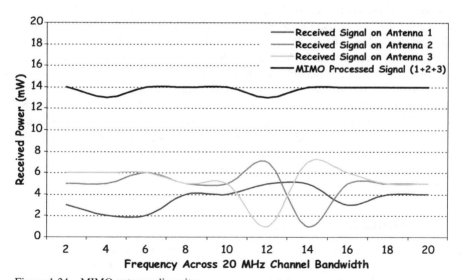

Figure 1.24 MIMO antenna diversity.

smaller set of receive antennas, fewer receiver chains are required. Depending on the number of receive antennas chosen, this may be done at the expense of a small sacrifice in capacity compared with the case when the full set of receive antennas are employed. Antenna selection is an exhaustive process of searching for the best pair of transmit and receive antenna combination that either gives the best SNR/ range (for diversity) or spectral efficiency/capacity (for spatial multiplexing).

The number of parallel spatial streams (also known as layers or codewords) is determined by the rank of the channel matrix (H). The rank of H indicates the number of linearly independent rows or columns. If the rank of a channel is r, at most r, spatial streams can be transmitted successfully. More generally, the rank of the MIMO channel matrix is governed by:

$$\text{Rank}(H) \leq \min(N_T, N_R), \tag{1.3}$$

where N_T and N_R are the number of transmit and receive antennas, respectively. Clearly, the spectral efficiency and channel capacity improve in proportion to the number of antennas used at both the transmitter and receiver. The pioneering research of Foschini [2] and Telatar [3] shows that in theory, MIMO capacity can increase linearly with min (N_T, N_R) or the number of antennas. This is in contrast to the logarithmic increase in capacity with SNR for SISO systems. If H is not full rank (i.e., rank deficient), some of its columns become linearly dependent (i.e., some columns can be expressed as a linear combination of other columns in the matrix). The information in these columns becomes redundant and does not contribute to the overall channel capacity. Thus, a rank-deficient MIMO channel matrix leads to low spectral efficiency and hence, low capacity. This is usually the case when signals arriving at the receiver are highly correlated in the fading characteristics. The correlation between the signal paths can be introduced by a small antenna spacing at the transmitter and receiver, a lack of scattering, or a strong LOS component. However, low fading correlation in the received signals does not guarantee high spectral efficiency. In spite of the low correlation, there are channels that still exhibit poor rank properties and hence, low capacity. Proper selection of a set of transmit antennas can avoid antennas that lead to linearly dependent columns. By redistributing transmit power to other antennas, a full rank channel matrix can be obtained.

Rich RF signal scattering ensures full MIMO spectral efficiency because independent paths are commonly present in such a multipath environment. The key is to efficiently map data symbols to various signals transmitted by individual antenna elements using a code matrix. In general, the rank and determinant of the code matrix determines diversity gain and coding gain, respectively. The diversity gain affects the bit error probability at high SNR and should therefore be maximized before coding gain.

1.6.4 Spatial Multiplexing

In spatial multiplexing, different data streams are sent simultaneously on different antennas to achieve a higher data rate. It is also known as Bell Laboratories layered space–time (BLAST) coding. Each antenna transmits an independently modulated

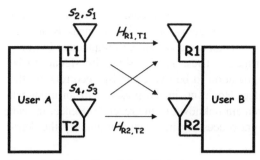

Figure 1.25   2 × 2 spatial multiplexing.

stream using a common channel bandwidth. Stream mapping is used if the number of data streams is less than the number of transmit antennas. The spatially multiplexed data streams are separated by the receive antennas using interference cancellation. Hence, the number of receive antennas should not be less than the number of transmit antennas. A 2 × 2 spatial multiplexing MIMO system is shown in Figure 1.25. This can be generalized to an $n \times m$ MIMO system. Data symbols s_1, s_2, s_3, and s_4 are mapped into a PSK or QAM constellation and applied independently to each OFDM subcarrier. The symbols are buffered to form the code matrix X that attempts to maximize the coding gain and channel capacity (data rate). In this case, the rate of X is 2 because four different symbols are transmitted in two symbol periods.

$$X = \begin{bmatrix} s_1 & s_2 \\ s_3 & s_4 \end{bmatrix}.$$

(1.4)

The general governing equation of matrices is given as

$$Y = HX + N,$$

(1.5)

where

Y = received signals, H = channel response

X = transmitted signals, N = channel noise.

If $H_{R1,T1} = 0.8$, $H_{R2,T1} = 0.2$, $H_{R1,T2} = 0.1$, $H_{R2,T2} = -0.8$, the channel matrix H for the 2 × 2 MIMO system becomes

$$H = \begin{bmatrix} 0.8 & 0.1 \\ 0.2 & -0.8 \end{bmatrix}.$$

In this case, the cross-coupled (self-interfering) signals are $H_{R2,T1}$ and $H_{R1,T2}$. A negative value for H indicates a signal phase reversal.

Full spatial diversity is not achieved with spatial multiplexing. The reduced diversity gain is due to a lack of redundancy across the antennas. This degrades the bit error performance and reduces the net data throughput at low SNR. In addition, the maximum data rate becomes limited due to path correlation, which increases as the number of transmit antennas increases. Finally, maintaining a consistent MCS is a challenge due to differing SNRs experienced by the receive antennas. Nonlinear receivers (e.g., maximum likelihood detectors) are complex to implement but may be needed to achieve better performance. These receivers require signals to be decoded jointly, allowing signal degradation to be overcome by decorrelating the streams using the channel knowledge. However, in the presence of an unknown interferer, the channel model may remain unknown to the receiver. Hence, instead of decorrelation across antennas, joint processing may actually amplify the detrimental impact of interference.

1.6.5   Space–Time Coding

Space–time coding (STC) was first introduced by Tarokh [4] from AT&T Research Labs in 1998. As the name implies, STC makes use of space (multiple antennas) and time domains in encoding/decoding data symbols. It combines channel coding (temporal diversity) with multiple antennas at the transmitter (spatial diversity) to improve transmission reliability and range. In general, STC optimizes diversity and range, whereas spatial multiplexing optimizes capacity. However, both methods require concurrent antenna transmissions in close proximity, as well as channel estimation at the transmitter or receiver. Recall that in spatial multiplexing, collocated antennas from the same device transmit different data symbols concurrently to increase the data rate. However, this is done at the expense of self-interference, which can be far more significant than weakened signal reflections resulting from multipath. Therefore, it is more difficult to recover the signal constellation in a spatially multiplexed MIMO system than a SISO system since there is a high probability of signal coupling in the data streams transmitted by antennas from the same device. Hence, spatial multiplexing demands better SNR than SISO to maintain the same BER.

   In STC, replica data symbols among multiple antenna signals are transmitted to maximize spatial diversity gain and minimize bit error rate. This redundancy improves the robustness and SNR/operating range even in fading scenarios, but may achieve a lower data throughput than a comparable spatial multiplexing system. Unlike spatial multiplexing, multiple antennas at the receiver are optional. This is an advantage in asymmetric links where high DL rates are crucial. It also simplifies receiver design with the use of fewer antennas. The high DL rates are achieved using higher-order modulations. This is in contrast to spatial multiplexing, which improves data rates using independent spatial streams. Well-constructed space–time codes achieve a fair amount of coding gain in addition to diversity gain. These codes are useful if the channel is unknown to the transmitter. They include space–time block codes, Trellis codes, and Turbo codes. If space–time block codes are used, then this gives rise to space–time block coding (STBC).

1.6.6   Alamouti Space–Time Coding

The Alamouti STC is a special case of STC using two transmit and one receive antennas (2×1). It is an example of a multiple input single output system (MISO)

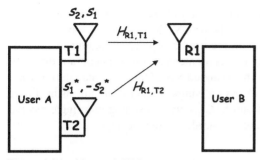

Figure 1.26   Alamouti STC.

that operates on pairs of consecutive OFDM symbols. In the first symbol interval, two OFDM symbols (namely s_1 from T1 and $-s_2^*$ from T2) are simultaneously transmitted from two antennas, where * denotes complex conjugation corresponding to a 180° reversal in phase. In the next symbol interval, signal s_2 is transmitted from T1, and signal s_1^* is transmitted from T2. This is illustrated in Figure 1.26. The code matrix becomes

$$X = \begin{bmatrix} s_1 & s_2 \\ -s_2^* & s_1^* \end{bmatrix}. \tag{1.6}$$

Since two data symbols s_1 and s_2 are transmitted in two symbol periods t_1 and t_2, there is no increase in capacity, and the rate is 1. The redundant symbols are added in space and time to improve diversity. Note that $XX^T = \alpha I_{2 \times 2}$, where X^T is transpose of X, α is a constant, and $I_{2 \times 2}$ is a 2×2 identity matrix. The receiver buffers the received symbols (r_1, r_2) for two symbol periods, which are given by:

$$\begin{bmatrix} r_1 \\ r_2 \end{bmatrix} = \begin{bmatrix} s_1 & s_2 \\ -s_2^* & s_1^* \end{bmatrix}\begin{bmatrix} h_1 \\ h_2 \end{bmatrix} + \begin{bmatrix} n_1 \\ n_2 \end{bmatrix} = \begin{bmatrix} s_1 h_1 + s_2 h_2 + n_1 \\ -s_2^* h_1 + s_1^* h_2 + n_2 \end{bmatrix}. \tag{1.7}$$

The receiver estimates the channel responses h_1 and h_2 via channel sounding. These responses correspond to $H_{R1,T1}$ and $H_{R1,T2}$, respectively, in Figure 1.26. Data preambles can be transmitted on alternate OFDM subcarriers from the two antennas. For example, the even subcarriers can be transmitted on antenna 1, whereas the odd subcarriers can be transmitted on antenna 2. Taking conjugates for r_2 (this operation does not introduce any correlation for the noise entities), we obtain:

$$\begin{bmatrix} r_1 \\ r_2^* \end{bmatrix} = \begin{bmatrix} h_1 & h_2 \\ h_2^* & -h_1^* \end{bmatrix}\begin{bmatrix} s_1 \\ s_2 \end{bmatrix} + \begin{bmatrix} n_1 \\ n_2^* \end{bmatrix}. \tag{1.8}$$

Thus, the recovered symbols s_1, s_2 are decoupled and can be decoded independently. This is an important feature in spatial stream recovery since the two spatial streams may encounter significantly different levels of signal attenuation and interference as they propagate through the wireless medium.

The Almouti STC achieves excellent performance even in highly correlated channels and is usable in the mobile mode. This is because the Alamouti STC employs orthogonal signals that work with a variety of channels. In addition, because only a single receive antenna is required, the best MCS can be chosen based on the prevailing SNR. If two receive antennas are used and placed a quarter of a wavelength apart, the independent fading on each antenna ensures maximum diversity due to the redundancy introduced by the Alamouti STC. Unlike spatial multiplexing, which requires joint multistream processing, single-stream decoding using linear processing is possible with the Alamouti STC receiver, resulting in a single receive chain. However, channel estimates are still needed at the receiver.

1.6.7 Beamforming MIMO Antenna Arrays

Conventional SISO and MIMO antenna types are designed to radiate laterally (not up or down) and this may not always be optimal in many deployments since the BS or AP is located at an elevated position compared with the user device in order to improve the signal coverage. For example, wireless APs are normally mounted above ceiling panels in enterprises. The use of active narrow-beam antennas offers the possibility of improved capacity at long distances and low power operation at short distances. Beamforming antenna arrays employ directional antennas that attempt to create LOS paths to the receivers so that the impact of multipath can be minimized. Dielectric lens and patch arrays on soft substrates can be used to produce low-cost antennas that are nearly omnidirectional in the azimuth direction but with narrow beams in elevation. Such antennas radiate energy only in a particular direction, providing gain along the intended direction and attenuation in the undesired directions. In doing so, a wireless coverage area can be broken down into smaller areas, each serviced by a directional antenna. This is equivalent to increasing the number of communication links without employing additional bandwidth, which increases system capacity. If only a single link is needed, the use of a directional antenna may reduce the transmit power and interference significantly, thus minimizing the overall energy consumption. Selection algorithms at the antennas assess the received signal strength and quality from different paths, giving the system a high probability of finding a path that is not corrupted by fading or interference. In addition, as can be seen in Figure 1.27, beamforming antennas are less dependent on correlated channels because a single channel estimate is sufficient for each link. This is in contrast to STC and spatial multiplexing, which require two or more channel estimates per link.

The use of beamforming antenna arrays is a promising technique to increase cell coverage and improve cell edge spectral efficiencies. Wireless links become implicitly distinct like in wired networks and can be physically separated and replicated using the same channel bandwidth. These arrays also provide a limited form of security by ensuring that the signal is directed at the legitimate station. Each pair of transmit–receive antennas may also use a different frequency channel to boost rates without causing self-interference. The antenna arrays can be made adaptive by using pilot signals on the UL and DL. When employed by the BS, such antennas can be used to track the locations of mobile users. Beamforming is normally applied on the DL and not on the UL. Since the antenna from the handset needs to be aimed at the correct antenna from the BS, it is difficult for beamforming systems to allow

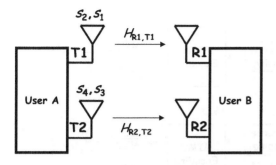

Figure 1.27   Beamforming antenna arrays.

new users to join a wireless cell or perform a handoff (or handover) when moving to a new cell. Thus, many handheld devices employ omnidirectional antennas for new user transmission. This implies that an access protocol is needed to resolve contention on the UL. A major drawback of beamforming is the so-called flashlight effect, where the measured channel quality becomes inaccurate due to the interaction of simultaneous active beams from neighboring arrays using the same frequency channel. Coordinated beamforming and switching has been suggested as a possible solution to this problem.

Beamforming antennas require more overheads in point-to-multipoint communications, since all arrays have to send the same information individually. For example, the same information has to be repeated periodically to all recipients or a specific group of recipients in broadcast or multicast operations. Although no additional bandwidth is required for this operation, the overall transmit power may be increased compared with unicast transmission. The need for more transmit power to service multiple links simultaneously is sometimes offset by the higher antenna gain that is available in the beamforming antennas. As the antenna gain is increased, the beamwidth becomes narrower. Very narrow antenna beamwidths—those associated with an antenna gain of 10 dBi or more—require automated antenna alignment. Beam-steering arrays can point the antenna beams in specific directions without manual intervention by dynamically adjusting the input voltage to each array to adaptively control the signal radiation angle. However, a beamforming protocol is needed to allow devices to discover each other, establish connections by aligning the beams, and coordinate operation in an efficient and interoperable manner. Such a protocol is critical to the performance of higher frequency systems (e.g., 60 GHz systems) since the beamwidths tend to be much narrower than lower frequency systems (e.g., 5 GHz systems) due to the higher gain. Once connected, antenna settings can be refined to maximize transmit and receive gains. As channels change, adjustments to the antenna settings can be made to optimize performance. This allows the highest data rates to be achieved, even with time-varying channels, such as those seen by mobile devices.

1.6.8   Downlink MIMO Architectures

Figure 1.28 illustrates two codeword options in DL MIMO implementation. In single codeword (SCW) or vertically encoded MIMO, only one encoder block or layer is

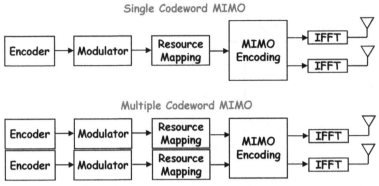

Figure 1.28   Single and multiple codeword MIMO.

required. Channel quality information (CQI) is used to indicate the quality of the DL channel as experienced by the user handset. For SCW, a single data stream is required, but spatial diversity is possible. Multiple codeword (MCW) or horizontally encoded MIMO employs multiple encoders or layers. In this case, CQI for multiple data streams are required. In addition, codeword-based successive interference cancellation to cancel out multiuser interference is needed at the receiver. For the same receiver type, the performance of the two MIMO implementations is almost the same.

1.6.9   Open-Loop and Closed-Loop MIMO

With open-loop transmit diversity, channel knowledge cannot be acquired at the transmitter due to the absence of an UL feedback path. Only CQI is available. Thus, open-loop MIMO is well suited for deployments that require high mobility speeds because UL feedback is restricted in these deployments. On the other hand, closed-loop transmit diversity allows channel state feedback to be sent by the handset when directed by the BS. Closed-loop MIMO requires CQI and channel state information (CSI) to be measured and sent back by the handset to the BS. This information is applied to a precoding procedure at the transmitter so that the BS can decide how best to modify the transmission, if needed. In this case, a robust signal format that uniquely identifies each transmitter is required before the signals are separated and demodulated. CSI is derived from channel sounding measurements. Examples of CSI are the channel matrix, the transmit rank indicator (RI), which indicates the number of useful spatial streams that can be spatially multiplexed based on the current channel conditions as experienced by the handset, and the precoding matrix indicator (PMI), which allows a handset to report to the BS its preferred precoding vector (normally selected from a codebook). With the additional information, closed-loop MIMO offers gains that may be comparable with using beamforming or adaptive MCS (Figure 1.29).

The handset can send quantized channel parameters to the BS to reduce the overheads. The PMI overhead can be measured by the feedback rate (in bits/ms/user) and the number of common DL pilots. Alternatively, both handset and BS can

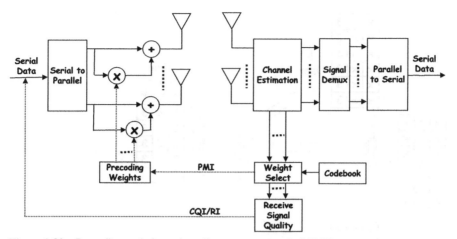

Figure 1.29   Precoding and channel quality measurement in MIMO.

predefine a set of codebooks, each containing an indexed set of coding matrices. For each channel rank, an index or value based on the codebook can be used, much like selecting a MCS from a table of MCSs. Thus, codebook-based closed-loop MIMO reduces the amount of feedback. Recall that the channel rank determines the number of simultaneous spatial streams that can be supported and should not exceed the number of transmit antennas.

1.6.10   Single-User and Multiuser MIMO

In single-user MIMO (SU-MIMO), only one user is communicating with the BS or AP. In multiuser MIMO (MU-MIMO), also known as collaborative or virtual MIMO, the BS or AP may communicate simultaneously with multiple users, thereby achieving higher aggregate rates compared with the single user case. User handsets listen to the pilot from the BS, which schedules the transmissions from different users to occur at the same time using the same channels. The overall network throughput is increased without the need for two or more transmitters at the user handset. However, the throughput of each user is not increased.

MU-MIMO is especially bandwidth efficient for the DL transmission due to the higher aggregation of traffic at the BS or AP. In addition, the BS or AP can efficiently allocate bandwidth resources based on channel feedback provided by the handset (e.g., transmit rank, PMI, CQI). Joint precoding can be centralized at the BS or AP, and cooperation between handsets is not required. MU-MIMO on the UL is challenging since the amount of cross-coupling needs to be estimated by the transmitting handsets, which may not be collocated. In addition, the BS or AP needs to estimate the channel responses of all independent spatial streams. Employing orthogonal UL transmissions (e.g., using the time or frequency domains) is a possible solution. A special case of such a solution is MU-SISO, where user handsets employ only a single antenna for transmission (Figure 1.30).

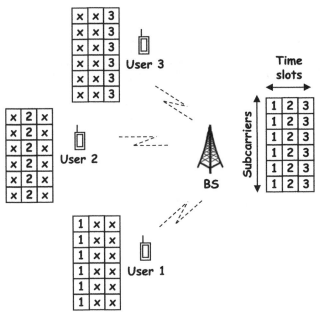

Figure 1.30   MU-SISO using nonoverlapping time slots.

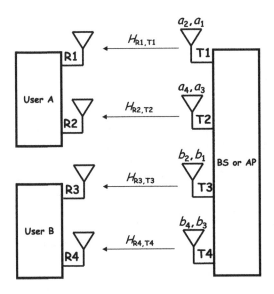

Figure 1.31   Downlink MU-MIMO transmission using beamforming antennas.

Since it is difficult to select the preferred spatial stream for different users using open-loop MIMO, closed-loop MIMO is recommended for MU-MIMO. Beamforming antenna arrays are capable of providing space division multiple access (SDMA) in MU-MIMO, where transmissions from multiple users can be divided in the spatial domain, and cross-coupling is significantly reduced. Figure 1.31 shows the operation of MU-MIMO using beamforming that is applied in the DL direction. Unlike DL beamforming, CSI may not be needed for UL beamforming. Note that MU-SISO results if only one transmit antenna is directed at the user handset.

1.7 INTERFERENCE

Interference is inherent in wireless transmission because it exists even when a single radio source is transmitting. Mitigating interference is a challenging problem in wireless networks because the transmitted signals cannot be shielded like in wired networks. In addition, the signal waveforms and transmission characteristics of the interfering sources may be unknown. Although transmit power control allows the radiation power to be utilized efficiently, this can be difficult to maintain in a fading or highly mobile environment. There are several forms of radio interference arising from known and unknown sources. The first category comprises self-interference caused by a single source (e.g., multipath propagation and multiantenna transmission) and multiuser interference caused by multiple sources sending information at the same time. In the latter case, because all users employ a known wireless transmission method, an access protocol can often be designed to resolve contention among the users, thereby transforming the multiuser interference into a single self-interfering source. The second category of interference may be generated randomly by radars (e.g., military and weather), malicious jammers, microwave ovens, and others. For licensed band operation (e.g., cellular networks), only known forms of interference may exist. For unlicensed band operation (e.g., wireless LANs), all interference types may exist. The carrier to interference plus noise ratio (CINR) is normally employed to characterize the channel quality in the presence of interference. CINR measures the fidelity of the desired carrier signal in the presence of interference, whereas SNR measures the fidelity of the demodulated signal. Since emerging 4G wireless systems are interference limited, CINR is widely used.

Unlike CDMA systems, where receivers attempt to cancel known interfering sources, 802.11 wireless LANs employ carrier sensing to avoid known and unknown interference sources and minimize contention. In this case, packet transmission is deferred when interference is detected. Unnecessary deferment may occur due to oversensitive sensing (e.g., sensing the transmission of a remote user), resulting in longer delays. In addition, the difficulty in achieving equal sensing and transmit ranges makes frequency reuse challenging for such networks.

1.7.1 Spatial Frequency Reuse

TDMA cellular users achieve orthogonality by transmitting at different times using different time slots assigned by the BS. An important feature of TDMA cellular systems is that the same set of frequency channels is reused after a far-enough distance (Figure 1.32). Frequency planning is therefore required, where a frequency channel is assigned to a BS in a specific cell or wireless coverage area. The bandwidth within a cell may be further subdivided into subbands, each served by a sector of the antenna at the BS. Hence, a hexagonal (honeycomb) cell is normally used to

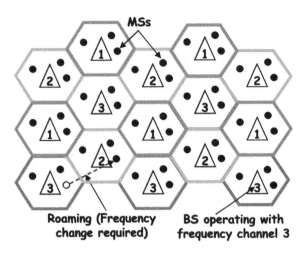

Figure 1.32 Three-cell frequency reuse in a sectorized TDMA cellular network.

represent six subbands in a single cell. When the user crosses a radio cell boundary, the assigned subband changes. Such spatial reutilization of bandwidth achieves a higher overall spectrum utilization by allowing multiple transmissions to take place simultaneously in different locations without causing excessive interference. A radio cell operating on one frequency channel is protected from the mutual interference arising in reused frequency channels by a surrounding ring of adjacent cells operating at other frequency channels. Ultimately, it is this mutual interference that limits the performance of TDMA cellular systems. In CDMA systems, the concept of frequency reuse translates to code reuse, where the same code is assigned to two users if they are far apart from one another (Figure 1.33).

A similar concept can be applied to indoor wireless LANs. However, because unlicensed frequency bands are employed in these networks, frequency planning is more complex and tedious due to the presence of known and unknown interfering sources that may change over time. With legacy 802.11 APs, manual site surveys are often conducted to calibrate the frequency channels according to the level of interference in each coverage area. However, newer 802.11 APs incorporate an ability to automatically select the best frequency channel based on channel measurement reports, thereby enabling plug-and-play operation. The channel may be switched dynamically as needed. The 802.11 user device will select the channel as dictated by the AP. In practice, channels that are spaced furthest apart (e.g., channels 1, 6, and 11 in the 2.4 GHz band) are used to reduce adjacent channel interference (caused by signal leakage from a neighboring channel).

The data rate selected by the AP (corresponding to a specific MCS) is typically based on the SNR of the wireless link at the receiver. A higher-order modulation, such as 64-QAM, is chosen if the SNR is high. The rate is continuously adapted to changing link conditions. If MIMO processing using spatial multiplexing is employed, rate selection is further complicated by the fact that multiple receive antennas on the same device may experience different SNRs. Since a single rate must be selected by all spatial streams originating from the same device, this implies

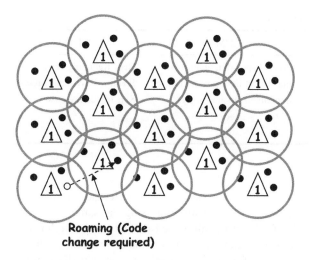

Roaming (Code
change required)

Figure 1.33   Spatial
frequency reuse in a
CDMA system.

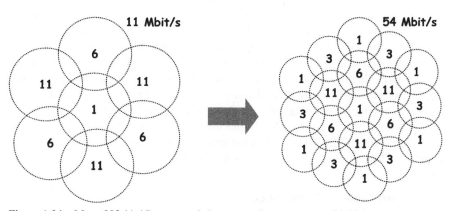

Figure 1.34   More 802.11 APs are needed to cover the same area with higher rates.

that the chosen rate is dictated by the lowest SNR from one of the receive antennas. Thus, the tradeoff between rate versus range is an important consideration in wireless LAN deployment. High data rates may only be relevant for small coverage areas. More 802.11 APs are therefore needed with denser coverage areas to carry more traffic (Figure 1.34). This implies greater cost, complexity, and interference with frequency reuse. In addition, more wires are needed to connect the APs.

1.7.2   Cochannel Interference

A common type of interference encountered in systems that employ frequency reuse is cochannel interference (CCI), which is caused by BSs or APs operating on

TABLE 1.4 CCI Mitigation Techniques

TDMA	CDMA	OFDMA
Control reuse patterns	Control spreading gain	Hard frequency reuse
Implement loose power control	Implement tight power control	Fractional frequency reuse
Implement guard times	Implement antenna arrays	Soft frequency reuse
Implement error control codes		
Implement antenna arrays		

the same channel. TDMA cellular users experience CCI interference from time-overlapping slots that originate from users transmitting using the same reuse frequencies in different radio cells. CCI can be minimized through proper frequency planning, but may not be completely eliminated. It is a major problem in direct sequence CDMA systems because the frequency channels are not spatially reused. In this case, CCI is a combination of multiaccess interference (MAI) due to transmission from multiple users, as well as ISI arising from multipath. The overall interference, which increases as the number of transmitting users increases, severely limits the capacity of these systems. It is important to realize that the performance of a spread spectrum receiver can vary widely even though the number of transmitting users is fixed. If the interfering users are located close to the receiver, MAI can be very large due to the higher receive power. Conversely, if the interfering users are far from the receiver, then MAI can be very small. This is known as the near/far effect, which can be fading induced. Thus, the geographic location of the receiver also influences the amount of MAI, and this changes as users move to different locations. Several CCI mitigation techniques commonly employed by TDMA, direct sequence CDMA, and OFDMA systems are listed in Table 1.4.

1.7.3 Multiuser Interference

Since MAI in direct sequence CDMA systems is highly structured, where spreading codes used by all users are known, interference cancellation techniques can be exploited. Examples include successive interference cancellation (SIC) and parallel interference cancellation (PIC) receivers that employ multiuser detection (MUD). These suboptimal linear detectors (usually implemented using a zero-forcing decorrelator) can completely eliminate MAI. They are an alternative to power control techniques that are used to overcome the near/far effect and channel fading, but the computational complexity may increase considerably with the number of users, especially when transmissions are conducted asynchronously. In addition to mitigating the near/far effect, MUD has the more fundamental potential of raising capacity by canceling multiuser interference. SIC appears to be simpler and requires less hardware than PIC. However, the processing delay presents the biggest drawback to SIC, since only a single user bit is decoded at each stage, and, thus, it takes at least x bit times to decode all users for each bit. In the case of PIC, it takes s bit-times (where s is the number of stages) to decode all users for each bit, and s is usually smaller than x. While increasing s improves the BER performance to a level similar to SIC, this may be achieved at the expense of excessive hardware complexity and processing delay. Both SIC and PIC schemes achieve better performance than the conventional detector. At the same time, they are simpler and less complex than the optimum detector. The processing speed of the decorrelator may limit the number of possible cancellations.

1.8   MOBILITY AND HANDOFF

Tracking a mobile user in a cellular environment creates additional complexity since it requires rules for handoff (or handover) and roaming. In addition, one must also cope with rapid fluctuations in the received signal power and traffic load variations that may change with user location and time. Handoff refers to the process of changing communication links so that acceptable channel quality and uninterrupted service can be maintained when users move across radio cell boundaries. Clearly, a handoff between two adjacent radio cells will change the number of active connections in each cell, and this in turn affects the traffic conditions and interference level in the cells. An effective handoff scheme must take into consideration three key factors:

- Propagation conditions
- Traffic load
- Switching and processing requirements.

It requires a neighboring BS to have a free channel with good signal quality and that the switchover be completed before any significant deterioration to the existing link occurs. The handoff process essentially comprises two stages namely,

- Channel quality evaluation and handoff initiation
- Allocation of bandwidth resources.

1.8.1   Intercell versus Intracell Handoff

When the user crosses two adjacent radio cells, an intercell handoff is required so that an acceptable channel quality for the connection can be maintained. Sometimes, an intracell handoff is desirable when the link with the serving BS is affected by excessive interference, while another link with the same BS (using a different antenna sector) can provide better quality.

1.8.2   Mobile-Initiated versus Network-Initiated Handoff

Network-controlled handoff (NCHO) has been widely used in first-generation analog cellular systems (e.g., AMPS). With NCHO, the channel quality is monitored only by BSs. The handoff decision is made under the centralized control of a mobile telephone switching office (MTSO). Typically, NCHO algorithms have handoff network delays in the order of several seconds, and have relatively infrequent updates of the channel quality from the alternate BSs. Mobile-assisted handoff (MAHO) is a decentralized strategy used in several digital cellular systems. The channel quality is measured by both the serving BS and the user. In TDMA systems, the user measures the signal strength during the intervals when it is not allocated a time slot. Channel quality measurements of alternate BSs are performed by the user. Link measurements at the serving BS are relayed to the user. The handoff decision is made by the user. The MAHO achieves the lowest delay (about 100 ms) and exhibits high reliability.

1.8.3 Forward versus Backward Handoff

Handoffs differ in the way a connection is transferred to a new link. Backward handoff algorithms initiate the handoff process through the currently serving BS. No access to the new link is made until resources have been allocated. The advantage of backward algorithms is that the signaling information is transmitted through an existing radio link, and, therefore, the establishment of a new signaling channel is not required during the initial stages of the handoff process. The disadvantage is that the algorithm may fail in conditions where the channel quality with the serving BS is rapidly deteriorating. This type of handoff is used mostly in TDMA cellular systems. Forward handoff algorithms activate the handoff process via a channel with the target (alternate) BS without relying on the current BS. Although the handoff process is faster, this is achieved at the expense of a reduction in handoff reliability. Handoffs can also be hard or soft. Hard handoffs release the radio link with the current BS at the same time when the new link with the target BS is established. Such handoffs are used in many TDMA cellular systems. Soft handoffs maintain radio links with at least two BSs. A link is dropped only when the signal level falls below a certain threshold. Soft handoffs are used in CDMA cellular systems.

1.9 CHANNEL ASSIGNMENT STRATEGIES

The assignment of frequency channels to BSs and to users is a fundamental operation of a mobile communication network. A classification of channel assignment schemes is shown in Figure 1.35. In fixed channel assignment, channels are assigned to cells for a relatively long period of time. This method obviously results in poor utilization of the available bandwidth when traffic patterns are bursty and change over time. Dynamic channel assignment represents the opposite extreme where channels are assigned to radio cells only when required. Such techniques can adapt to traffic load changes in real time but suffer from increased network management overhead. Between the extremes of fixed and dynamic channel assignment lie flexible channel assignment, channel borrowing, and hybrid channel assignment. Flexible channel assignment is essentially fixed assignment that is altered regularly according to predicted changes in the traffic load. Borrowing strategies can also be considered to be a variant of fixed channel assignment where channels not in use in their allocated cell can be temporarily transferred to congested cells on a

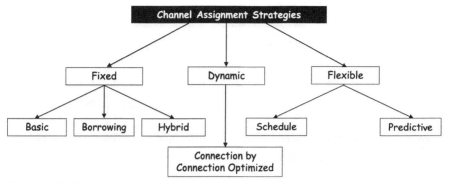

Figure 1.35   Channel assignment classification.

connection-by-connection basis. Hybrid assignment allocates a fraction of the channels according to fixed assignment and the rest according to dynamic assignment.

1.9.1   Medium Access Control Protocols

For wireless networks, sharing of bandwidth is essential because radio spectrum is
not only expensive but also inherently limited. For example, Verizon recently spent
$3.6 billion to purchase just 20 MHz of the advanced wireless services (AWS)
spectrum licenses from major U.S. cable companies to expand its 4G wireless services. This is in contrast to wired networks, where bandwidth can be increased
arbitrarily by adding extra cables. However, the broadcast nature of the wireless link
poses a difficult problem for multiuser access in that the success of a transmission
is no longer independent of other transmissions. To make a transmission successful,
interference must be avoided or at least controlled. Otherwise, multiple simultaneous
transmissions may lead to collisions and corrupted signals. This sharing task is made
harder when disparate traffic types created by multimedia applications are to be
transported across the network. A multiple access or MAC protocol is required to
resolve access contention among users and transform a broadcast wireless network
into a logical point-to-point network. The domains that contention resolution can be
achieved include time, frequency, code, space, or some combination. Note that in
general, packet reception consumes less resources than packet transmission because
there is no access contention in packet reception.

Defining a MAC protocol essentially implies the specification of a set of rules
to be followed by each member of the user population in order to share a common
bandwidth resource in a cooperative manner. The overall communications channel
can be divided into subchannels, which are then assigned to contending users. Typically, there are more users than the available subchannels, but only a fraction of all
users have packets to transmit at any given time. The central problem, therefore, is
to locate the users with data to send in order for these users to share the channel
efficiently. In general, MAC protocols can be categorized under contention, reservation, polling, and fixed allocation (Figure 1.36). A comparison of these protocols is
shown in Table 1.5. At one extreme, where no control is enforced, two or more users

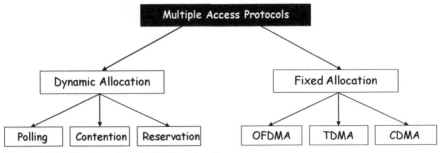

Figure 1.36   Classification of MAC protocols.

TABLE 1.5   Comparison of MAC Protocols

MAC protocol	Collisions	Control overhead	Idle time
Contention	Yes	No	No
Reservation	No	Yes	No
Polling	No	Yes	No
Fixed allocation	No	No	Yes

may transmit at the same time and conflicts may occur. These uncontrolled (contention) schemes are very easy to implement, but pay the price in the form of wasted bandwidth due to collisions. The other extreme is represented by a rigid system of static control (fixed allocation), where each user is permanently assigned a portion of the total bandwidth for its exclusive use. Such control places hard limits on the number of users sharing a given bandwidth. The class of dynamic control protocols (reservation and polling) dedicates a small part of the bandwidth for control, and this control information is used to determine the identity of the users with data to transmit. While the performance of contention and reservation techniques is dependent on the combined traffic from all users in the network, the performance of static allocation and polling schemes is strongly influenced by the traffic requirements of each individual user.

The multiple access capability of spread spectrum systems is distinctly different from narrowband systems. Not only can simultaneous transmissions be tolerated, the number of such transmissions should be large in order to justify the bandwidth spreading and achieve high network capacity. Thus, as long as different receivers are involved, MAC protocols in spread spectrum systems are normally designed to have multiple transmissions taking place at the same time. In this case, interference may sometimes dominate over noise as an error-producing mechanism.

1.9.2   Signal Duplexing Techniques

The duplexing scheme is usually described along with a particular MAC scheme. Half-duplex transmission is usually employed in radio transceivers due to cost and implementation considerations. For example, 802.11 transmission is half-duplex because either the AP is transmitting or one of the users within its coverage area is transmitting. Two half-duplex transceivers can be employed to achieve full-duplex transmission, each operating on an independent frequency band, one for the transmission and the other for reception. This introduces technical difficulties (e.g., saturation of the receiving antenna by the transmitter) and additional cost (e.g., separate radio hardware needed for each frequency band). An alternative is to employ duplexing methods, such as frequency division duplex (FDD) and time division duplex (TDD). These methods are commonly employed in cellular standards. The amount of spectrum required for both FDD and TDD is similar. The difference lies in the fact that FDD employs two bands of spectrum separated by a certain minimum bandwidth (guard band), while TDD requires only one band of frequencies for UL and DL transmissions. TDD has an advantage here, since it may be easier to find a single band of unassigned frequencies than two bands of frequencies separated by the required bandwidth.

TDD transceivers are half-duplex because the UL and DL bursts alternate on the same frequency channel but do not occur simultaneously. This is achieved by allocating a set of time slots to the UL and DL frames, with the ratio determining the relative UL and DL bandwidth requirements. This separation produces a delay, which is inherent in any time division techniques (e.g., TDMA) that employ a fixed frame structure. The DL frame is always transmitted first so that users at different locations can perform ranging operations on the UL to compensate for different signal propagation delays. The UL frame is transmitted after the DL frame. Sufficient turnaround time must be provided to allow the half-duplex transceivers to switch from transmit to receive and vice-versa.

In FDD, the UL and DL frames operate on nonoverlapping frequency channels, and can therefore transmit simultaneously (i.e., UL/DL frames can be coincident in time). FDD requires some frequency separation between UL/DL channel pairs to mitigate self-interference, but unlike TDD, no time gaps are required. Both TDD/FDD alternatives support adaptive burst profiles where MCS options may be dynamically assigned on a burst-by-burst basis. An important advantage of TDD over FDD is the ability to support asymmetric traffic loads since transmission time in TDD can be apportioned flexibly between the UL and DL. FDD systems may transmit at a longer range than TDD systems (or equivalently, requires lower transmit power). FDD also avoids the turnaround delay associated with TDD. Unlike TDD, a FDD BS may not use the UL channel information to enhance DL transmission.

Since FDD employs different frequencies in the forward and reverse directions of transmission, diversity antennas have to be employed at both the BS and the user. TDD, however, benefits from antenna diversity without the need for multiple antennas. This is attributed to the fact that for a given frequency, the signal attenuation in a radio channel is typically reciprocal. For example, a TDD system can have the BS select the best signal from its antenna when receiving a specific signal from a user. When the BS next transmits to that user, it can employ the same antenna with the same power budget. Note that while the channel is reciprocal in attenuation, this may not apply to the interference. Thus, dual antennas may still be needed for a TDD system.

Another antenna-related design consideration when selecting a duplex scheme is whether a duplexer is required. A duplexer adds weight and cost to a radio transceiver, and can place a limit on the minimum size of a handset. This is because dual channels imply more complex receivers than a single-channel system, where all receivers operate at the same frequency. TDD is a burst mode transmission scheme. During the transmission, the receiver is deactivated. Thus, TDD systems are capable of providing bidirectional antenna diversity gain without employing duplexers. However, in terms of equipment utilization, a TDD transceiver effectively remains idle half of the time. Although duplexers are employed by most FDD systems, they are not required in FDD systems employing TDMA since the transmit and receive time slots occur at different times. A simple RF switch performs the function of the duplexer, but is less complex, smaller in size, and cheaper. Such a switch connects the antenna to the transmitter when a transmit burst is required and to the receiver for the incoming signal.

Since FDD uses different frequencies for each direction of transmission, interference is not possible even if the timing on the two frequencies is not synchronized. However, on each TDD link, precise synchronization is required. Otherwise, overlapping transmit and receive bursts will result in a reduction of overall system capacity. Each user is synchronized to a common clock time reference, allowing transmissions by each user to arrive at the intended receiver at a time agreed upon by all users. Ranging estimates of the forward plus reverse propagation times are also required to allow each handset to compensate for differences between its distance to the BS. This is fundamental for establishing new connections, since the handset can normally detect the BS signal (transmitted at higher power and from an elevated location), but the BS may not be able to detect the signal from the power-limited handset.

1.9.3 Orthogonal Frequency Division Multiple Access

Orthogonal frequency division multiple access (OFDMA) has become a key enabling air interface for 4G wireless standards. Like OFDM, OFDMA is mitigates multipath fading and achieves efficient spectral utilization. Unlike OFDM, where all subcarriers within the channel are assigned to same user, OFDMA allows BSs to assign a subset of subcarriers to different users for a predetermined amount of time (Figure 1.37). This procedure results in more granular bandwidth allocation and can be used to optimize the total system throughput under specific constraints (e.g., packet delay, transmit power, etc.). In addition, user contention on the UL is avoided, whereas a separate MAC protocol is required with OFDM. For instance, the 802.11 standard employs OFDM together with carrier sense multiple access (CSMA) to resolve access contention, which is not required in OFDMA. Like CDMA, an OFDMA system achieves a frequency reuse of 1. In this case, different sets of subcarriers may be assigned to different users within a cell or to the same user moving across cell boundaries. Thus, all wireless cells employ the same frequency band but may employ different sets of subcarriers, much like CDMA networks using the same frequency band but different spreading codes. This means that unlike conventional TDMA cellular deployment that requires different frequency bands for neighboring

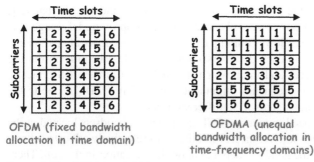

OFDM (fixed bandwidth
allocation in time domain)

OFDMA (unequal
bandwidth allocation in
time–frequency domains)

Figure 1.37   OFDM and OFDMA bandwidth allocation.

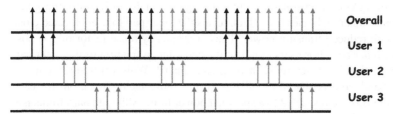

Figure 1.38   Grouped subcarrier allocation.

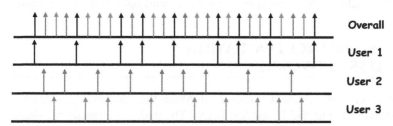

Figure 1.39   Permutated (distributed) subcarrier allocation.

cells, OFDMA allows overlapped cells to share the same frequency resource, which leads to a more flexible deployment, since frequency planning is not required. More importantly, fast handoffs in high-mobility communications can be achieved using the same frequency band in adjacent cells.

There are two general methods where OFDMA subcarriers can be allocated to different users. These methods provide frequency diversity that minimizes multipath interference. Figure 1.38 shows the grouped subcarrier allocation. Subcarriers may interact at the cell edge where coverage areas overlap. In this case, a group of subcarriers may collide. To further improve frequency diversity, permutated (distributed) subcarrier allocation allows non-contiguous subcarriers to vary from cell to cell (Figure 1.39). In this case, one or more subcarriers may collide at the cell edge.

OFDMA minimizes the transmit power from the user handset. Alternatively, it can allow more transmit power per subcarrier, thereby improving UL range performance. OFDMA requires a lower training overhead and subcarrier allocation can be dynamic (i.e., can vary in each data burst). OFDMA is less helpful on the DL because the power is shared, and the power per subcarrier from the BS typically remains constant. Time and frequency alignment is crucial in OFDMA due to the need to maintain orthogonal signals from different users.

OFDMA networks need to manage CCI. In particular, handsets located at the cell edge can be prone such interference. A solution is to selectively allocate subcarriers, rates, and power to the handsets depending on their location in the cell. There are three major frequency reuse patterns for mitigating intercell interference: hard frequency reuse, fractional frequency reuse (FFR), and soft frequency reuse. Hard frequency reuse splits the overall bandwidth into a number of distinct subbands so that neighboring cells transmit on noninterfering subbands. FFR splits the given bandwidth into inner and outer parts. It allocates the inner part to users located close to the BS with reduced transmit power. Thus, a frequency reuse factor of 1 is possible for the inner part. For users closer to the cell edge, a fraction of the outer part is employed with a frequency reuse factor greater than 1. With soft frequency reuse, the overall bandwidth is shared by all BSs (i.e., a reuse factor of 1), but for each subcarrier, the BSs are restricted to a specified power limit. Although hard frequency reuse is simple to implement, it suffers from reduced spectral efficiency. On the other hand, soft frequency reuse achieves high spectral efficiency but requires the coordination of multiple BSs, as well as accurate channel reports and interference measurements at the cell edge. FFR is a compromise between hard and soft frequency reuse.

1.10 PERFORMANCE EVALUATION OF WIRELESS NETWORKS

We present several measurement studies to illustrate the performance of single-antenna and multiantenna 802.11 wireless LANs. The studies are based on a single wireless connection between an AP and a laptop, both capable of supporting up to two spatial antenna streams. Figure 1.40 shows that at short distances, the rate for 802.11n is about 2.4 times greater than 802.11a/g. This is because 802.11n employs spatially multiplexed MIMO as opposed to SISO in 802.11a/g. However, at longer distances, the rates for 802.11a/g (54 Mbit/s maximum rate) are actually higher than 802.11n (maximum rate is set to 130 Mbit/s for this experiment). In all cases, the same channel bandwidth of 20 MHz is chosen. Thus, the gains for using multiple spatial streams may be relevant for short distances. At longer distances, single antenna systems achieve better throughput performance. The key factor that degrades MIMO performance at longer distances is the need for link adaptation when independent antenna branches collocated on the same device experience varying signal levels.

While mobility issues assume lesser importance in wireless LANs than in cellular systems, adjusting the MCS to accommodate different operating distances is a difficult challenge due to varying channel and interference conditions. As shown

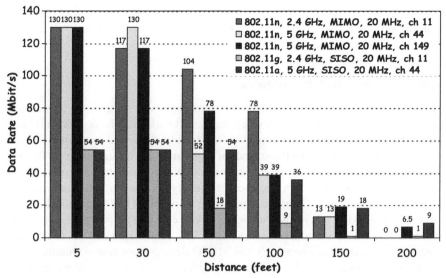

Figure 1.40   PHY data rate versus range (enterprise environment).

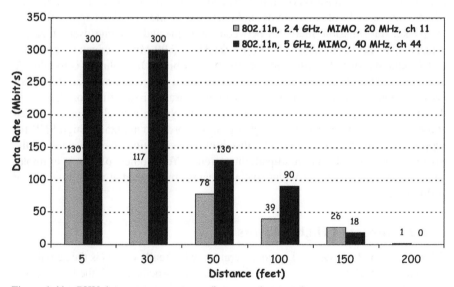

Figure 1.41   PHY data rate versus range (home environment).

in Figure 1.40, the rate of a 5 GHz radio can be better or worse than a 2.4 GHz radio due to different levels of interference in a dense office environment. However, a 2.4 GHz radio typically achieves a longer range than a 5 GHz radio for the same transmit power. This is exemplified in Figure 1.41, which shows the range performance in a typical single-family home. In this case, despite using a wider channel bandwidth of 40 MHz, the operational range for the 5 GHz radio is shorter than the

2.4 GHz radio. Similarly, the use of MIMO may only be effective at short distances (e.g., at 100 ft or less). A higher transmit power can improve range (e.g., using channel 149 instead of channel 44 in the 5 GHz band).

1.10.1 Impact of Link Adaptation

The use of a broad and dynamic range of transmission rates within a single coverage area and frequency channel may severely impact network performance. For instance, a few users transmitting continuously at low rates at the fringe areas of reception may degrade the overall data throughput of high-rate users located closer to the AP. The problem exists in both forward and reverse directions of transmission, and is due to data packets from high rate users frequently waiting for lower rate packets to complete transmission. This is caused by the native 802.11 MAC protocol, which is based on CSMA, where user devices or APs can initiate packet transmission whenever the wireless medium is sensed to be idle. Similarly, a mixture of low rate legacy SISO devices and new high rate MIMO devices operating within the same distance and channel may also lead to this problem. As an example, suppose a low-rate user is transmitting 500-byte packets at 1 Mbit/s (4 ms packet transmission time), whereas a high rate user is transmitting 500-byte packets at 130 Mbit/s (0.031 ms packet transmission time). The packets are transmitted continuously, which is typical of video streaming and peer-to-peer file download applications. If the packet transmission alternates between the high- and low-rate users, then the actual data rate for both users is 4000/(4 + 0.031) or 0.99 Mbit/s, which is very close to 1 Mbit/s, but over 130 times slower than 130 Mbit/s. One solution is to allocate additional airtime for the high rate users to transmit or receive a larger number of packets. This can be achieved by preallocating the transmission time for all users in the same coverage area using time slots. Another solution is to allow users located at the fringe areas to transmit at a higher power in order to maintain a high SNR for supporting a higher rate. However, this high-power option may be done at the expense of greater CCI and multipath interference. Yet another solution is to implement MU-MIMO, where different spatial streams are directed at users transmitting at different rates.

1.10.2 Impact of Higher Layers

While the PHY data rate is a key consideration, what really counts is the actual data throughput, which takes into account the overheads associated with the lower layers (i.e., PHY and MAC). The transmission control protocol (TCP) is a bidirectional network protocol that incorporates congestion and flow control using a sliding window mechanism to minimize network congestion. In doing so, device buffer overflows leading to packet loss are avoided. Thus, the end-to-end TCP throughput is highly dependent on how fast the window rotates. The maximum TCP throughput is affected by the roundtrip propagation delay and the congestion window size. The maximum throughput can be estimated by the ratio of the maximum window size over the roundtrip time. For example, if the maximum window size is 65,535 bytes (or 64 KB) and the roundtrip time is 0.5 second, then the maximum TCP throughput becomes limited to about 1 Mbit/s. The roundtrip time is dependent on the span of the network. Clearly, it is important to match the maximum TCP throughput with the available PHY and MAC rates. Unlike TCP, the user datagram protocol (UDP) is not subjected to such a throughput limit, since it is a unidirectional protocol.

Since TCP or UDP are designed for wired networks, there are several performance issues to consider when applying these protocols on wireless networks. This is because wireless networks encounter higher error rates, lower bandwidth, and more frequent signal outages than wired networks. For example, the performance of TCP over a wireless link may be affected by TCP's congestion avoidance algorithm, which is controlled by the sender. The algorithm is optimized for networks that experience low packet loss. Hence, TCP makes the implicit assumption that packet losses are a result of network congestion. However, pauses during handoffs (when users move among different wireless coverage areas) can be perceived as periods of heavy losses by the transport layer, causing retransmission timeouts. In addition, excessive retransmissions over the wireless link can lead to retransmission timeouts at the TCP layer. In both cases, TCP reacts by drastically reducing the current transmission rate. First, TCP reduces the transmit window size to restrict the amount of data flowing through the network. Second, it activates the slow-start mechanism that limits the increase in the data rate to the same rate that ACKs are received. The overall effects of the slow start algorithm is to increase the time required to reach the maximum throughput, both at the beginning of a TCP session and whenever a data segment is lost or corrupted. Finally, TCP resets the retransmission timer to a backoff interval that doubles with each consecutive timeout. These measures are effective in restricting the impact of congestion on the network. However, TCP takes a long time to recover from a transmission rate reduction, resulting in severe throughput degradation. Fortunately, such problems can be mitigated using a lossless link layer that provides local retransmissions using per-packet ACKs. In this case, a transmitter resends a packet after a timeout if it does not receive an ACK from the receiver. The 802.11 standard adopts this approach so that the wireless link becomes loss-free with a reduced effective bandwidth. As a result, most of the packet losses seen by the TCP sender are caused by network congestion due to buffer overflows at the user handset, AP, or other interconnected devices along the network path. Like TCP, UDP can also operate reliably over wireless networks that employ per-packet ACKs at the link layer, including 802.11 networks.

The hypertext transfer protocol (HTTP), based on the native TCP standard, is one of the most enduring network protocols of all time, providing a bidirectional mechanism to support Web browsing and more recently, high-quality audio/video transport. In addition, HTTP has been widely used to enable remote management and configuration of different Internet access devices, such as modems, routers, gateways, set-top boxes, and voice-over-IP (VoIP) phones. Figure 1.42 shows that the HTTP throughput can be reduced by as much as 50% of the PHY rates when streaming a 1080p high-definition (HD) video. The PHY rates are selected based on the received signal strength and not on the bandwidth demanded by the application. While the overheads of the 802.11n PHY and MAC can be substantial, optimizing the HTTP window size (and hence the HTTP throughout) is also important in ensuring that the PHY rate utilization is maximized. Figure 1.43 shows the impact of MAC layer ACKs on the HTTP throughput. As can be seen, when local packet ACKs are disabled, more end-to-end retransmissions will be handled by the HTTP layer.

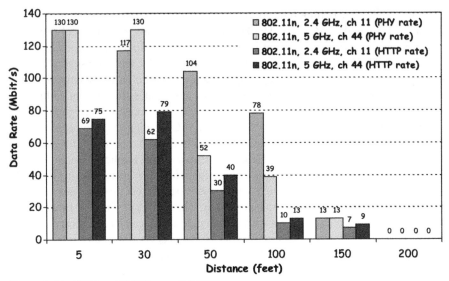

Figure 1.42 PHY and HTTP rates (MIMO).

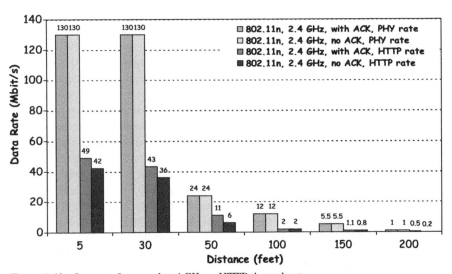

Figure 1.43 Impact of per-packet ACK on HTTP throughput.

This in turn reduces the transmit congestion window size and hence the available throughput.

1.10.3 Impact of Number of Antennas

The number of spatial streams supported by a device normally dictates the minimum number of antennas. For example, a three-stream device requires three or more antennas. The number of transmit antennas typically does not exceed the number of spatial streams. However, if there are more antennas than the number of spatial streams, the best antennas (e.g., antennas placed furthest apart) can be chosen among the available set of antennas to enhance the performance of transmission and reception. Suppose an 802.11n AP employs up to three spatially multiplexed streams, but the user device can only demultiplex up to two spatial streams. In this case, the maximum data rate of the wireless system will be limited to two streams. However, because there are three available antennas at the AP, it may select the two best antennas to transmit to the user device. Similarly, the AP may choose the two best antennas to receive transmissions from the device. Thus, the system performance can be better than an 802.11n AP using only two antennas.

1.10.4 Impact of Centralized Control

In 802.11 networks, data packets are transmitted asynchronously at the highest available data rate by the AP or user device whenever an idle wireless medium is detected. The AP provides a wireless interface for the user device to connect to the wired network but does not control the transmission from the device (i.e., APs and user devices are peers). In longer-range wireless access networks, including cellular networks, user handset transmission is tightly controlled by the BS. These networks also incur a greater signal propagation delay than 802.11 wireless LANs due to the longer operating range. Since TCP requires ACKs, both UL and DL transmissions are needed. Consider a file download application. The rate of TCP packets received by a user handset on the DL is inversely proportional to the request-grant cycle (RGC) duration (i.e., the turnaround time to receive a bandwidth grant from the BS) plus the TCP ACK transmit time. Suppose the RGC duration is between 3 and 5 ms, and each 40-byte UL TCP ACK takes a further 1.5 ms to send. Thus, the maximum packet rate in packets per second (pps) for the UL ranges from 154 to 222 pps, or equivalently, 49 to 71 Kbit/s. Assuming the maximum-length (1500-byte) TCP data packets on the DL, the same packet rate gives a maximum DL rate of between 1.848 and 2.664 Mbit/s. Thus, even if the UL rate is low, the DL can still achieve high TCP rates. The highest bandwidth utilization on a 1 Mbit/s UL and 10 Mbit/s DL becomes 7.1% and 26.64%, respectively. This implies that up to about five simultaneous TCP connections (running at the peak rate) can be supported on the DL. However, if the UL rate slows down to between 49 and 71 Kbit/s, then only 1 maximum rate DL TCP connection can be supported. Note that since UDP does not require ACKs, it requires only a unidirectional connection, and hence, the maximum data rate for the DL can be higher. For example, assuming the same RGC duration (3–5 ms) and a file download application, the highest UDP rate for the DL ranges from 2.4 to 4 Mbit/s, which corresponds to 200–333 pps.

1.11 OUTDOOR DEPLOYMENT CONSIDERATIONS

The very high frequency (VHF) band ranges from 30 to 300 MHz, and the the ultra-high frequency (UHF) band ranges from 300 MHz to 3 GHz. These bands are usually too high a frequency for ionospheric refraction or reflection of the radio signal back

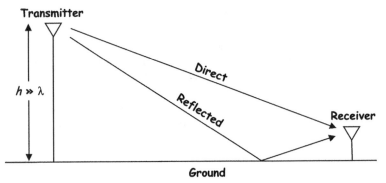

Figure 1.44   LOS signal propagation in VHF/UHF bands.

to earth. Therefore, LOS signal propagation is more appropriate for these bands. In this case, the transmitting antenna is elevated several wavelengths or more above the ground level, and the received signal is a summation of the direct and ground-reflected components (Figure 1.44). The available bandwidth is good for high-quality FM radio and TV channels, but the range of roughly 100 km restricts its use to local coverage. Propagation problems in these bands include reflection from ground and buildings, tropospheric refraction and scattering, diffraction over hilltops and buildings, and multipath scattering caused by buildings and trees. In urban settings, more physical obstacles are usually encountered, leading to signal reflections, whereas in rural areas, a larger cell coverage can be expected.

The earth bulges 12 ft every 18 miles (125 ft with buildings and trees). This becomes worse for shorter wavelength (or higher frequency) operation. To compensate for the Earth's curvature, 60% of the center of the first Fresnel zone (essentially an ellipsoid corresponding to the antenna radiation pattern with a circular aperture) should be kept free of obstacles for LOS operation. Signals traveling within the first Fresnel zone will add constructively at the receiver. Obstacles normally occur below the LOS path in outdoor transmission. Professional installation of outdoor antennas is recommended for LOS operation. Mobility may cause large-scale signal fading and the Doppler effect. The following sections provide simulation studies to illustrate the impact of fixed and mobile path loss models (based on Reference 5), single and multicarrier operation, and the modulation and operating frequency on wireless network performance.

1.11.1   Fixed Access Path Loss Model

The free space model can be represented as

$$P_l = -20 \log\left(\frac{\lambda}{4\pi d}\right),$$

(1.9)

where

d = distance between users (m)

λ = wavelength of radio signal (c/f_c).

The suburban model can be represented as

$$P_l = 20\log\left(\frac{4\pi \times 100}{\lambda}\right) + 10\left(a - bh_1 + \frac{c}{h_1}\right)\log\left(\frac{d}{100}\right) + 6\log\left(\frac{f_c}{2 \times 10^9}\right) - 20\log\left(\frac{h_2}{2}\right),$$

(1.10)

where

h_1 = BS height

h_2 = Handset height

a, b, c = Terrain dependent parameters.

Typical terrain values for fixed access are:

Terrain A (hilly-high tree density): $a = 4.6$, $b = 0.0075$, $c = 12.6$

Terrain B (moderate): $a = 4.0$, $b = 0.0065$, $c = 17.1$

Terrain C (flat-moderate tree density): $a = 3.6$, $b = 0.005$, $c = 20.0$.

As can be seen from Figure 1.45, for the same transmit power, bandwidth, and data rate, the suburban B model incurs higher signal attenuation than the free space model, significantly reducing the operating range. Single carrier transmission is used in this case.

1.11.2 Mobile Access Path Loss Models

The pedestrian model can be represented as

$$P_l = 40\log\left(\frac{d}{1000}\right) + 30\log\left(\frac{f_c}{10^6}\right) + 49.$$

(1.11)

The vehicular model can be represented as

$$P_l = 40 \times (1 - 0.04h_1) + \log\left(\frac{d}{10^3}\right) - 18\log(h_1) + 21\log\left(\frac{f_c}{10^6}\right) + 80.$$

(1.12)

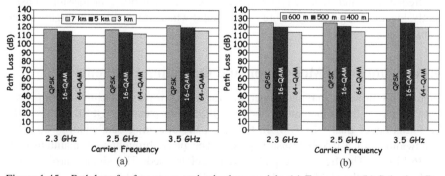

Figure 1.45   Path loss for free space and suburban models. (a) Free space. (b) Suburban B.

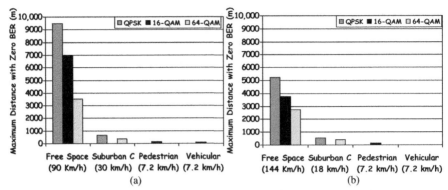

Figure 1.46   Maximum distance performance for different channel models. (a) Path loss only. (b) Path loss plus multipath model.

Figure 1.46 shows the performance of the mobile access models when compared with the fixed access models. The performance of these models is also evaluated in the presence of a multipath channel model. In this case, there is a fixed transmitter communicating with a stationary or moving station. The configurations are OFDM transmission, 2.5 GHz carrier frequency, 20 MHz bandwidth, and 0.5 W transmit power. The suburban C path loss model is chosen. The multipath channel model has an rms delay spread of 400 ns, giving a coherence bandwidth of 2.5 MHz. The Doppler spread at 2 m/s, and 2.5 GHz is 16.67 Hz. The peak throughput is 0.8 Mbit/s for the vehicular channel model and 1.136 Mbit/s for all other channel models. As expected, the maximum distance for error-free transmission using the vehicular model is much shorter than the rest, due to the higher signal attenuation. When multipath is taken into account, the performance of all models degrades.

1.11.3   Single Carrier and Multicarrier OFDM Comparison

The delay performance of single carrier and multicarrier transmission is shown in Figure 1.47. The free space path loss model is employed with a 20 MHz bandwidth, 0.5 W transmit power, and a peak throughput of 1.136 Mbit/s is maintained for all frequencies and distances. As can be seen, multicarrier transmission consistently incurs a longer delay (more than double) when compared with single-carrier transmission.

1.11.4   Impact of Modulation and Operating Frequency

Figure 1.48 shows the maximum operating range for error-free transmission improves with a more robust modulation, such as QPSK. For example, QPSK consistently performs better than 64-QAM. In this case, the operating bandwidth is 20 MHz bandwidth with a 0.5 W transmit power. The peak throughput is maintained at 1.136 Mbit/s. The free space path loss model is used. The single carrier mode performs better than OFDM mode due to the absence of a multipath model. However,

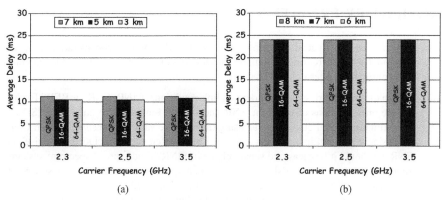

Figure 1.47   Single-carrier and multicarrier delay performance. (a) Single carrier. (b) OFDM.

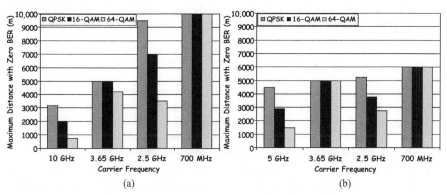

Figure 1.48   Single-carrier and multicarrier delay performance. (a) Single carrier. (b) OFDM.

using a lower operating frequency, such as the 700 MHz band, removes the differences in the range performance regardless of modulation.

REFERENCES

[1] Apple Computer, et al., *Report and Recommendations of the Millimeter Wave Communications Working Group to the Federal Communications Commission*, 1996, http://transition.fcc.gov/oet/dockets/et94-124/etiquette.pdf.

[2] G. J. Foschini, "Layered Space-Time Architecture for Wireless Communication in a Fading Environment when using Multi-element Antennas," *Bell Labs Technical Journal*, pp. 41–59, Autumn 1996.

[3] I. E. Telatar, "Capacity of Multi-antenna Gaussian Channels," *European Transactions on Telecommunications*, Vol. 10, pp. 585–595, 1999.

[4] V. Tarokh, N. Seshadri, and A. R. Calderbank, "Space-Time Codes for High Data Rate Wireless Communication: Performance Criterion and Code Construction," *IEEE Transactions on Information Theory*, Vol. 44, pp. 744–765, 1998.

[5] OPNET WiMAX Specialized Model, 2007, http://www.opnet.com/WiMAX.

HOMEWORK PROBLEMS

1.1. Why are equally spaced subcarriers used in OFDM? Which consideration is more important when increasing the number of OFDM subcarriers: to increase the data rate or to mitigate a larger delay spread?

1.2. Which of these metrics are dependent on link conditions: CINR, SNR, or BER? Evaluate the pros and cons of using the SNR and signal-to-interference-noise ratio (SINR) for estimating signal quality versus using the absolute value of the signal power. Consider single and multiple antenna systems. Can interference be included as "noise" in SNR, or should it be measured separately as in SINR? Since spread spectrum systems suppress interference, is SINR relevant for such systems? How can one measure SNR or channel quality without including interference? Hint: Will it be possible to measure the channel impulse response of known and unknown interference sources?

1.3. The Rician channel matrix can be represented by the following equation, where H_d is the direct LOS component channel matrix, H_s is the scattered component channel matrix (i.e., the Rayleigh channel matrix), P_d is the power of the direct component, and P_s is the power of the scattered component.

$$H = \sqrt{\frac{K}{K+1}} H_d + \sqrt{\frac{1}{K+1}} H_s \quad \text{where } K = \frac{P_d}{P_s}.$$

If P_d and P_s are normalized, and $P_d + P_s = 1$, derive a simplified equation for H based on P_s. Explain whether Rician fading (with a strong LOS path) occurs more frequently in an indoor environment than in an outdoor environment. How will such fading impact the performance of STC, spatial multiplexing, and beamforming MIMO systems? Which of these systems are more suited for non-LOS deployments?

1.4. Can carrier sensing be used to evaluate channel quality? If the transmission from one 802.11 user interferes with another 802.11 user, can the transmission be classified as an unknown interfering source since the content of the information generated by the source is random and unknown? Is carrier sensing needed for devices operating with beamforming antennas and 60 GHz frequencies? How will carrier sensing performance be affected if a station chooses channel 1 and another station uses channels 1 and 2? If two or more transmit antennas are used on the same device, must carrier sensing be performed on all antennas?

1.5. It can be shown that transmitting the shortest Ethernet packet (i.e., 64 bytes) achieves the lowest throughput because this incurs the most processing overheads compared with transmitting longer packets. Explain why this is so and whether it applies to both wired and wireless networks. The maximum Ethernet packet length for wireless networks is typically restricted to 500 bytes, whereas wired Ethernet networks are limited to 1500-byte packets.

1.6. Different equalizer types have their individual merits, but the important considerations are how much power they consume and the amount of training overheads. These considerations are related to the complexity in design. A key reason why OFDM is used in many wireless standards is to reduce the symbol rate to mitigate ISI, thereby simplifying the design of equalizers, possibly eliminating them in some instances. Recently, there has been interest in frequency domain equalizers. Single-carrier OFDM (SC-OFDM) is normally used in conjunction with frequency domain equalization. Frequency domain equalization in a SC-OFDM system is the frequency domain analog of a conventional linear time domain equalizer. For channels with severe delay spread, it is simpler than the corresponding time domain equalizer because of the FFT operations and the simple channel inversion operation. This technique can achieve better performance in bit error rate performance and PAPR compared with conventional OFDM. Explain how SC-OFDM can be used on the UL transmission to simplify the design of the user handset and shift the complexity of signal processing to the BS.

1.7. 4G wireless systems are designed to support a frequency reuse of 1 to enable the efficient use of valuable radio spectrum. However, this gives rise to more interference at the cell edge, where MSs and BSs must employ higher transmit power. Consider a point to multipoint network, with one BS managing several MSs. Transmissions between the BS and MSs are realized by means of OFDMA. Orthogonal subcarrier allocation on the UL will eradicate intracell interference between different MSs. However, since MSs A and B in the overlapped region can receive signals from two adjacent BSs, subcarriers may collide when the two BSs transmit at the same time on the DL. This is because a BS may not coordinate the allocation of subcarriers for a neighboring cell, which uses the same set of subcarriers. This reduces the success rate of handoffs and compromises the data throughput of the system. In addition, packet retransmissions will require more power from the user device and create higher CCI. Consider a new subcarrier allocation scheme to resolve collisions in the overlapped cell region. When there is a collision, the MSs will select the next subchannel with half the number of subcarriers allocated by the BS for the current subchannel. The subcarriers in the subchannel are selected in a random fashion. Explain whether this approach reduces the collision probability of the retransmission and the waiting time.

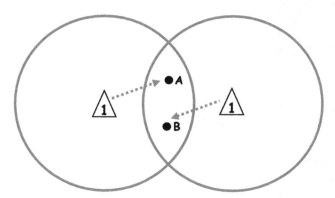

1.8. OFDM employs FEC to correct symbol errors at the receiver. The FEC is embedded in each transmitted packet, and this constitutes a recurring overhead. Consider a different method of FEC, where a dedicated FEC packet is transmitted after a group of N packets. Do you think this method will incur less overheads? What are the pros and cons? Can such a method be used to replace the need for data compression and reduce storage requirements? How can you modify this method to service real-time traffic involving voice and video?

1.9. In a wired LAN, only physically connected stations can hear each other. These stations are therefore authorized stations. However, this may not apply to wireless LANs. If a wireless LAN is connected to a wired LAN, how will this compromise the security of the wired LAN? Evaluate whether it is easier (and faster) to detect a denial of service (DoS) attack on a wired network than on a wireless network. Such attacks can be initiated by malicious users constantly transmitting information on the network.

	t_1	t_2	t_3	t_4
T1	s_1	s_2	s_5	s_6
T2	$-s_2^*$	s_1^*	$-s_7^*$	$-s_8^*$
T3	s_3	s_4	s_7	s_8
T4	$-s_4^*$	s_3^*	s_5^*	s_6^*

Figure A   Code matrix for four transmit antennas.

	t_1	t_2	t_3	t_4
T1	s_1	s_2	0	0
T2	$-s_2^*$	s_1^*	0	0
T3	0	0	s_3	s_4
T4	0	0	$-s_4^*$	s_3^*

Figure B   Code matrix for four transmit antennas.

	t_1	t_2
T1	s_1	s_2
T2	$-s_2^*$	s_1^*
T3	s_3	s_4
T4	s_5	s_6

Figure C   Code matrix for four transmit antennas.

	t_1	t_2
T1	s_1	s_2
T2	$-s_2^*$	s_1^*
T3	s_3	s_4
T4	$-s_4^*$	s_3^*

Figure D   Code matrix for four transmit antennas.

	t_1	t_2
T1	s_1	s_2
T2	$-s_2^*$	s_1^*
T3	s_3	s_4

Figure E   Code matrix for three transmit antennas.

1.10. (a) The code matrices in Figures A and B employ four transmit antennas with four symbol periods. "0" indicates no transmission. Find the rates. (b) The code matrices in Figures C and D employ four transmit antennas with two symbol periods. Find the rates. (c) The code matrix in Figure E employs three transmit antennas with two symbol periods. Find the rate. From the code matrices of Figures A–E, list the code matrices that allow all the received symbols to be decoded independently.

1.11. Suppose a 3×3 MIMO system (three transmitters) achieves a similar performance as a 4×2 MIMO system (four transmitters), which system would you choose? What are the advantages of combining MIMO with multiband OFDM where different antennas employ different frequency channels? What antenna spacing would you choose for STC, spatial multiplexing, and beamforming MIMO: 1/4 or 1/2 wavelength spacing?

1.12. Consider an Alamouti STC system with two transmit and two receive antennas. The second receive antenna is used for additional diversity. Express the receive symbols (r_1, r_2) as a function of the transmitted symbols (s_1, s_2). Repeat for the 2×2 spatial multiplexing and 2×2 beamforming MIMO systems. Evaluate the number of channel estimates for 2×2 spatial multiplexing, STC, and beamforming. If $h_1 = h_2$, derive the receive symbols for the three MIMO types. What is the rank of the channel matrix? Note that this corresponds to the case of correlated channels. Repeat for the case when the channels are reciprocal. Is antenna beamforming useful for 2×2 and 2×1 Alamouti STC? Will MU-MIMO perform better using spatial multiplexing, STC, or beamforming? How can Alamouti STC be employed in UL and DL MU-MIMO?

1.13. More robust modulations (e.g., QPSK, 16-QAM) tend to achieve the full spatial multiplexing gain almost all the time. However, 64-QAM tends to achieve less. Is this related to the operating range, with longer-range lower-order modulations providing a richer multipath scattering environment? Or is this due to receive antennas experiencing different SNRs? Are beamforming systems affected by this problem? Should unequal transmit power be allocated to the spatial streams? Strictly speaking, in MIMO, the total transmit power across all antennas should be the same as an equivalent SISO transmission.

1.14. What are the advantages and disadvantages for deploying a 2.4 GHz network with four surrounding cells instead of six in first ring? Assume all cells have the same area of coverage.

1.15. Identify the wireless cells in Figure 1.34 that will experience cochannel or adjacent interference in the 2.4 GHz band. Assume that the transmit and receive ranges are confined within each circle (i.e., transmitter power and receiver sensitivity are about the same). How will the interference pattern change if the transmit range is two times shorter than the receive range? In practice, the sensing range is longer than the transmit range because the receiver need not decode a signal when sensing the presence of interfering signals. What are the drawbacks for having a more sensitive sensing range? Note that as users roam between cells served by different APs, they will not be affected by the sensitivity of the carrier sensing mechanism since the relative signal strengths are evaluated, rather than the absolute signal values. Smaller cells may lead to less contention within each cell. Evaluate the advantages and disadvantages of using small cells with packet forwarding in each cell to improve the coverage area and network capacity.

1.16. In the figure below, analyze whether cochannel and/or adjacent channel interference are present in each 20 MHz channel of the 2.4 GHz band. For 802.11 APs operating on channels 3, 7, and 11, the operating transmit power is above the interference power. However, the AP on channel 1 is operating below the interference power. Rearrange the APs (without changing the transmit power) such that all APs operate at an optimum level above the interference power. The operating channels remain the same: 1, 3, 7, and 11.

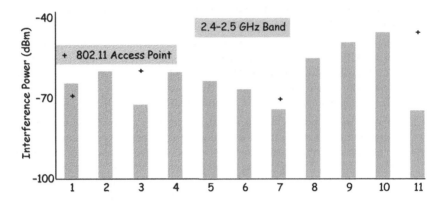

1.17. Show that 25 MHz is the minimum bandwidth to separate two 20 MHz channels in the 2.4 GHz band. For this bandwidth separation, show that carrier 6 is closest to carrier 1. Repeat for the case when 30 MHz is the minimum separation and confirm that carriers 1 and 7 are the closest. Determine the minimum separation to separate two 40 MHz channels and identify the appropriate carriers.

1.18. Are MIMO systems interference or noise-limited systems? How would you quantify the performance of MIMO systems using E_b/N_0? How will MIMO perform in an environment with a strong LOS signal, such as in an outdoor environment? Joint processing at the receiver for multiple spatial streams together with unknown interfering sources may have a detrimental impact. Explain whether channel sounding procedures can model the interference.

1.19. By implementing a MAC protocol to control channel access, only one user is transmitting at any one time in a wireless cell. Therefore, multiuser interference is reduced, although it may not be completely eliminated because of possible CCI. In this case, if there is a rich scattering environment with no LOS, STC MIMO performs well. Explain whether STC MIMO will perform well when a LOS path and rich scattering are both present.

1.20. Given an interference-prone environment in the 2.4 GHz band, which MIMO method will you use with OFDM: STC or spatial multiplexing? Will either method achieve higher data rates and/or better range performance than an OFDM system with no MIMO capability? Both OFDM and spread spectrum systems are able to mitigate the impact of fading. OFDM achieves this by using multiple subcarriers. Spread spectrum systems employ a wider signal bandwidth so that fading affects only a small portion of the bandwidth. In addition, spread spectrum systems are able to spread the power of unknown interference sources over a wide bandwidth, making them appear as low-power background noise. Like OFDM, the effective data symbol rate is reduced due to the need for bandwidth spreading. Explain whether OFDM is able to mitigate unknown interference sources, such as spurious emissions from microwave ovens in the 2.4 GHz band. Can MIMO systems help reduce the impact of these interference sources? What can you conclude about the effectiveness of MIMO-OFDM in mitigating frequency-selective fading, multiuser interference (including CCI), and unknown interference sources?

1.21. For wireless access networks covering several miles, are licensed bands preferred over unlicensed bands? For wireless communications in a large convention hall or an airport terminal with many active devices, would you expect a licensed spectrum, low bandwidth channel (e.g., 5 MHz channel in the 2.3 GHz band) to provide a higher bit rate than an unlicensed spectrum, higher bandwidth channel (e.g., 20 MHz channel in the 2.4 GHz band)? Evaluate whether carrier sensing and MIMO (requiring channel estimates) are useful for operation in these bands. How will the performance of open and closed loop MIMO systems differ for licensed and unlicensed bands?

1.22. When spatially multiplexing two data streams over two transmit antennas, would you recommend the use of one common channel accessible by both antennas (e.g., a single 40 MHz channel) or two independent channels (e.g., two 20 MHz channels), each channel assigned to an antenna? If one of the two channels experiences low SNR, whereas the other channel maintains high SNR, how will this impact the spatial multiplexing gains? Will it be better off to transmit a single stream on a single channel using one antenna? Suppose, instead of low SNR, one of the channels is rank deficient. Will single antenna transmission be more desirable in this case? Note that these antennas are collocated on the same device. Explain whether the flashlight effect will occur in a spatially multiplexed MIMO system.

1.23. Which system performs better: using OFDM to service two different users simultaneously with two 20 MHz channels or using OFDMA to service these two users asynchronously using a 40 MHz channel? Can OFDMA achieve better performance than MU-MIMO? Justify your reasoning in terms of the number of simultaneous users supported, multipath mitigation, and the achievable data rate or throughput. Hint: OFDMA may allow all channels of say the 2.4 or 5 GHz bands to be used simultaneously.

1.24. Beamforming enhances the bit rate performance, energy efficiency, and security of the spatial streams, and can help locate and identify user devices. Beamforming also allows the rate and quality of each stream to be adapted to link conditions even in multicast operations. Can beamforming be used to reduce multiuser interference in MU-MIMO with concurrent transmissions arising from the BS and directed to different users? Will beamforming be useful for the UL transmission? If beamforming is used on the DL but not on the UL, how will this impact the overall performance on the UL? If beamforming is used on the UL, is a MAC protocol still needed? Note that if beamforming is possible on the UL, acknowledgment or channel feedback sent by the handsets can be used to improve link performance. This is in contrast to wireless systems that employ broadcast antennas to send information to multiple devices simultaneously. In this case, acknowledgments or channel estimates that are concurrently sent by two or more handsets will result in collisions.

1.25. Compare the use of SIC in MU-MIMO and CDMA systems. Are CDMA systems likely to support more concurrent transmissions than MU-MIMO systems? Since there is no bandwidth spreading in MU-MIMO, will this imply that it may be more prone to interference than CDMA systems?

1.26. 60 GHz systems experience high signal attenuation but virtually no multipath or CCI. These systems employ channel bandwidths that are more than a 100 times wider than 2.4 and 5 GHz systems (2.16 GHz vs. 20 MHz). However, a wider bandwidth may potentially imply more interference sources, which may not be modeled accurately. Would you recommend a single antenna system with rate fallback or a two-antenna system with no rate fallback? If the two-antenna system is chosen, identify the specific type of MIMO system (i.e., STC, spatial multiplexing, or beamforming). Would you consider spatial multiplexing and beamforming to be equivalent at 60 GHz? Repeat for the case when the 700 MHz band is employed.

1.27. Multipath propagation is not always a bad phenomenon since it reduces the probability of complete phase cancellation at the receiver. This diversity property is sometimes exploited by radio receivers to improve reception. However, it can also result in higher amplitude variation in the received signal. These effects are shown in the figure below. For simplicity, it is assumed that the received signal comprises the sum of the direct LOS signal and the reflected signals (each with a random amplitude and phase). Whether the sum of the signal components cancel or reinforce each other strongly depends on the difference in their phase angles. If this phase difference is near 180°, then the net result is a deep fade in the received signal. As can be seen, these fades are separated by about half a wavelength. At 2.4 GHz, this is approximately 0.06 m. If a user moves at 10 km/h (pedestrian speed) in this frequency band, fades can be expected every 22 ms. Clearly, the rate of fading is proportional to the velocity of the user motion. Compare the multipath phenomenon with the use of subcarriers in OFDM. What can you say about the use of more subcarriers per channel? How will this impact the use of multiple antennas? Note that in all cases, the frequency of the received signal remains unaffected by the phase or amplitude distortions of the reflected signals.

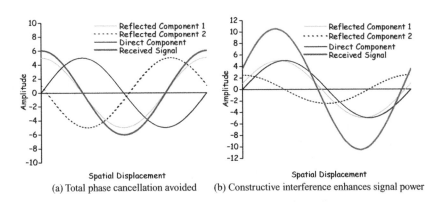

(a) Total phase cancellation avoided　(b) Constructive interference enhances signal power

1.28. For a home networking environment where network connectivity is required between a wireless router and multiple consumer devices, such as HDTV, game console, and smartphone, would you recommend the use of STC, spatial multiplexing, or beamforming MIMO antennas? Justify your answer. What are the benefits of employing MU-MIMO in such an environment?

1.29. First-generation wireless LAN products employ the 900 MHz ISM unlicensed band, which ranges from 902 to 928 MHz. The allowed power output is up to 1 W. FCC section 15.249 allows 50 mV/m of electrical field strength at 3 m, which corresponds to an EIRP of −1.23 dBm. The harmonics limit is one hundredth of the fundamental level, 500 μV/m, corresponding to an EIRP of −41.23 dBm. Evaluate the pros and cons of using this band compared with the 2.4 and 5 GHz bands.

1.30. Consider a MIMO system where the transmit and receive antennas are chosen based on the receive power level. How will such a system perform compare with a system that chooses the antennas based on the channel response? Consider another system whether the number of transmitting antennas is always less than the total number of available antennas at the transmitter. This minimizes the number of receiver chains to be implemented. The transmitting antennas are selected such that the SNR is maximized at the receiver. How will such a system perform compared with a system where all available antennas at the transmitter are employed, regardless of SNR? Explain whether switched antenna diversity at the receiver will perform better than the case when a single antenna is transmitting at the source and both receive antennas at the destination are turned on and the individual signals from both antennas are weighted (in a manner that is proportional to the signal strength or amplitude) and then added. This latter process, which amplifies the stronger signal and attenuates the weaker signal, is also known as maximum ratio combining. It is an example of a single input multiple output (SIMO) system.

1.31. The table below shows the product specifications of a legacy 802.11 wireless LANs. Clearly, a longer operating range leads to a higher delay spread, which reduces the data rate. What is the connection between the bit error rate and the packet error rate? Can a higher receiver sensitivity reduce the transmit power?

Bit error rate	$<10^{-8}$
Range in an open environment (100 bytes of user data)	1200 ft (2 Mbit/s)
	1400 ft (1 Mbit/s)
Receiver sensitivity	−90 dBm (2 Mbit/s)
	−93 dBm (1 Mbit/s)
Delay spread for a packet error rate of <1%	400 ns (2 Mbit/s)
	500 ns (1 Mbit/s)
Transmit power	15 dBm (32 mW)
Power consumption	Doze mode (9 mA)
	Receive mode (230 mA)
	Transmit mode (330 mA)

1.32. The figure below shows the range performance of a phased antenna array 802.11 switch operating in the 2.4 and 5 GHz bands. The customer premise equipment (CPE) uses a regular omnidirectional antenna and communicates with a central switch that can be located several miles away. Such a switch is useful for reducing the number of APs in office buildings and campus deployments. The antennas of the switch focus multiple narrow beams on the user devices. Since the same transmit power is used for the UL and DL, explain the differences in the operating range for the indoor and outdoor environments. Is carrier sensing useful for these systems? Why is the DL performance better than the UL in some cases but achieves the same performance at the maximum and shorter ranges? How would the performance of such a system differ from the case when the CPE uses beamforming arrays?

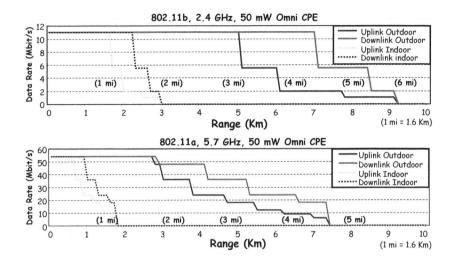

1.33. In the following multicarrier transmission method, a symbol (16-QAM) is split into four subsymbols (QPSK). The symbol interval remains constant. Comment on the bandwidth requirements of this scheme. Suppose four 16-QAM symbols are transmitted instead, using a symbol interval that is four times longer than the original symbol period (i.e., symbol rate is reduced four times). Comment on the bandwidth requirements of this scheme. What can you conclude on the tradeoff(s) between the number of subcarriers (representing lower speed symbols) and the spacing between the subcarriers?

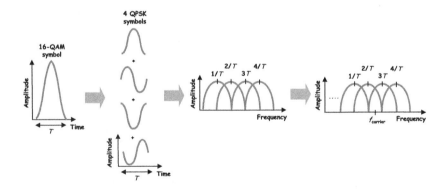

1.34. Beamforming antenna arrays may be combined with spatial multiplexing to achieve better range while maintaining high capacity. Explain whether STC, including Alamouti STC, can be implemented using beamforming antennas, and describe the potential benefits and drawbacks. Note that in this case, the number of receive antennas can be less than the number of transmit antennas. However, independent fading on the two paths may not be sufficient.

1.35. HomePlug power line communications technology employs 84 OFDM subcarriers ranging from 4 to 21 MHz. It enhances the OFDM basic functionality by dynamically comparing each subcarrier with the current characteristics of the power line. Specific subcarriers that experience high attenuation or line noise are dropped. Since noise spikes can be random and can occur at any time, FEC is required to reconstruct any bits that may be corrupted. Explain whether a similar OFDM method can be adopted for wireless networks.

1.36. The public Internet is a lossy network where any router along the path can drop packets if it becomes congested. Depending on the distance of the path, losses of 2–3% are typical, although over 10% can also be experienced. Consider the following FEC scheme that adds an error recovery packet to N packets that are sent, minimizing retransmissions and reducing the associated delay. The FEC packet contains information that can be used to reconstruct the missing packets. The method can made adaptive to minimize overheads. For example, if no losses are detected for a connection, FEC is disabled. Explain whether such a method will perform well in an error-prone wireless link.

1.37. A lot of Web content is text-based and therefore highly compressible. This will reduce the number of packets that need to be transported from the application to the user device, which will reduce the response time. All popular Web browsers support compression. Through the use of the accept encoding request in the HTTP header, the client can indicate it can receive compressed content. The compression algorithm (e.g., gzip and deflate) can be specified in the same header. Compare the effectiveness of such a method with the packet-level FEC method (previous problem) when applied to an error-prone wireless link.

1.38. OFDMA is useful for resolving contention on the UL channel. Since the DL is a broadcast channel from a single source (i.e., the BS), discuss the merits of using OFDM versus OFDMA on the DL.

1.39. Beamforming antenna arrays may extend the coverage of wireless systems. However, a broadcast synchronization channel (also known as beacon or pilot channel) is normally needed to ensure all handsets in different locations are synchronized to the BS or AP. The synchronization channel requires omnidirectional transmission so that new and existing users can easily connect with the BS or AP. However, if the user moves out of synchronization range, the connection is lost even if the data channel (using beamforming) is within range of the BS or AP. Explain how this problem can be solved to improve the coverage performance of a wireless system using beamforming antennas. Suppose the 60 GHz band is used for data transmission. Discuss the merits of using a 2.4 or 5 GHz channel for synchronization.

1.40. Energy efficiency determines the sustainability of wireless communications. Design a wireless network that will consume the least power to achieve reasonable data rate and operating range. Justify your solution in terms of the antenna method used, the frequency band, the possibility of imperfect estimation of time-varying channel statistics, and the modulation and error control methods. Consider the tradeoffs in using smaller cells, which require more cells to serve the same number of users in the same coverage area, hence requiring more packet transmissions (possibly accompanied by more handoffs) compared with an equivalent single cell operation. Although deployment cost may increase with more cells, lower power BSs, APs, and handsets can be used.

1.41. Layer-free networks enable flexibility in accessing and responding to a variety of measured wireless channel and transmission statistics. Explain whether such an approach will lead to incompatibility issues in wireless communications (one of the major objectives of the Open Systems Interconnect (OSI) layered model is to standard-ize the structure of network layers so that transmitting and receiving devices from different vendors can seamlessly communicate). Justify your reasoning from the per-spective of licensed and unlicensed frequency operation. Explain whether the CSMA protocol can be considered as a layer-free protocol since it uses interference information to transmit packets. Would you classify pure broadcast networks (e.g., satellite, GPS, and FM radio) as layer-free networks too?

1.42. Would you consider it odd that many CDMA cellular networks are deployed using licensed bands and yet interference cancellation is required in the CDMA receivers? Explain whether interference cancellation will work well in unlicensed frequency bands where unknown interference sources may be present. Is CSMA useful when used in conjunction with CDMA in unlicensed band operation?

1.43. Explain whether interference avoidance multiple access methods (e.g., CSMA) perform better or worse than interference driven methods (e.g., CDMA). Justify your answer in terms of the level of channel reliability (e.g., poor or good BER) and the level of packet losses (e.g., buffer overflow caused by excessive retransmissions).

1.44. Cellular networks are better at supporting mobile users but suffer from low efficiency due to the need to deal with unpredictable changes in the SNR resulting from fast movements. These SNR variations make it difficult to select the appropriate MCS. Describe the flaw(s) in this argument.

1.45. In a multipath environment, explain whether STC MIMO is spectrally more efficient than a spread spectrum system employing a single spreading code. Justify your answer for the single user case. Suppose a CDMA wireless handset uses different spreading codes to support concurrent transmissions from multiple antennas. In this case, how will the spectral efficiency compare with an equivalent STC system using the same number of transmit antennas as the number of spreading codes? Note that although both STC MIMO and spread spectrum systems generally perform well in a multipath envi-ronment, there is an upper limit to the interference they can tolerate.

1.46. Explain whether STC and spatial multiplexing can be implemented in the UL of a MU-MIMO system where two handsets are transmitting simultaneously, each equipped with a single antenna. Thus, each transmitter resides on a different device, whereas the receiver resides on the same device, which is a prerequisite for MIMO communications. The advantage of such a system is the lower transmit complexity than a single device transmitting using two antennas. However, reference pilot signals from different trans-mitters have to be orthogonal (unique) in order to recover the channel coefficients, but precoding is not needed for the pilot signals. Will such a system lead to lower transmit power and more efficient energy usage for each transmitting handset?

1.47. The effects of multipath change with distance and operating frequency. Optical infrared systems employ baseband or unmodulated wireless transmission. Explain whether these systems will suffer from multipath fading.

1.48. In general, the effect of delay spread is to cause the smearing of individual symbols in the case where the symbol rate is sufficiently low, or to further cause time-dispersive fading and ISI if the symbol rate is high. ISI is a form of self-interference that increases the error rate in digital transmission, an impairment that cannot be overcome simply by improving the SNR. This is because increasing the signal power in turn increases the self-interference. For a data rate of 2 Mbit/s with 2 bits/symbol (i.e., using QPSK), the symbol rate is 1 Msymbol/s. Compute the rms delay spread that will cause adjacent symbols to overlap by 0.1 symbol. Note that at higher symbol rates or larger delay spreads, the difference in delay among the various signal reflections arriving at the receiver can be a significant fraction of the symbol interval. Normally, a delay spread of more than half a symbol interval results in indistinguishable symbols and a sharp rise in the error rate.

1.49. TCP is designed to operate over a variety of networks using an "hourglass" network architecture, serving different PHY and MAC layers (e.g., low- and high-speed networks, wireless and wireline networks) over a common IP framework. TCP file downloads perform better over links with a low bandwidth-delay product. Thus, a higher download rate demands a lower delay. TCP is also sensitive to packet loss. Hence loss rates may be restricted to 10^{-5} to 10^{-7}. This is in contrast to VoIP flows that can tolerate delays of about 100 ms and packet losses of up to 1%. Explain whether the poor performance of TCP over wireless networks is due to the lower rates and/or the higher error rates. One of the remedies to recover from lost data segments quickly is to incorporate fast retransmit and fast recovery into the TCP protocol. The random early discard is a method that forces faster retransmits rather than timeouts. It selectively penalizes high throughput TCP sessions. Explain whether this method will work well with wireless networks. Repeat for the case when the sender is made aware of the existence of wireless links so that it knows some packet losses are not due to congestion. The sender can then avoid invoking congestion control on noncongestion-related losses. The buffer (queue) length is sometimes used as a metric to indicate network congestion. Explain whether this metric applies to a network that comprises an interconnection of wireline and wireless networks.

1.50. An early wireless LAN product employs sectorized horn antennas operating in the 18–19 GHz band (see figure below). The overall transmission coverage is split into six sectors, each covering a 60° segment. A similar method is widely adopted in cellular systems where each sector of the BS is used to service a number of subscribers. How does omnidirectional MU-MIMO transmission and beamforming compare with this approach? Explain your reasoning in terms of the number of simultaneous stations supported, the effectiveness of multipath mitigation, the need for intracell handoff, and the achievable data rate.

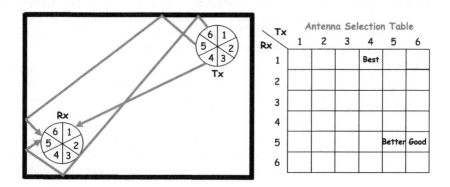

1.51. Many 802.11 APs allow the user to select a channel based on current interference conditions. For example, the user may evaluate active channels employed by several nearby APs and choose the channel that is least likely to encounter interference. Other routers go one step further and allow the user to fix the data rate (i.e., the MCS). Discuss the consequences of fixing the channel and data rate in an 802.11 wireless LAN. Note that fixing the MCS is effectively cutting off SNR feedback from the receiving device, much like UDP transmission. Since 802.11 employs asynchronous transmission at the maximum available rate with no centralized control, will TCP rates become similar to UDP rates?

1.52. In general, a more efficient code rate is normally chosen by a higher-order modulation. Since a higher-order modulation requires a higher SNR, will this imply that a more efficient code rate (e.g., 5/6) is more effective in correcting errors than a less efficient code rate (e.g., 1/2)?

1.53. A European hotel operates a 2.4 GHz 802.11 network using channels 1, 6, and 11 to serve international guests. It is suffering from high packet loss (5–35%), high packet latency (5–6 seconds), and spotty coverage. Evaluate the following solutions to improve the performance and coverage of the system:

- Switching some of the APs from 2.4 to 5 GHz band
- Employing four channels (i.e., 1, 5, 9, and 13) in the 2.4 GHz band
- Placing a smartphone near an AP with good signal strength and then relaying the 802.11 network connection over Bluetooth (also operating at 2.4 GHz) to the laptop
- Decreasing the AP receiver sensitivity (by reducing the coverage area from large to small)
- Increasing the minimum data rate from 1 to 2 Mbit/s
- Decreasing the transmit power from 20 dBm (100 mW) to 10 dBm (10 mW).

1.54. A big advantage of wireless networks operating at 5 GHz or lower is the ability to operate indoors as well as outdoors. Some measurements were conducted using the same hardware configurations to compare the performance of 802.11 in indoor and outdoor deployments. A wireless data rate of 60 Mbit/s at 300 m was obtained for an outdoor network, whereas a lower rate of 34 Mbit/s at a shorter distance of 93 m was obtained for an indoor network. Explain the substantial difference in the rate and range for the outdoor and indoor networks.

1.55. Etiquette transmission allows incompatible systems to co-exist. For example, the 802.11, Bluetooth, and Zigbee wireless standards may all co-exist in the 2.4 GHz band by sensing the wireless medium before data transmission. Are standards in etiquette transmission required to allow these systems to interoperate?

1.56. Having a reasonable span for a network is crucial for the success of any aspiring network technology. Some lessons can be learned from the approach taken by the wired Gigabit Ethernet Task Group more than 10 years ago. When the group realized that increasing the Ethernet clock rate by tenfold (which reduces the time needed to transmit an Ethernet frame by a factor of 10) directly shrinks the network span from 200 m (100Base-T) to 20 m, the IEEE 802.3z committee developed a mechanism to preserve the 200 m (660 ft) span. Evaluate the options for increasing the range performance of 60 GHz wireless systems without compromising the high data rate that is available in these systems.

1.57. Link adaptation using a set of MCSs may lead to fluctuations in wireless data rates and performance. Describe an antenna technology that can be used to maintain a fixed MCS even as the operating range or fading conditions change.

1.58. In a MU-MIMO network, is it possible for different users to transmit on the same time–frequency resource? If this is possible, spectrum efficiency is enhanced with more degrees of freedom than the case when a single user transmits using multiple spatial streams. In a cellular network, large number of users may potentially send data at the same time. How can the BS partition the users into different groups to reduce processing requirements, and how can spatial interference be reduced with precoding?

1.59. Cellular systems employ ranging operations to estimate the distance between the user handset and BS. To do this, the time of arrival or the power of the signal can be measured. Which metric is more accurate? Since more powerful error control is typically used for longer distance transmission, how will this impact ranging accuracy?

1.60. In general, for a 2 × 2 antenna system, 4 channel estimates are required for spatial multiplexing—2 for Alamouti STC and 1 for each beamforming link. Explain how these channel estimates affect the performance of the system. What is the channel rank for the Alamouti STC? ∎

CHAPTER 2

IEEE 802.11 STANDARD

The IEEE 802.11 standard has enjoyed unprecedented success over the past decade. 802.11 chipset shipments topped 1 billion in 2011, a threefold increase compared with 2010. The recent surge in demand for 802.11 chipsets is driven by a large array of sleek personal devices that require high-speed wireless connectivity. A decade ago, laptops replaced desktop computers as the main driver for 802.11. Smartphones and tablets are now beginning to replace laptops as the key driver for 802.11. The 802.11 standard is the most popular wireless standard to date and possibly the easiest to read. It is also known by the trade name of Wireless Fidelity (or Wi-Fi). The 802.11 wireless LAN is a cost-effective alternative to a wired Ethernet LAN in connecting end-user devices. Unlike wired networks, bandwidth can be reused in 802.11 wireless networks. 802.11 has become an increasingly important extension of cellular service. Cellular providers rely on 802.11 to offload voice, data, and video traffic from their crowded and expensive licensed spectrum to a wireline broadband connection. This also benefits users, who do not have to pay more when crossing the bandwidth caps enforced by these providers. Similarly, wireline providers with no cellular service are using 802.11 to expand their broadband service offerings in homes and public areas. This chapter provides a detailed coverage of 802.11, with an in-depth review of the key PHY and MAC technologies.

2.1 802.11 DEPLOYMENTS AND APPLICATIONS

802.11 deployments have grown significantly in indoor and outdoor access networks, including enterprise wireless LANs, home wireless networks, and large-scale public hotspots. 802.11 technology has also been deployed in trains, airplanes, parking meters, utility meters, smart-grid meters, and in innovative applications, such as water sprinkler controllers, RFID tags, and sensors.

Currently, there are about 750,000 802.11 hotspots in over 150 countries. During the Christmas period in 2009, Google provided free 802.11 hotspots in 47 U.S. airports and on every Virgin America flight. AT&T operates the largest 802.11 network in the United States. There are over 30,000 AT&T hotspots, with key locations that include Starbucks cafes and MacDonald's restaurants. AT&T also provides access to 190,000 hotspots worldwide. Its U.S. hotspot connections in public areas

Broadband Wireless Multimedia Networks, First Edition. Benny Bing.
© 2013 John Wiley & Sons, Inc. Published 2013 by John Wiley & Sons, Inc.

totaled more than 1.2 billion in 2011 (or over 3 million connections per day), four times higher than 2010. More impressively, AT&T's 802.11 hotspot subscribers are growing at a rate that is 25 times faster than its cellular subscribers. Cablevision, a major cable operator in the United States, spent more than $300 million to deploy an 802.11 public network with more than 10,000 access points (APs) spread across its New York, New Jersey, and Connecticut markets. This network covers coffee shops, train stations, parks, and beaches, and is complemented by 7,000 hotspots in central business districts affiliated with Cablevision customers. Like AT&T's hotspot service, the wireless service is free to Cablevision's broadband subscribers. Over 500,000 Cablevision customers have used the network, and 60% of those customers connect to it more than 10 times a month. The service offers 15 Mbit/s inbound and 4 Mbit/s outbound rates, and Cablevision launched a dedicated digital cable channel designed to help its customers find indoor and outdoor 802.11 APs.

The hotspot concept extends to a bigger scale and becomes cooperative with 802.11 community networks. For example, community hotspot builder FON (http://www.fon.com/en) offers free 802.11 gateways for users to share their broadband Internet connections with others in exchange for being able to use other users' connections for free when they are away from home. Subscribers may also make money by selling access to nonsubscribers. Unlike 802.11 community networks that are operated by residential users, the scale of such a network is no longer restricted to a single community in the neighborhood but is available worldwide. Using 802.11 in overseas hotspots is cheaper than roaming on 3G. iPassConnect offers its iPhone and iPod Touch customers access to iPass' 100,000 hotspots around the world via a single login. 802.11 public hotspots and community networks illustrate a key strength of the technology: pervasive high-speed wireless access can be achieved without costly base stations or customer premise equipment. The success of 802.11 public hotspots is a remarkable turnaround compared with 10 years ago when the company that first brought 802.11 hotspot service to Starbucks cafes went bankrupt in 2001. Unlike ad hoc hotspot providers in the past, the current success of public hotspots can be attributed to service providers ensuring consistent quality of service. The Wi-Fi Alliance (WFA) [1] plans to streamline the 802.11 hotspot capability with an interoperable platform called Wi-Fi Certified Passpoint, allowing devices to automatically discover and authenticate Passpoint hotspots.

802.11 services are available in major airlines such as American Airlines, Delta, and AirTran Airways. The typical cost is $9.95 on flights of 3 hours or less and $12.95 for longer flights. Not to be outdone, train operators are also providing similar services. The San Francisco Bay Area Rapid Transit District (BART) signed a 20-year contract with Wi-Fi Rail to provide 802.11 services throughout the BART transit system and on all commuter trains. Major U.S. cable operator Comcast offers free 802.11 service for its Internet customers on 120 commuter train stations in New Jersey. In addition, customers can access thousands of outdoor and indoor 802.11 hotspots for no additional fee in Philadelphia, New Jersey, and Delaware (see http://www.comcast.com/wifi). The National Express East Coast offers 802.11 services in 43 trains that ply a 950-km route between London and Inverness (Scotland). Deutsche Bahn offers 802.11 services in trains plying Dortmund, Cologne, Stuttgart, Munich, Frankfurt, and Hamburg. The Eurostar offers 802.11 services on most of its train stations in major cities in the United Kingdom and Western Europe. 802.11 services on trains are also planned.

Municipal 802.11 networks have evolved into more reasonable business models to include local government tenants and academic institutions. This approach is in contrast to the failed municipal businesses of Earthlink and MetroFi in early 2008. Silicon Valley Power purchased 802.11 assets from MetroFi in September 2008. The nonprofit electric utility is using these assets to help Santa Clara customers reduce power bills. Meters are read every 15 minutes and usage information is immediately relayed to the customer to reduce energy consumption during peak times or when rates are high. The system also offers free outdoor 802.11 service to the surrounding neighborhood.

2.2 802.11 TODAY

802.11 data rates have climbed from 11 to 300 Mbit/s within a decade. As a result, over 80% of smartphone users prefer data transfer using 802.11 than 3G or even 4G. Security is no longer an issue with Wi-Fi protected access (WPA). Since 2002, no practical attacks have been mounted against WPA. 802.11 networks have the potential to displace wired Ethernet LANs, especially in connecting user devices. In fact, many small businesses no longer use wired LANs. Five billion 802.11 chipsets have been shipped since 2000. In contrast, 370, 300, and 120 million chipsets were shipped in 2010, 2007, and 2005, respectively. The dramatic increase in shipments between 2010 and 2011 is driven by the demand for 802.11n and the rapid adoption of handheld personal devices, such as smartphones and tablets.

Due to the excellent work done by the WFA, APs, home gateways, adapters, and chipsets from different vendors are able to interoperate. This allows users to access different 802.11 networks without switching devices. Carriers can authenticate a user in a public 802.11 hotspot, secure the link, and coordinate with roaming partners for seamless handoffs between APs. Thus, 802.11-enabled personal devices can be used virtually anywhere: from office to public spaces to the home and even in different countries. 802.11 provides a unified network device management for building large-scale networks, including outdoor mesh networks supporting diverse user devices. It is a great example demonstrating the convergence of wireless, computing, and consumer electronics (CE). For example, Intel has embedded 802.11 in all its microprocessor chips since 2003. Many software operating systems can detect 802.11 networks automatically. In addition, 802.11 is the first wireless network interface to penetrate popular consumer devices, such as Internet-enabled HDTVs, digital cameras, Blu-ray players, and game consoles. This penetration has enabled multimedia content to be delivered seamlessly with the touch of a button. 802.11's penetration in the CE space is unlikely to be emulated by 3G/4G cellular and 802.16. When 802.11 is combined with portable personal devices, truly cordless communications can be realized. In this case, no power cord or network cable is needed since many new laptops, tablets, and smartphones can last over 8 hours with a single battery charge. In a nutshell, 802.11 access has become universally ubiquitous, as shown in Figure 2.1.

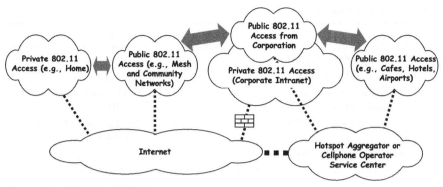

Figure 2.1   Ubiquitous 802.11 access.

2.3   IEEE 802.11 STANDARD

The IEEE 802.11 standard specifies wireless connectivity for fixed, portable, and moving stations in a geographically limited area. The standard is established by the IEEE 802.11 Working Group (WG) [2], which was formed in late 1989 and sponsored by the 802 Local and Metropolitan Area Networks Standards Committee (LMSC) of the IEEE Computer Society. Because the 802.11 transmission mode is packet based involving Ethernet frames (with 48-bit MAC addresses), hence it is sometimes known as wireless Ethernet. The IEEE 802.11 WG is subdivided into a number of task groups (TGs), each focused on establishing a specific amendment. The main standard is supplemented by alphabetical amendments. The alphabets of the amendments are issued according to the date that the TG was established. For example, TG a was formed before TG b and TG e was formed before TG i, and so on. An amendment can be ratified at a later date than the formation of the TG. For instance, 802.11e was ratified later than 802.11i. Currently, all amendments are based on single mandatory medium access control (MAC) protocol. The main standard is updated and consolidated every 4 years: 802.11ma (March 2007) and 802.11mb (March 2011). 802.11ma and 802.11mb are also known as 802.11-2007 and 802.11-2011, respectively. Amendments issued before the main standard are included in the main standard. For example, the 802.11a, b, d, e, g, h, i, and j amendments (all issued prior to 3/2007) are included in 802.11-2007, whereas 802.11n (issued on 11/2009) is not included, and needs to be studied in conjunction with 802.11-2007. However, 802.11n is included in 802.11-2011. The 802.11 standard and ratified amendments can be downloaded free of charge from Reference [3]. However, draft amendments are available to members only. Some of the key amendments are shown in Table 2.1. To avoid confusion, no TG was named 802.11l, 802.11o, 802.11x, 802.11ab, or 802.11ag.

The evolution of the 802.11 data rates (up till 802.11g) is shown in Table 2.2. The popular physical layer (PHY) transmission methods are orthogonal frequency-division multiplexing (OFDM), complementary code keying (CCK), and direct-sequence spread spectrum (DSSS). Frequency-hopped spread spectrum (FHSS) and

TABLE 2.1 802.11 Amendments

Ratified amendments		Amendments under development	
a	54 Mbit/s 5 GHz PHY (9/1999)	m	Standard maintenance
b	11 Mbit/s 2.4 GHz PHY (9/1999)	aa	Video transport streams
d	Compliance in multiple regulatory domains (6/2001)	ac	Very high throughput under 6 GHz
e	Quality of service (9/2005)	ad	Very high throughput at 60 GHz
g	54 Mbit/s 2.4 GHz PHY (6/2003)	ae	QoS for management frames
h	Spectrum and transmit power management at 5 GHz (9/2003)	af	Wireless LANs for TV whitespace
i	MAC security enhancements (6/2004)		
j	4.9–5 GHz operation in Japan (9/2004)		
k	Radio resource measurement (3/2008)		
n	Enhancements for higher throughput (11/2009)		
p	Wireless access in vehicular environments (7/2010)		
r	Fast basic service set transition (6/2008)		
s	Mesh networking (9/2011)		
u	Interworking with external networks (8/2011)		
v	Wireless network management (2/2011)		
w	Protected management frames (9/2009)		
y	3.65–3.7 GHz operation in the United States 6/2008)		
z	Extensions to direct link setup (9/2010)		

packet binary convolutional coding (PBCC) are optional PHY methods. PBCC is also known as high-rate DSSS. Note that spread spectrum is not used as a multiple access technique in 802.11 wireless LANs. Rather, it is used to protect data signals against the effects of multipath propagation and other signal impairments. 802.11a (approved on Sept 1999) was the first wireless standard to adopt OFDM and the first 802.11 amendment to employ the unlicensed 5 GHz band. Subsequently, the 2.4 GHz 802.11g OFDM PHY was ratified with the same data rates as 802.11a (i.e., 6 to 54 Mbit/s). Many 802.11g chipset vendors implement the highest data rate of 54 Mbit/s, even though this rate is optional. The DSSS/CCK rates in 802.11b are mandatory in 802.11g, and fall back to these lower rates are possible. Since 802.11g encompasses the earlier 802.11b DSSS amendment, it is common to refer the 802.11b/g combo as 802.11g. Similarly, 802.11n includes 802.11a/g. As can be seen in Figure 2.2, there is near exponential growth in the maximum available data rate since 1997. 802.11 really took off with the 802.11b amendment, providing a maximum data rate of 11 Mbit/s in the 2.4 GHz band. This rate was increased to 54 Mbit/s with 802.11a/g and then 300 Mbit/s with 802.11n (extensible to

TABLE 2.2 Evolution of Radio-Based 802.11 Data Rates

Data Rate (Mbit/s)	DSSS 802.11 (1997) 2.4 GHz		802.11a (1999) 5 GHz		802.11b (1999) 2.4 GHz		802.11g (2003) 2.4 GHz	
	Mandatory	Optional	Mandatory	Optional	Mandatory	Optional	Mandatory	Optional
1	DSSS	FHSS			DSSS		DSSS	
2	DSSS	FHSS			DSSS		DSSS	
5.5					CCK	PBCC	CCK	PBCC
6			OFDM				OFDM	CCK-OFDM
9				OFDM				CCK-OFDM, OFDM
11					CCK	PBCC	CCK	CCK-OFDM, PBCC
12			OFDM				OFDM	CCK-OFDM
18				OFDM				CCK-OFDM, OFDM
22								PBCC
24			OFDM				OFDM	CCK-OFDM
33								PBCC
36				OFDM				CCK-OFDM, OFDM
48				OFDM				CCK-OFDM, OFDM
54				OFDM				CCK-OFDM, OFDM

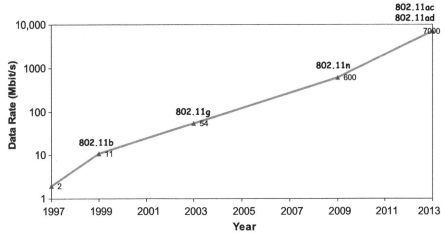

Figure 2.2 802.11 data rate evolution.

600 Mbit/s). The 1 Gbit/s barrier has been broken with the 802.11ac and 802.11ad amendments, which are expected to be ratified in 2013.

2.4 IEEE 802.11 NETWORK ARCHITECTURE

The basic service set (BSS) is the basic building block of an 802.11 wireless LAN. A BSS consists of two or more 802.11 devices, which may include an AP and one or more stations (STAs). STAs are user devices, such as laptops, smartphones, or tablets. Information is exchanged between 802.11 devices using frames. Members of a BSS communicate in a wireless coverage area called the basic service area (BSA). A BSS can operate in an independent (ad hoc) or infrastructure mode. An independent BSS comprises only one BSS with STAs (no AP), as shown in Figure 2.3. An independent BSS is also known as an IBSS in the 802.11 standard. This should be distinguished from an infrastructure BSS, which must have an AP. Each BSS has a unique identification called the BSSID. This corresponds to a MAC address of an AP in an infrastructure BSS and to a randomly generated MAC address in an independent BSS. Typically, an infrastructure network comprises one or more APs (hence one or more BSSs) that are interconnected via a wired backbone network (also known as the distribution system or DS). The DS can also be a point-to-point 802.11 wireless mesh network. Wireless stations in a BSS can access the DS via the AP. The DS allows wireless stations to access fixed network resources via a portal. The portal is essentially a layer-2 bridge (often residing within an AP) whose main function is to filter MAC addresses so that data frames targeted for a station that lies outside the current BSS can be forwarded to a neighboring AP. The antenna attached to the AP is usually mounted high in enterprise networks but may also be placed anywhere that is practical as long as the desired radio coverage is obtained.

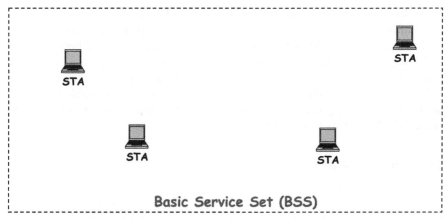

Figure 2.3   A BSS in an independent network.

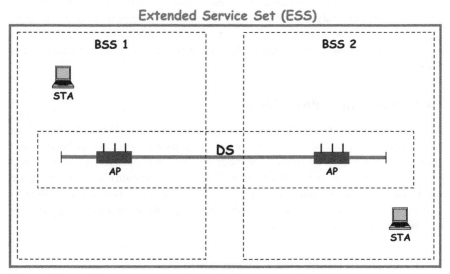

Figure 2.4   An ESS in an infrastructure network.

An extended service set (ESS) comprises a group of BSSs linked together by the DS, as shown in Figure 2.4. An ESS offers the possibility of mobile services where wireless stations can roam seamlessly from one BSS to an adjacent BSS without breaking the network connection. Thus, an ESS is effectively an infrastructure network with two or more interconnected APs. Like the BSS, an ESS has a unique identifier called the ESSID, which is a prerequisite for roaming. Unlike the BSSID, an ESSID involves the DS (i.e., wired network). Members of an ESS communicate in a wireless coverage area called the Extended Service Area (ESA). This information can be obtained from the service set identity (SSID) field of the beacon frame that is transmitted periodically by the AP. The length of the SSID varies from 0 to 32 bytes (or octets) and can be generated randomly to ensure no conflicts with neighboring networks. Multiple SSIDs on one AP correspond to different virtual LANs with different security levels. For example, guest users may gain restricted access to the network without interfering with intracompany traffic. Broadcast SSIDs are typically used in hotspots. They facilitate roaming and enable new security methods involving network authentication. To simplify home network management, a single SSID is usually defined.

2.4.1   Joining a BSS

The first step is authentication between the station and the AP. This is handled by exchanging request/response management frames. The second step is association between the station and the AP for infrastructure wireless LANs or between stations in the BSS for independent wireless LANs. The station can only be associated with one AP or BSS at any one time. Disassociation is a disconnect declaration from either an AP or a station. Association is necessary but not sufficient to support mobility. To support mobility and roaming, a reassociation function must be used in conjunction with the association function. Reassociation enables an established association to be transferred from one AP to another and is always initiated by the station.

2.4.2   Association Procedures

Association enables the establishment of a wireless link between a station and an AP in an infrastructure network. The station that joins a network is capable of transmitting and receiving data frames only after association is completed. The four-step process in establishing a new connection (initiated by a station) is shown in Figure 2.5. In this case, the station associates with AP B because it is nearer (resulting in a stronger SNR). The association request may originate from a new user logging into the network for the first time.

2.4.3   Disassociation and Reassociation

Disassociation and reassociation are dynamic processes. This is because stations may power on, power off, move within range or go out of range. These processes are handled by the request and response management frames. To illustrate how both processes interact in a roaming situation, consider the case when the station moves to a new location and decides that the link to its current AP is poor. The station scans for another AP or uses information from previous scans. If the new AP is found, the station sends a reassociation request to the new AP (Figure 2.6). If the reassociation response is successful, the station becomes connected to the new AP. Otherwise, the station scans for another AP. When the AP accepts the reassociation request, it indicates reassociation to the DS. Information on the DS is then updated. The old AP is notified of the change through the DS via a disassociation request and response. A disassociation notification can also be issued by a station when it powers down.

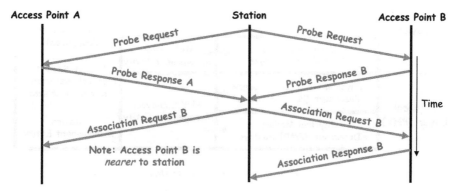

Figure 2.5   Association procedures.

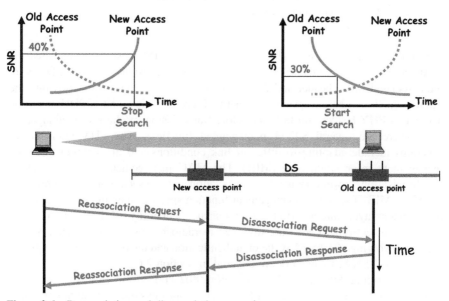

Figure 2.6   Reassociation and disassociation procedures.

2.5   IEEE 802.11 BASIC REFERENCE MODEL

The basic reference model is shown in Figure 2.7, which is divided into two sublayers. The physical medium dependent (PMD) sublayer deals with the characteristics of the wireless medium. It defines a method for transmitting and receiving data through the medium, such as modulation and coding. The PHY convergence procedure (PLCP) sublayer specifies the mapping of the MAC protocol data units (MPDUs) into the PLCP service data units (PSDUs) suitable for the PMD sublayer. It allows a common MAC protocol to be defined. It also performs carrier sensing for the MAC layer. The hierarchy of data flow at the 802.11 transmit station

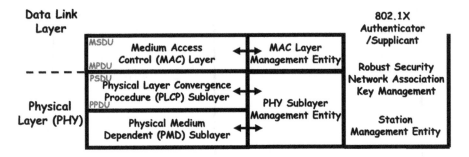

MSDU: MAC Service Data Unit

MPDU: MAC Protocol Data Unit

PSDU: PLCP Service Data Unit

PPDU: PLCP Protocol Data Unit

Figure 2.7   Basic reference model for 802.11.

can be described as follows. The highest data unit at the MAC layer is called the MAC Service Data Unit (MSDU), which can be encapsulated in a MPDU with the associated MAC header and frame check bits. The MPDU is mapped into the PSDU, which can be encapsulated in a PLCP protocol data unit (PPDU) with the associated PHY header and tail bits before transmission by the PMD sublayer.

There are several 802.11 management functions. The PHY management functions include adapting to different link conditions and maintaining the PHY management information base (MIB). The MAC management functions include synchronization, power management, association and reassociation, and maintaining the MAC MIB. The station management function specifies how the PHY and MAC management layers interact. Both station management entity (SME) and MAC layer management entity (MLME) are involved in radio resource management. 802.11 security involving 802.1X MAC layer authentication and the robust security network association (RSNA) framework are covered in Section 2.6.

The original 802.11 PHY specification (802.11-1997) allows three transmission options, namely DSSS and FHSS radio transmission, and omnidirectional optical transmission using diffuse infrared (DFIR). DSSS is mandatory in the original 802.11 standard, whereas FHSS and DFIR are optional. The evolution of the 802.11 radio PHYs is shown in Figure 2.8.

2.5.1   OFDM PHY

OFDM has been chosen due to its excellent performance in highly dispersive wireless channels. OFDM also allows considerable flexibility in the choice of different modulation methods. The OFDM channel bandwidth of 20 MHz is chosen as a compromise between having high data rates per channel and having a reasonable number of channels in the allocated spectrum. Each 20 MHz channel is typically occupied by a number of overlapped OFDM subcarriers in 802.11. For instance, as shown in Figure 2.9, there are 52 OFDM subcarriers (each subcarrier occupying a

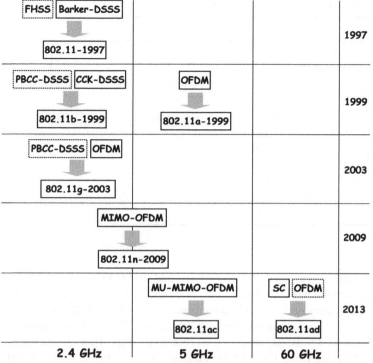

Figure 2.8   Evolution of 802.11 radio PHY (optional PHY in dotted boxes).

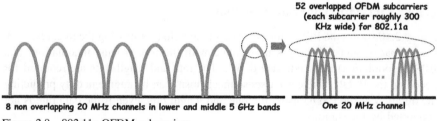

Figure 2.9   802.11a OFDM subcarriers.

bandwidth of roughly 300 KHz) in 802.11a, which operates in the 5 GHz band. This also applies to 802.11g in the 2.4 GHz band.

The main OFDM specifications for 802.11 a/g are listed in Table 2.3. Out of the 52 subcarriers in each channel, 48 subcarriers carry user data, while the remaining 4 subcarriers are pilots that facilitate synchronization and phase tracking for coherent demodulation. Each subcarrier serves as a communication link between the AP and the stations. The 800 ns guard interval (GI) is sufficient to enable good performance on channels with delay spreads of up to 250 ns. An optional shorter GI of 400 ns may be used in small indoor deployments. Binary phase-shift keying

TABLE 2.3 802.11 a/g OFDM PHY Specifications

Parameters	Specification
Mandatory data rates (Mbit/s)	6, 12, 24
Optional data rates (Mbit/s)	9, 18, 36, 48, 54
Number of subcarriers	52 (48 for data, 4 for pilots)
Sampling rate	20 Msample/s
Guard interval	800 ns (16 time samples), 400 ns is optional
Channel spacing	20 MHz
Signal bandwidth	16.6 MHz
Modulation for subcarrier	BPSK, QPSK, 16-QAM, 64-QAM
Bit-interleaved convolutional coding	Constraint length = 7, rate = 1/2, 2/3, 3/4

(BPSK), quadrature phase-shift keying (QPSK), and 16-QAM are the supported subcarrier modulation schemes, with 64-QAM as an option. Error control is provided by convolutional coding with rate 1/2 and a constraint length of 7. The three code rates of 1/2, 2/3, and 3/4 are obtained by code puncturing. The symbol duration for a short GI (400 ns) is 3.6 µs, and for a long GI (800 ns) is 4 µs. Thus, the maximum data rate for 802.11a/g (long GI, 20 MHz channel) is given by 48 data subcarriers × 1/(4 µs) symbols per second × 6 bits/symbol (i.e., 64-QAM) × 3/4 code rate or 54 Mbit/s.

2.5.2 OFDM PLCP Frame Format

Figure 2.10 shows the PPDU format. The signal field (comprising the rate, reserved, length, parity, and tail bits) is transmitted with the most robust modulation (i.e., BPSK modulation) at a code rate of 1/2. The rate and length bits are required for decoding the PSDU. The carrier sense mechanism can be augmented by predicting the duration of the PSDU using the rate and length bits, even if the data rate is not supported by the station.

2.5.3 Medium Access Control

The MAC protocol allows multiple stations to access the shared wireless medium. The original 802.11 standard defines a mandatory access mechanism called distributed coordination function (DCF). DCF employs carrier sense multiple access with collision avoidance (CSMA/CA) and per-frame acknowledgment (ACK) for contention-based multiple access. Contention-free service is provided by an optional MAC protocol called point coordination function (PCF), which is essentially a polling access method. PCF can allow priority data to pass before normal data. It achieves higher throughput and lower latency than DCF when there are many stations with data to send. PCF relies on the asynchronous access service provided by

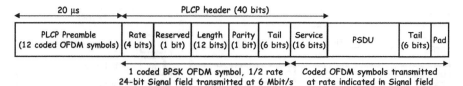

Figure 2.10   PPDU frame format.

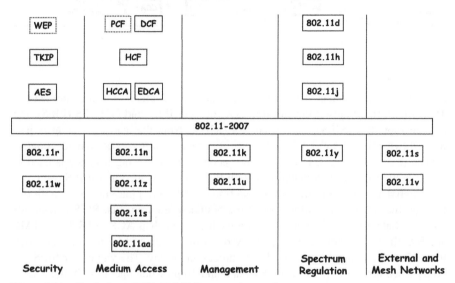

Figure 2.11   Evolution of 802.11 MAC protocol, security, management, spectrum regulation, and internetworking (optional features in dotted boxes).

the DCF. Hence, PCF is residing on top of DCF. DCF assumes less importance in high-rate amendments, such as 802.11n. This is because the backoff mechanism (that resolves contention when two or more frames collide) may incur higher overheads than sending a reservation frame. Nevertheless, the DCF performs functions usually associated with higher layer protocols. These functions include fragmentation, error recovery, mobility management, power conservation, and encryption. These additional functions allow the MAC sublayer to conceal the unique characteristics of physical wireless layer from higher layers. The evolution of the 802.11 MAC protocol is illustrated in Figure 2.11. The DCF and PCF MAC protocols have since been extended and combined into an integrated MAC protocol called the hybrid coordination function (HCF).

2.5.4   Interframe Space Definitions

The interframe space (IFS) and slot time definitions are shown in Figure 2.12. These time gaps are independent of the wireless data rate. Carrier sensing, also known as

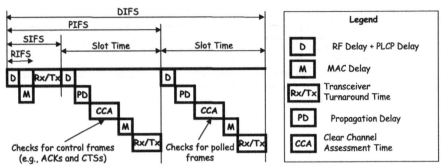

Figure 2.12   Interframe space (IFS) definitions.

clear channel assessment (CCA), is used to arbitrate contention in medium access. This is performed during the slot time, DCF IFS (DIFS), and PCF IFS (PIFS), but not the short IFS (SIFS) or reduced IFS (RIFS). Note that carrier sensing is performed twice for DIFS. The first instance of carrier sensing is used to detect control frames (e.g., ACK and CTS), while the second carrier sensing instance is used to detect polling frames (activated only with PCF). DIFS is the longest IFS used as a minimum delay between successive data frames. PIFS is an intermediate-length IFS used by the AP to poll stations with time-bounded requirements. SIFS is used for all immediate response actions (e.g., control frames, such as ACK, RTS, CTS, BAR, and BA). It is a function of receiver delay, delay in decoding PLCP preamble, header, transceiver turnaround time, and MAC processing delay. The values for SIFS are 10 μs when operating in the 2.4 GHz band and 16 μs when operating in the 5 GHz band. The one-way signal propagation delay is included in the slot time. The slot time is

- 20 μs (long) or 9 μs (short) when operating in the 2.4 GHz band
- 9 μs when operating in the 5 GHz band.

The slot time is used primarily for backoff purposes, as described in the next section. It is the sum of the carrier sensing time, transceiver turnaround time, signal propagation delay, and MAC processing delay. Only RIFS, SIFS, and the slot time are defined in the 802.11 standard. DIFS and PIFS are derived from these definitions. As can be seen from Figure 2.12, DIFS = SIFS + 2 × slot time, whereas PIFS = SIFS + slot time.

802.11n introduces the RIFS of 2 μs to separate multiple 802.11n PPDUs from a single station. Because RIFS is the shortest IFS interval among all IFS intervals, it minimizes overhead and improves network efficiency. RIFS may not be used between 802.11n PPDUs with different receiver MAC addresses. Another backoff interval called arbitration IFS (AIFS) can be used to prioritize different traffic types. AIFS was introduced by the 802.11e amendment. A longer AIFS corresponds to lower priority level (Figure 2.13). Up to 8 AIFS intervals (with increments of one slot time) have been defined to support eight traffic classes. The shortest interval corresponds to DIFS for backward compatibility.

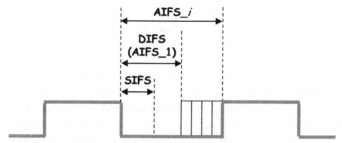

Figure 2.13   Traffic differentiation using AIFS.

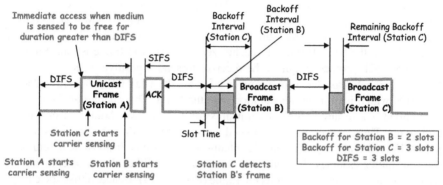

Figure 2.14   Broadcast and unicast DCF transmission.

2.5.5   Distributed Coordination Function

According to the DCF, when a station decides to transmit a new data frame, it must first sense the medium. If the medium is sensed to be idle for a duration greater than the DIFS, the frame is transmitted immediately. Otherwise, the transmission is deferred, and the backoff process is initiated. The backoff process is a collision avoidance technique that reduces the high probability of a collision immediately after a successful frame transmission, including repeat collisions. This is achieved by introducing a random time interval between successive frame transmissions, effectively separating the total number of active stations into smaller groups, each using a different time slot. Unlike wired Ethernet, the 802.11 backoff process is also initiated immediately after a successful frame transmission to prevent a station from dominating channel transmission time.

If the medium is detected to be busy, the station must delay until the end of the DIFS interval and further wait for a random number of time slots (collectively known as the backoff interval) before attempting transmission (Figure 2.14). Specifically, a station generates a backoff interval that is uniformly distributed between zero and a maximum contention window. This backoff interval is used to initialize the backoff timer. The timer is decremented only when the medium is idle but is suspended when another station is transmitting. The decrementing period is the slot time. Thus, carrier sensing is cooperative because it allows the station to listen to the medium before transmitting and so it will not disrupt or interfere with other ongoing transmissions. This allows the coexistence of 802.11 and non-802.11 devices from different vendors using the same unlicensed bands (e.g., 2.4 and 5 GHz bands).

At each slot time, carrier sensing is performed to determine if there is activity on the medium. If the medium is idle for the duration of the slot, the backoff interval is decremented by one slot. If a busy medium is detected, the backoff procedure is suspended and the backoff timer will not decrement for that slot. In this case, when the medium becomes idle again for a period greater than the DIFS, the backoff procedure continues decrementing from the slot that was previously disrupted. This implies that the new backoff interval is now shorter than the initial interval. Hence, a data frame that was delayed while performing the backoff procedure has a higher probability of being transmitted earlier than a newly arrived data frame. This ensures some measure of fairness in medium access. The process is repeated until the backoff interval reaches zero and the data frame is transmitted.

The DCF protocol requires the receiver to send a positive ACK back to the transmitter if a unicast data frame is received correctly. The ACK is transmitted after the SIFS, which is of a shorter duration than the DIFS. This enables an ACK to be transmitted before any new frame transmission (i.e., ACKs should not encounter collisions). Thus, a receiving station need not sense the medium before transmitting the ACK. ACKs also signify end of contention-free periods. If no ACK is returned, the transmitter assumes the data frame is corrupted (either due to a collision or transmission error) and retransmits the frame (Figure 2.15). Hence, unlike the carrier sense multiple access with collision detection (CSMA/CD) protocol used in legacy wired Ethernet, the CSMA/CA mechanism infers collisions only after the entire data frame is transmitted. Without collision detection, frames transmitted using CSMA need not be of a certain minimum length. In addition, the sender cannot distinguish between a collision and a transmission error. Note that backoff intervals are only increased when retransmissions occur (Figure 2.16). In this case, the repeat collision is a result of the two stations selecting the same initial backoff interval (two slots).

Broadcast frames cannot be acknowledged due to the possibility of colliding ACKs. To reduce the possibility of repeated collisions, after each unsuccessful transmission attempt, the contention window is doubled until a predefined threshold

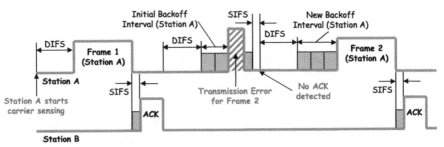

Figure 2.15 DCF retransmission.

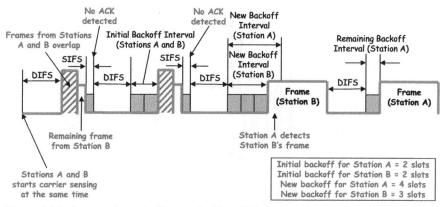

Figure 2.16   Increase in backoff interval with collisions.

is reached. Thus, the backoff interval increases exponentially up to a maximum contention window.

The DCF protocol ensures that stations cannot dominate network usage. This is achieved by forcing stations with multiple data frames to activate the backoff procedure, thereby giving another station a chance to transmit (Figure 2.17). However, the backoff procedure is not used when the station decides to transmit a new data frame and the medium has been free for a duration greater than DIFS.

2.5.6   Virtual Sensing

Hidden collisions may arise when a station or AP is able to successfully receive frames from two or more stations, but these stations cannot receive signals from each other. Although such collisions are mostly unintentional, they can be significant in large-scale 802.11 networks operating in the unlicensed bands. For example, in Figure 2.18, stations A and B may transmit to station C simultaneously because stations A and B cannot hear each other (i.e., they are out of range). In this case, it is possible for one of the stations to sense an idle medium even when the other station is transmitting. This results in a hidden collision at the receiving station (station C). Note that stations A and B are unaware of the collision at station C (i.e., they are hidden from each other). The problem is due to the failure of the carrier-sensing mechanism. Since an AP is able to communicate with all stations within the wireless subnet and nonoverlapping frequency channels are used by adjacent APs, it may appear that such collisions can be avoided. However, it turns out that hidden collisions are also very common in office deployments with wall partitions, as illustrated in Figure 2.19. Because stations A and B are separated by wall partitions, they may not be able to sense each other's transmission. However, both stations are within range of the AP. If stations A and B transmit to the AP simultaneously, a hidden collision results. Since APs are normally available in each residential home or apartment, they are either not shared or shared by only a few users. Thus, hidden collisions are less likely to arise in these deployments.

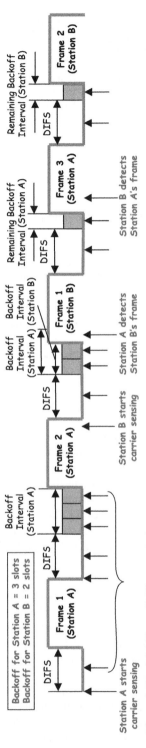

Figure 2.17 Multiframe DCF transmission.

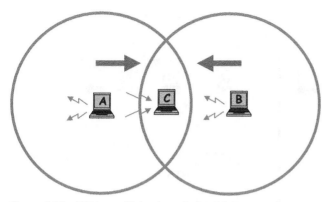

Figure 2.18   Hidden collision in an independent network.

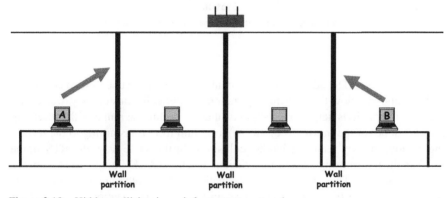

Figure 2.19   Hidden collision in an infrastructure network.

802.11 attempts to solve the hidden collision problem by using a virtual carrier sensing mechanism. The mechanism distributes reservation information by announcing the impending use of the wireless medium. With virtual sensing, a sender and a recipient can notify all in-range devices that the link is in use. The basic idea is a four-way handshake where short control frames called request-to-send (RTS) and clear-to-send (CTS) are exchanged prior to the transmission of the data frame (Figure 2.20). The RTS frame is issued by the sending station while the CTS frame is issued by the destination station to grant the source station permission to transmit. On correct receipt of the data frame, the destination station sends an ACK frame to the source station. The RTS and CTS frames contain the destination addresses as well as a duration field defining the time interval between the transmission of the data frame and the returning ACK frame.

It is clear that with virtual sensing, collisions may occur during the transmission of an RTS frame; hence, the problem of hidden collisions is only partially solved. The RTS frame may overlap either a RTS or a CTS frame from another station. Although the vulnerable collision period for RTS is twice the duration of the frame, this can be much shorter compared with the vulnerable period for a data

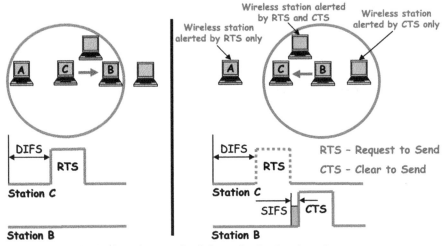

Figure 2.20   Distribution of reservation information in virtual sensing.

frame. The short RTS and CTS frames minimize the overhead due to such collisions, and in addition, allow the transmitting station to infer collisions quickly. Moreover, the CTS frame alerts neighboring stations (that are within the range of the receiving station but not of the transmitting station) to refrain from transmitting to the receiving station, thereby reducing hidden collisions. In the same way, the RTS frame protects the transmitting area from collisions when the ACK frame is sent from the receiving station. Thus, reservation information is distributed among stations surrounding both the transmitting and receiving stations (i.e., stations that can hear either the transmitter or the receiver or both).

Virtual sensing transforms DCF to a distributed form of reservation MAC protocol. All nontransmitting stations that successfully decode the duration field in the RTS and CTS frames store the medium reservation information in a Net Allocation Vector (NAV), as shown in Figure 2.21. For these stations, the NAV is used in conjunction with carrier sensing to detect the availability of the medium. Hence, these stations will defer transmission if the NAV is nonzero or if carrier sensing indicates that the medium is busy. Note that unlike the backoff timer, the NAV timer is continuously decremented regardless of whether the medium is sensed busy or idle. Since stations can hear either the transmitter or the receiver refrain from transmitting until the NAV expires, the probability of hidden collisions is reduced. Like the ACK mechanism, virtual sensing cannot be applied to data frames with broadcast and multicast addressing due to a high probability of collision among a potentially large number of CTS frames.

Due to the large overhead involved, the mechanism is not always justified, particularly for short data frames. Hence, 802.11 allows short data frames to be transmitted without virtual sensing. This is controlled by a parameter called the RTS threshold. Only frames of lengths above the RTS threshold are transmitted using virtual sensing. It was first reported in Reference [4] that performing carrier sensing

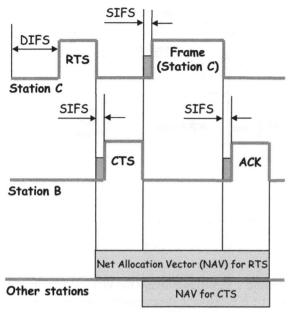

Figure 2.21   Frame transmission using 802.11 virtual sensing.

(via DIFS) before sending an RTS may not be justified. As we shall see later, in 802.11n, this requirement is waived and DIFS is replaced by SIFS. The use (activation) of virtual sensing is optional, but the mechanism must be implemented so that all stations can decode and interpret RTS and CTS frames, and backoff for the appropriate time interval as specified in the NAV.

It is important to note that at long operating distances, RTS/CTS virtual sensing becomes ineffective. Contending stations close to AP may not detect the CTS issued by a remote station, and this blind spot probability increases with distance. In addition, the CTS and RTS header durations are short compared with the data frame transmission time; hence, there is a probability that both RTS and CTS may be missed. This probability may increase with distance due to presence of more contending stations.

2.5.7   Point Coordination Function

The native DCF employs contention-based access, which may not be suitable for real-time traffic due to the random backoff delay needed to avoid or resolve collisions. The optional PCF may be used to support time-bounded services. PCF employs a centralized access scheme where stations in a wireless coverage area are allowed to transmit only when polled by an AP (acting as a central controller), which has priority access to the medium. Thus, unlike DCF, stations and APs are no longer peers, and transmission becomes scheduled and not asynchronous. To allow other stations with asynchronous data to access the medium, the 802.11 MAC protocol alternates between DCF and PCF, with PCF having higher priority access

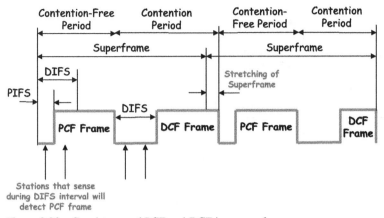

Figure 2.22   Coexistence of PCF and DCF in a superframe.

(Figure 2.22). This is achieved using a superframe where the PCF is active in the contention-free period, while the DCF is active in the contention period.

A beacon frame notifies all stations in the wireless coverage area to refrain from transmission for the duration of the contention-free period (unless specifically polled by the AP). The contention-free period can be variable in length within each superframe without incurring any additional overhead. At the beginning of the superframe, if the medium is free, the PCF gains control over the medium. If the medium is busy, then the PCF defers until the end of the data frame or after an ACK has been received. Since the PIFS is of shorter duration compared with DIFS, PCF can gain control of the medium immediately after the completion of a busy period.

Note that the contention period may be variable, and this causes the contention-free period to start at different times. Similarly, a data frame may start near the end of the contention period, thereby stretching the superframe and causing the contention-free period to start at different times. Thus, the repetition interval of the contention-free period can vary depending on the network load. Another important observation is that hidden collisions are eliminated by the PCF because transmit permissions are initiated by well-sited APs (as opposed to stations initiating transmission independently using CSMA). If the APs employ nonoverlapping frequency channels, then the simultaneous transmission of transmit permissions do not interfere even when multiple permits are received by stations located in an overlapped region between two APs.

2.5.8   Hybrid Coordination Function

The HCF provides a combination of contention and contention-free channel access. The HCF contention access is called enhanced distributed channel access (EDCA), whereas HCF controlled channel access (HCCA) supports contention-free polling with deterministic delays. EDCA is effectively an extension of DCF. Unlike DCF, EDCA allows the contention backoff intervals to be prioritized using the AIFS.

EDCA supports 8 user priorities (UPs), which are identical to the IEEE 802.1d priority tags. HCCA is essentially the PCF. It allows the AP to poll stations during the
contention-free period and supports reservation requests called transmission opportunities (TXOPs). A TXOP allows the station to send multiple data frames within a
specified NAV period. The traffic specification (TSPEC) allows a station to signal
its traffic requirements to the AP. The hybrid coordinator (i.e., the AP) places the
station on a polling list whenever data is available for transmission, thereby allowing
efficient polling. The station maintains separate queues for different traffic types as
indicated in the quality of service (QoS) field.

2.5.9   Synchronization

Beacon frames are critical in maintaining network synchronization. They convey
timing information and important network parameters, as shown in Figure 2.23.
They also enable new stations to join the network. In an infrastructure wireless LAN,
the AP periodically transmits beacon frames. Receiving stations adjust local clocks
the moment the beacon is received. Synchronization becomes a distributed function
for IBSS networks. When the station wishes to discover an existing BSS within
range (either after power-up, power-save mode, or entering a BSS), it can obtain
synchronization information by passive scanning (the station waits for the beacon
frame from the AP) or active scanning (the station tries to locate an AP by transmitting a Probe Request and then listens for a Probe Response from the AP).

2.5.10   Transmit Opportunity Scheduling

The TXOP allows a station to transmit a number of data frames that fit a specified
duration. This is defined by a starting time and a maximum duration, and is obtained
by successfully contending for medium. It also facilitates block acknowledgment
(BA) of multiple data frames in 802.11n. The scheduled service interval (SI) is
calculated in two steps as follows:

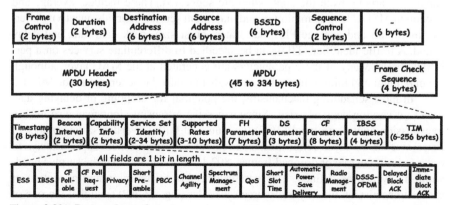

Figure 2.23   Beacon frame format.

- First, the scheduler calculates the minimum interval of all maximum SIs for all admitted traffic streams.
- Second, the scheduler chooses a value lower than the minimum interval obtained in the first step that is a submultiple of the beacon interval. For example, if the beacon interval is 100 ms and the minimum of all maximum SIs is 60 ms, the scheduler chooses 50 ms as the scheduled SI.

To compute the TXOP duration for an admitted stream, the scheduler uses the following parameters:

- Average data rate (m) and average MSDU size (L) obtained from TSPEC
- Scheduled SI (s)
- Minimum PHY data rate (R)
- Maximum MSDU length of 2304 bytes (M)
- Overheads in time units (O).

R is the minimum PHY rate negotiated in the TSPEC for a specific station. The TXOP duration for that station is computed as follows. First, the scheduler calculates the number of MSDUs that arrive at the average data rate during the SI:

$$N = sm/L. \tag{2.1}$$

Second, the scheduler calculates the TXOP duration as the maximum of the time to send N data frames at R and the time to send a maximum length MSDU at R. Overheads (O) due to the MPDU header, SIFS, and ACK are included in both time intervals.

$$TXOP = \max(NL/R + O, M/R + O). \tag{2.2}$$

2.5.11 Traffic Specification Construction

The TSPEC is constructed at the SME. The 802.11 standard does not provide requirements on how TSPECs are generated (i.e., this is vendor-specific). However, typical parameters arising from the application are suggested. These include: average and maximum MSDU length, minimum and maximum service intervals, inactivity interval, minimum and average PHY data rates, burst size, peak data rate, and delay bound. A traffic identifier (TID) can be specified to differentiate services on a per-MSDU basis. Many of these parameters are generated at the MAC layer, although the maximum and minimum service intervals may also be generated in the MLME.

In most scheduling mechanisms, the potential for dropping or delaying data frames at varying degrees exists. Since excessive retransmissions may also lead to transmission control protocol (TCP) timeouts and buffer overflows, the number of successive MAC layer retransmissions should be dimensioned appropriately. In order to limit the rate of dropped frames, 802.11 recommends that the surplus bandwidth allowance be defined. This parameter refers to the minimum allocation of excess transmission time by a resource scheduler (e.g., an AP) such that the rate of frame dropping becomes bounded. This takes into account the expected frame error probability in addition to the frame drop probability inherent with the wireless medium. Three different cases are considered:

- Single-frame transmission (e.g., control frame transmission)
- Multiple-frame transmission (e.g., Web download)
- Multiple-frame streaming (e.g., voice and video streaming).

In single-frame transmission, if the wireless medium causes independent errors from frame to frame and the error probability is invariant for all frames of the same length, then the probability of a dropped frame after n retransmissions is $p_{\text{drop}} = (p_e)^{n+1}$, where p_e is frame error probability in a single transmission attempt. If $p_e = 0.1$ and $p_{\text{drop}} = 10^{-8}$, then $n = 7$. In other words, the scheduler has to accommodate seven TXOPs. In multiple frame transmission, suppose N frames need to be transmitted as a group and the frame error probabilities are Bernoulli distributed. The Bernoulli distribution is a discrete probability distribution, which takes value 1 with success probability p and value 0 with failure probability of $1 - p$. Then it can be shown that p_{drop} is given as:

$$p_{\text{drop}} = \sum_{k=N_{\text{ex}}}^{N+N_{\text{ex}}} \binom{N+N_{\text{ex}}}{k} p_e^{N_{\text{ex}}} (1 - p_e)^{N+N_{\text{ex}}-k}.$$

$$(2.3)$$

If $N = 100$, $p_e = 0.1$, $p_{\text{drop}} = 10^{-8}$, then $N_{\text{ex}} = 38$ (Figure 2.24). The surplus bandwidth allowance, $S = N_{\text{allocated}}/N_{\text{payload}} = (N + N_{\text{ex}})/N = 1.38$. If $N = 100,000$, $p_e = 0.1$, $p_{\text{drop}} = 10^{-8}$, then $N_{\text{ex}} \sim 12,000$, $S \sim 1.12$. The surplus bandwidth decreases as N increases. For an infinite stream of frames with no delay constraints, S approaches

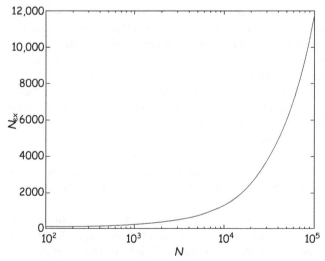

Figure 2.24 Computing surplus bandwidth with $p_e = 0.1$ and $p_{\text{drop}} = 10^{-8}$.

the lower bound of 1.111. If BA is employed, the probability of failing to receive a BA must be taken into account.

The TXOP computation in the previous section assumes an error-free wireless medium. In order to compute the appropriate medium transmission time (C_{tx}), the bandwidth requirements of the application and the expected error performance of the wireless medium must be considered. The application requirements may be captured adequately by the average MSDU length (L) and the average data rate (m). The wireless medium requirements may be captured by the surplus bandwidth allowance (S) and the minimum PHY rate (R). In general,

$$C_{tx} = S \times \text{MSDU rate} \times \text{MSDU exchange time}, \tag{2.4}$$

where MSDU rate = m/L and MSDU exchange time = $L/R + O$

2.5.12 Radio Resource Measurement

Radio resource measurement (RRM) collects network information for higher 802.11 layers. RRMs are exchanged in pairs of requests and reports, providing information about beacons, measurement pilots, summary of received frames, noise histograms, station statistics, location configuration information, neighbor report, link measurements, access delay, and so on. RRM can be supported via a driver or firmware upgrade (no hardware upgrade) and offers some form of automated radio management. For example, transmit power and frequency channel adjustment allow APs to change coverage area depending on number of stations, deal with changes in traffic demands (e.g., heavy usage in a conference room), and compensate for failed/disconnected APs.

RRM can be combined with dynamic channel selection (DCS) to allow the BSS to move to another channel with better channel characteristics. This preserves QoS metrics and enables automatic frequency planning. It provides a mechanism for overlapping but different BSSs (e.g., in adjacent apartments) to avoid one another, and helps mitigate the impact of malicious APs. RRM can also be combined with transmit power control (TPC) to enable stations to communicate at the minimum power. This reduces interference between BSSs using the same channel, thereby improving the QoS level for all BSSs.

Since APs may collect channel statistics from the wireless station, this facilitates automatic site surveys and reports. Such self-organizing APs eliminate the need for time-consuming site surveys in dense or large-scale wireless LANs. RRM employs the received channel power indicator (RCPI) that measures the absolute received power of the 802.11 signal at the antenna output. This is in contrast to the received signal strength indicator (RSSI), an optional parameter that measures the energy difference between the beginning and end of the 802.11 PPDU header at the receiver front end.

In many cases, performance metrics (e.g., overall throughput) can be optimized if opportunistic algorithms are employed. However, opportunistic RRM techniques always favor advantaged users who have good link conditions and/or low interference levels. The problem becomes worse when wireless stations have low mobility. In this case, the link conditions become slowly varying (or even static) and may lead to long-term unfairness. Performance fairness can be included as one of the QoS requirements (e.g., as a condition on the minimum throughput per user). Fair RRM schemes may penalize advantaged users. Hence, there should be a tradeoff between overall system performance and fairness requirements.

2.5.13   Station Power Management

The 802.11 standard defines three power modes that are incorporated into the MAC protocol. These modes are listed below:

- **Transmit:** Transmitter is activated.
- **Awake:** Receiver is activated.
- **Doze:** Transceiver is not able to transmit or receive.

The MAC protocol allows the mobile station to switch from full power (active) mode to low power (sleep) mode during a time interval defined by the AP without losing information. The actual power consumption is not defined in the standard and is implementation dependent. The AP keeps an updated record of stations currently operating in the power saving mode. It then buffers the data frames addressed to these stations until the stations specifically request for the frame or when they resume communication with the AP. If the station moves to a different AP, frames are forwarded across the wired LAN to the station. The power save mode limits the station's ability to transmit and receive data. Additionally, the station may miss incoming traffic if the AP buffer overflows. This can happen if the sleep duration is set to a value much greater than the beacon interval.

The AP periodically sends beacon frames with a traffic indication map (TIM), as shown in Figure 2.25. The beacon announces which power-saving stations have unicast traffic cached at the AP. The beacon transmission may be delayed by an ongoing data transmission. In the power-save mode, the wireless station regularly wakes up prior to any TIM broadcasts from an AP. The station need not check every TIM broadcast. If there are frames intended for the station, it shifts from doze to awake mode. The station then sends a polling message to the AP to retrieve these frames. To maintain synchronization, a timer continues to run as a station sleeps. Synchronization allows extreme low-power operation. The availability of broadcast, multicast, and unicast messages are indicated via the delivery TIM (DTIM), with broadcast and multicast messages sent first. The DTIM interval has a longer duration (a multiple of the TIM interval). For an independent wireless LAN, since no AP is available, beacon transmission becomes a distributed responsibility. Since all frames that cache at the AP are eventually transmitted to the station, the power save benefit may not be achieved if a station operating the sleep mode is regularly receiving traffic. In this case, the buffer resources at the AP must be properly dimensioned to prevent overflows.

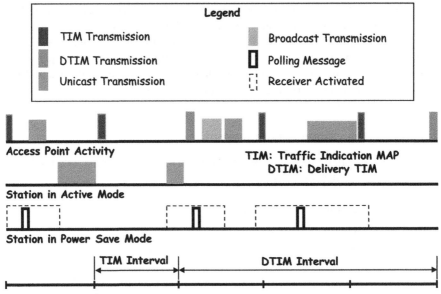

Figure 2.25 Power management procedures.

2.6 IEEE 802.11 SECURITY

By default, 802.11 data frames are transmitted without encryption, which implies that any 802.11-compliant station can potentially eavesdrop traffic within range. Privacy ensures eavesdroppers cannot read the network traffic. This is typically achieved using frame-by-frame encryption by the transmitting station and decryption by receiving station. Authentication restricts the ability to send and receive on the network, allowing only preapproved stations to join a network. This is a more crucial function than privacy and should be the first step that a station must pass when attempting to connect to an 802.11 wireless LAN. It is handled by exchanging 802.11 management frames. If authentication is open, then any 802.11-compliant station can be authenticated, allowing a proprietary authentication protocol to be used. If authentication is based on a shared key, then the station must prove that it knows the shared key to be authenticated.

Although the original 802.11 standard includes a basic security provision, this was enhanced with the 802.11i amendment. 802.11i removes several vulnerabilities associated with the earlier security mechanism. 802.11i provides stronger mutual (two-way) authentication mechanisms. These include preauthentication and key caching (to support roaming) using the IEEE 802.1X standard. Key management algorithms cover key hierarchy, cryptographic key establishment, and more frequent key and IV rotations. 802.11i enhances data encapsulation and data transfer using AES-CCMP encryption and TKIP encryption. Other technical highlights include support for independent BSS and infrastructure BSS, and cipher and authentication negotiation. 802.11i focuses on protecting the data frame. The

802.11w amendment extends this capability by protecting 802.11 management frames against eavesdropping and forgery.

2.6.1 Wired Equivalent Privacy

As the name implies, Wired Equivalent Privacy (WEP) is designed to provide a level of security comparable with a typical wired LAN operating without encryption. It is an optional provision for security in the original 802.11 standard. WEP uses an encryption and device authentication stream cipher based on the Rivest Cipher 4 (RC4) encryption, a version of the Rivest, Shamir, and Adleman (RSA) algorithm. The original version (WEP1) employs static 40-bit secret keys that can be preshared. A later version (WEP2) provides an upgrade to 104-bit keys. The use of static keys simplifies the set up of small networks (e.g., home networks), but is more vulnerable to hackers. The steps for encryption and decryption are shown in Figure 2.26. The secret key is concatenated with a known 24-bit initialization vector (IV). This results in a 64-bit (or 128-bit) seed to input to a pseudorandom generator. The generator outputs the key sequence (which is longer than the secret key) that is equal to the longest frame length. The key sequence is then combined with the user data (integrity protected by an integrity check value or ICV) before the data frame is transmitted. Like the frame check sequence (FCS), ICV is calculated using the cyclic redundancy check (CRC) method, but unlike the FCS, is used to authenticate the message (i.e., ICV will detect any modifications to the encrypted data message). Because the ICV is encrypted, it is difficult for attacker to modify the ICV so as to match the modifications to the encrypted data field. Each station can be assigned up to four WEP encryption keys. WEP protects the MPDU data payload only (if it exists) and not the PHY or MPDU header.

To operate in a connectionless mode, every data frame is appended with a 24-bit IV. This restarts the pseudorandom generator for each frame. Unlike the secret key, the IV changes periodically to make it difficult for attackers to discover the secret key using the key sequence. A common 40-bit (or 104-bit) shared key is used to authenticate, encrypt, and decrypt data. Only stations with a valid shared key will be allowed to be associated to the AP. If a station receives a data frame that is not

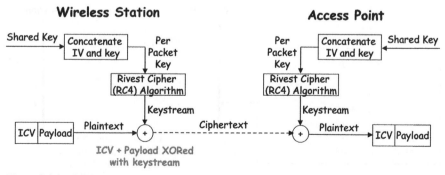

Figure 2.26   WEP encryption.

encrypted with the appropriate key, the frame is discarded and never delivered to the intended receiver. The key can be delivered to participating stations via a secure channel that is independent of the 802.11 network. It may also be manually delivered to the legitimate user. The shared key is secret but usually fixed. This is in contrast to the IV, which is known but changes. However, because the 24-bit IV is attached to the message unencrypted, the receiver (and attacker) knows which IV to use and it is therefore prone to attacks.

WEP's shared key authentication provides access control. A standard challenge–response procedure is used in conjunction with the WEP encryption features to authenticate the station. However, WEP provides one-way authentication only, as shown in Figure 2.27. In this case, the AP challenges a station. The one-way authentication allows a rogue AP (e.g., illegitimate APs plugged into a legitimate 802.11 network) to masquerade as a legitimate AP. This problem can be solved by employing two-way authentication, which requires a third component called an authentication server (not defined in WEP).

WEP provides only basic protection to the transmitted information. A good hacker can break the 40-bit WEP key in a few hours using commoditized attack or sniffing tools, compromising confidentiality of data. The weak encryption and one-way authentication are prone to man-in-the-middle (MITM) attacks where an intruder interposes himself between two other wireless devices. In this case, the intruder can determine the challenge and challenge response, compromising access control. Intrusions are also possible with lost or stolen devices. The static key makes it labor intensive to configure and manage keys for hundreds or thousands of wireless stations. This may be required for large enterprises or university campuses. The associate and disassociate messages are not authenticated. An attacker can continuously send streams of associate and disassociate messages, preventing legitimate stations from accessing the wireless LAN in a denial of service (DoS) attack. In home wireless networks, the weak encryption may lead to stream theft, allowing an intruder to gain access to content another subscriber is paying for. Because WEP does not define a process for allocating and distributing encryption keys, this makes it difficult to determine who is and who is not to be given permission to receive a particular data stream.

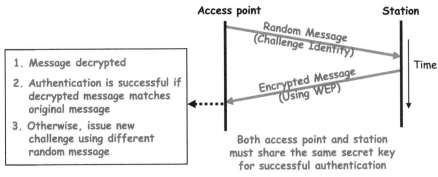

Figure 2.27   WEP authentication.

2.6.2   Robust Security Network Association

The RSNA framework defines enhancements to WEP and its 802.11 authentication scheme. RSNA includes:

- Stronger authentication mechanisms, such as key management algorithms
- Cryptographic key establishment
- An enhanced data encapsulation and data transfer mechanism with CCMP (TKIP is an option for backward compatibility).

The two main authentication components are the IEEE 802.1X port access entity (PAE) and authentication server (AS). The AS is external to the 802.11 network. With wireless connectivity, access to a layer 2 network cannot be assumed to be via the same physical port of entry. This creates a need to identify the user that is attempting to gain access to a given port. 802.1X provides a solution by employing the Extensible Authentication Protocol (EAP) over LANs (EAPOL) to authenticate stations with the AS. The EAP messages may function over Ethernet or 802.11. However, 802.1X does not specify a mandatory EAP method to implement. The 802.1X PAE is present on all wireless stations and APs, and controls data forwarding to and from the MAC layer. The AP implements an authenticator PAE based on an EAP authenticator. The station implements a supplicant PAE based on an EAP peer. The AS authenticates the user, station, and AP (contrast WEP, which only provides station authentication) and plays an important role in key distribution. The AS communicates with the 802.1X supplicant via the 802.1X authenticator, enabling each station to be authenticated to the AS and vice versa (Figure 2.28). Note that the AS is able to detect rogue APs since it knows all valid APs.

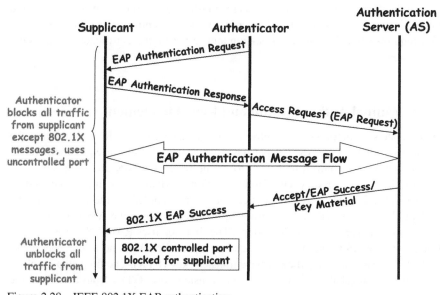

Figure 2.28   IEEE 802.1X EAP authentication.

When the authenticator first detects the station, it forces the port into an unauthorized state and forwards only 802.1X traffic. The authenticator then replies with an EAP request identity message to obtain the identity of the station. The station's EAP response message containing the station's identity is forwarded to the AS. On receiving an EAP success message from the AS, the station is authenticated, and all data traffic from the station is forwarded. The success message contains the session keys, which are used by the authenticator to encrypt the EAP key message. This is sent to the station immediately following the success message. With this information, both the authenticator and the station can program their encryption keys dynamically, making them hard to crack. At logoff, the station will send an EAP logoff message, and the authenticator will change the state of the station port to unauthorized. RSNA depends on the EAP to support mutual authentication of the AS and the wireless station. EAP allows intermediate devices connected to the supplicant and the AS to be transparent with respect to the authentication. Thus, the EAP framework can be extended to wired LANs, enabling an enterprise to use a single security architecture for all access methods (wireless and wired) and allowing seamless compatibility when deploying heterogeneous wireless networks.

EAP allows multiple user-authentication methods to be encapsulated under a unified security protocol. This in turn provides flexibility in the choice of authenticating users. More importantly, EAP allows several authentication methods to be multiplexed in sequence (i.e., serial authentication). This capability can be exploited in new security approaches that allow user to authenticate the network before revealing its identity. It also allows a mobile user to quickly reauthenticate the network and maintain connections as it roams across different networks. EAP is compatible with many popular operating systems of mobile computing devices. As an example, the EAP transport layer security (EAP-TLS), defined in RFC 5216, can be used to provide certificate-based mutual authentication (Figure 2.29). If the EAP-TLS authentication is successful, a secret master key is known to the station and AS. This key is subsequently delivered to the authenticator by the AS. EAP-TLS requires prior distribution of station and server digital certificates via a secure connection.

2.6.3   Mutual Authentication and Key Management

Two private symmetric keys are introduced. The master key (MK) authenticates the wireless station with the AS. The pairwise master key (PMK) is used by the wireless station and AP for controlling access to the network. The MK and PMK are valid only for the current session. The wireless station to AP authentication is handled by 802.1X EAP. The AP to AS authentication can be handled by the Remote Access Dial-In User Service (RADIUS) protocol and assumes that the AS is trusted. Key management is performed in three steps. The first step uses RADIUS to forward the PMK from the AS to the AP. The second step uses the PMK and a four-way handshake to derive and verify the pairwise transient key (PTK). The final step employs a group key handshake to send the group transient key (GTK) from the AP to the wireless station.

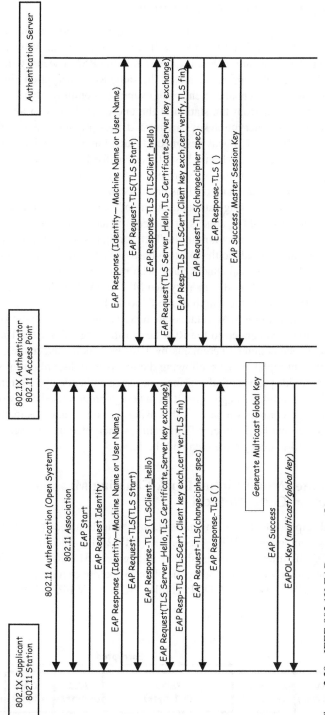

Figure 2.29   IEEE 802.1X EAP message flow.

113

2.6.4 Temporal Key Integrity Protocol

Temporal key integrity protocol (TKIP) is a primary component of Wi-Fi Protected Access (WPA) version 1.0 (WPA1). It was defined by the WFA before the full 802.11i standard was ratified and is an option in the RSNA security framework. TKIP runs WEP as a subcomponent. It uses the same RC4 encryption algorithm as WEP but contains several enhancements. For example, it employs dynamic encryption keys that disappear from the station at the end of each session and extended 48-bit IVs. Because the encryption keys are rotated, they provide stronger protection than static keys in WEP. TKIP was designed to patch known flaws identified in the WEP algorithm. It adds a per-frame hash function (for key construction) and IV sequencing rules. The sequencing rules protect against replay or spoofing. The addition of temporal key derivation algorithms prevents encryption abuse and key reuse. The addition of a keyed message authentication code called message integrity code (MIC) protects against MITM attacks or forgery, including forgery of the FCS of each data frame. A critical constraint in the TKIP design is that it must only require software or firmware upgrade to the existing base of 802.11 devices that implement WEP. Since 2002, no practical attacks have been mounted against TKIP.

TKIP adds four fields to the unencrypted MPDU: IV, EIV, MIC, and ICV. The extended IV (EIV) extends the length of the IV to 48 bits (Figure 2.30). The 48-bit IV, RC4 key, and transmitter MAC address are mixed to produce a 128-bit key (recall that the IV is simply appended to the RC4 key in the original WEP). The IV also serves as a sequence counter (increments by 1 for each new data frame), which is designed to prevent replay attacks, allowing the receiver to discard out of sequence frames. TKIP provides substantially stronger message authentication. It retains ICV for backward compatibility but prevents it from being compromised easily by adding the MIC. The MIC computes a 64-bit value using the source and destination MAC addresses and the data field contents. This value is encrypted using a separate RC4 key from that used to encrypt the data and the ICV.

2.6.5 Counter-Mode Cipher Block Chaining Message Authentication Code Protocol

Counter-mode cipher block chaining message authentication code protocol (CCMP) is the primary component of WPA Version 2.0 (WPA2). It is specified in IETF RFC 3610. The CCMP frame format is shown in Figure 2.31. CCMP provides both authentication and privacy by encoding the plaintext before encrypting it. It removes the use of the ICV in WEP and TKIP. Unlike TKIP and WEP, CCMP requires an

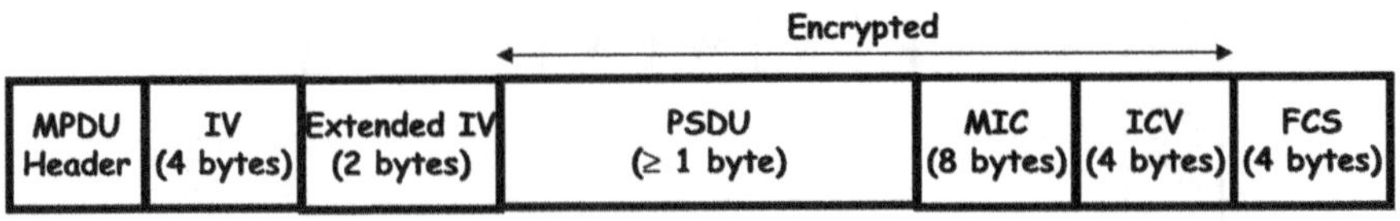

Figure 2.30   TKIP frame format.

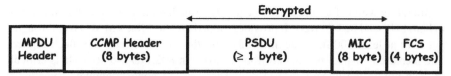

Figure 2.31   CCMP frame format.

AS for mutual authentication and eliminates the need for key rotation. Like TKIP, CCMP uses 128-bit temporal encryption keys derived from a master key obtained during the 802.1X negotiation. CCMP employs the Advanced Encryption Standard (AES) symmetric encryption, which is much stronger than the RC4 algorithm. AES is defined in Federal Information Processing Standard (FIPS) publication 197. It is a symmetric block cipher. Although AES supports key lengths of 128, 192, and 256 bits, the key length of 128 bits is sufficient to provide security and requires less processing time than a longer key length. The input to the encryption and decryption algorithms is a 128-bit block of plain text, comprising a matrix of 4 × 4 bytes, which is called a state array. The first row of the state array is not shifted, the second row is shifted by 1 byte, the third row by 2 bytes, and the fourth row by 3 bytes. Each column of bytes is then mapped into a new column of bytes. A similar process is applied to the 128-bit key. A byte-by-byte exclusive-OR operation is then performed on the state array using the key. The entire process (state array and key) is repeated 10 times to produce the final result. AES encryption requires hardware upgrades to legacy 802.11 devices. Since 2007, virtually all 802.11 products support AES-CCMP, and WPA2 is required for WFA interoperability certification. Thus, AES should be turned on as it operates efficiently on hardware.

2.6.6   Protection of Management Frames

By default, 802.11 management frames are not encrypted. An attacker can discover the layout of the network via unprotected unicast management frames, pinpoint the location of devices, and launch more successful DoS attacks against the network. 802.11w protects these frames against forgery and provides confidentiality. By employing TKIP or AES, an attacker cannot encrypt a management frame using a valid key, and a station can safely ignore these unencrypted frames. Because broadcast management frames are less common and do not contain sensitive information, 802.11w only protects these frames against forgery. By employing the MIC, which is appended to the nonsecure management frame, the attacker cannot forge broadcast management frames. A broadcast deauthentication attack causes a station to disassociate from the AP and lose its connection. Such broadcast management frames can be further protected using a pair of related one-time keys for the AP and the station.

2.7   IEEE 802.11N AMENDMENT

The 802.11n TG was formed on September 11, 2003. There were 254 members when the amendment was ratified on November 30, 2009. However, draft-n devices were proliferating well before the amendment was ratified. 802.11n defines modifications to both PHY and MAC layers. It provides more than 100 Mbit/s MAC layer throughput over at least 100 ft. To achieve this aim, the raw data rates are much higher. The increased channel bandwidth from 20 MHz in 802.11a/g to 40 MHz is critical in determining the maximum data rate. There are 12 nonoverlapping channels of 40 MHz in the 5 GHz band. In contrast, there are two overlapping 40 MHz channels in the 2.4 GHz band, but only one nonoverlapping 40 MHz channel is available in North America. 802.11n provides full interoperability with legacy 802.11 equipment. For example, a 2.4 GHz 802.11n device must support all mandatory 802.11g rates, while a 5 GHz 802.11n device must support all mandatory 802.11a rates. Unlike contention-based protocols, reservation protocols are useful for stations that carry a high amount of traffic (e.g., HD video traffic). Native 802.11n employs CSMA, but there are provisions to make it more reservation-based using virtual sensing, allowing a station to transmit multiple data frames in a single TXOP.

2.7.1   Data Rates and Dual Band Operation

802.11n operates on 2.4 and 5 GHz unlicensed bands with legacy 20 MHz and new 40 MHz channel bandwidths. The 40 MHz channel is optional. Only OFDM is permitted. It supports a maximum of four transmit and four receive antennas (i.e., 4×4 system), and up to four concurrent spatial streams are permitted. All spatial streams originate from a single station (i.e., a single connection). The GI can be 800 ns (regular) or 400 ns (short). The 400 ns GI is optional. With two spatial streams and a 40 MHz channel, a short GI gives a PHY data rate of 300 Mbit/s. The most common MIMO configuration supporting two spatial streams is 2×2, achieving an aggregate rate of up to 300 Mbit/s. Newer 802.11n home gateways currently support three spatial streams (3×3 MIMO) with an aggregated PHY rate of 450 Mbit/s. The 802.11n PHY data rates (in Mbit/s), as well as the modulation and coding schemes (MCSs) for one and two spatial streams, are listed in Table 2.4. The data rates for a 20 MHz channel and 800 ns GI are mandatory. However, WFA certification requires that only 802.11n APs implement two spatial streams. If an 802.11n station supports only one spatial stream, the AP will switch to the one spatial stream mode since it knows the capabilities of the station. Hence, the 802.11n AP must be configured to activate one or two spatial streams. The total number of data subcarriers in a 20 MHz channel in 802.11n is 56, which is four more than 802.11a/g. This explains why the 58.5 Mbit/s rate for MCS 6 is slightly higher than 802.11a/g (54 Mbit/s). When the channel bandwidth is doubled to 40 MHz, the number of data subcarriers is slightly more than doubled, going from 52 to 108.

The optional 802.11n PHY data rates (in Mbit/s) for three and four spatial streams are listed in Table 2.5. The maximum data rate of 600 Mbit/s gives a spectral efficiency of 15 bit/s/Hz. Figure 2.32 shows that the maximum 802.11n data rates increase linearly as the number of spatial streams increases. The increase is greater for 40 MHz channels when compared with 20 MHz channels. Among the optional data rates, the 40 MHz/400 ns GI combination is the most popular. Dual band 802.11n devices offer either selectable or concurrent operation using the 2.4 and 5 GHz bands. For selectable dual band devices, the user must select one of the

TABLE 2.4   MCSs and PHY Rates in 802.11n (1 and 2 Spatial Streams)

MCS index	Code rate	Modulation	Number of spatial streams	Data rate (Mbit/s)			
				20 MHz		40 MHz	
				800 ns GI[a]	400 ns GI	800 ns GI	400 ns GI
0	1/2	BPSK	1	6.5	7.2	13.5	15
1	1/2	QPSK	1	13	14.4	27	30
2	3/4	QPSK	1	19.5	21.7	40.5	45
3	1/2	16-QAM	1	26	28.9	54	60
4	3/4	16-QAM	1	39	43.3	81	90
5	2/3	64-QAM	1	52	57.8	108	120
6	3/4	64-QAM	1	58.5	65	121.5	135
7	5/6	64-QAM	1	65	72.2	135	150
8	1/2	BPSK	2	13	14.4	27	30
9	1/2	QPSK	2	26	28.9	54	60
10	3/4	QPSK	2	39	43.3	81	90
11	1/2	16-QAM	2	52	57.8	108	120
12	3/4	16-QAM	2	78	86.7	162	180
13	2/3	64-QAM	2	104	115.6	216	240
14	3/4	64-QAM	2	117	130	243	270
15	5/6	64-QAM	2	130	144.4	270	300

[a]Mandatory rates.

frequency bands. Concurrent devices operate in both 2.4 and 5 GHz bands at the same time. In this case, when one band is transmitting, the other band is receiving, thus achieving full-duplex communications.

2.7.2   Error Control

Like legacy 802.11a/g systems, 802.11n employs binary convolutional code (BCC) for error control, which is mandatory. The low-density parity-check (LDPC) coding is optional. The LDPC option defines a systematic block code with possible block lengths of 324, 432, 486, 540, 648, 864, 972, 1080, 1296, 1458, and 1620. It provides additional gain over BCC but uses the same code rates as BCC (i.e., 1/2, 2/3, 3/4, and 5/6). The desired code rate is obtained by puncturing. The parity check matrices are sparse and highly structured to facilitate both encoding and decoding.

2.7.3   High-Throughput Station

An 802.11n station is known as a high-throughput (HT) station. The PHY features that distinguish an HT station from a non-HT station include:

TABLE 2.5   Optional MCSs and PHY Rates in 802.11n (3 and 4 Spatial Streams)

MCS index	Code rate	Modulation	Number of spatial streams	Data rate (Mbit/s) 20 MHz 800 ns GI	20 MHz 400 ns GI	40 MHz 800 ns GI	40 MHz 400 ns GI
16	1/2	BPSK	3	19.5	21.7	40.5	45
17	1/2	QPSK	3	39	43.3	81	90
18	3/4	QPSK	3	58.5	65	121.5	135
19	1/2	16-QAM	3	78	86.7	162	180
20	3/4	16-QAM	3	117	130	243	270
21	2/3	64-QAM	3	156	173.3	324	360
22	3/4	64-QAM	3	175.5	195	364.5	405
23	5/6	64-QAM	3	195	216.7	405	450
24	1/2	BPSK	4	26	28.9	54	60
25	1/2	QPSK	4	52	57.8	108	120
26	3/4	QPSK	4	78	86.7	162	180
27	1/2	16-QAM	4	104	115.6	216	240
28	3/4	16-QAM	4	156	173.3	324	360
29	2/3	64-QAM	4	208	231.1	432	480
30	3/4	64-QAM	4	234	260	486	540
31	5/6	64-QAM	4	260	288.9	540	600

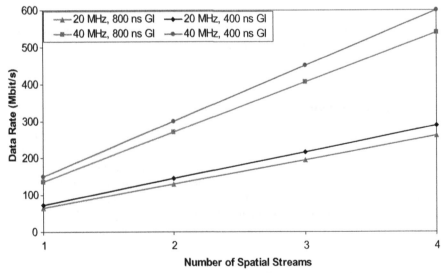

Figure 2.32   Maximum data rate versus number of spatial streams in 802.11n.

- MIMO operation (both transmitter and receiver use multiple antennas)
- Spatial multiplexing (SM) (data streams are transmitted on multiple spatial streams via multiple antennas at transmitter and receiver)
- Spatial mapping (including transmit beamforming) (the mapping of one or more data streams for transmission over multiple spatial streams)
- Space–time block coding (STBC)
- LDPC coding
- Antenna selection (ASEL).

Three PPDU formats are allowed, namely non-HT format, HT-mixed format, and HT-greenfield format, as shown in Figure 2.33. The PPDUs may be transmitted with 20 or 40 MHz bandwidths. The HT station MAC features include:

- Frame aggregation
- BA
- Power save multipoll (PSMP) operation (allocates resources by providing a time schedule for an AP and its stations to access the wireless medium)
- Reverse direction (RD) (reserves wireless medium for transmission and response using the reverse direction protocol—an optional feature that improves the efficiency and latency for bidirectional network traffic, such as TCP SEND followed by TCP ACK)
- Coexistence mechanisms to support non-HT stations.

The Service field contains 16 bits, with bit positions 0 to 6 transmitted first. They are used to synchronize the descrambler in the receiver. The remaining 9 bits (7 to 15) are set to zero and reserved for future use. The PPDU Tail contains six bits of zero to return the error control encoder to the zero state. This improves the error probability of the error control decoder, which relies on future bits when decoding. The number of bits in the Data and Tail fields is a multiple of the number of coded bits in an OFDM symbol (48, 96, 192, or 288 bits). To achieve that, pad bits are inserted if necessary.

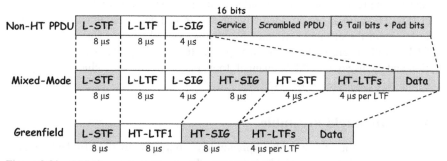

Figure 2.33   PPDUs supported by the 802.11n HT station.

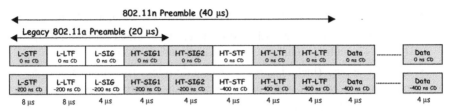

Figure 2.34   802.11n mixed mode preamble for two spatial streams.

2.7.4   Mixed Mode Preamble

The mixed mode preamble is needed for compatibility with IEEE 802.11a/g. It starts with the 802.11a/g preamble. The 802.11n mixed mode preamble for two spatial streams is shown in Figure 2.34. The legacy short training field (L-STF) is identical to 802.11a/g except that different transmitters use different cyclic delays (CDs). This also applies to the legacy long training field (L-LTF). The STFs from different transmitters have low cross-correlation. For example, a CD of −400 ns (or a cyclic advance of 400 ns) minimizes correlation between two different transmitted short symbols. The L-STF uses a CD of only −200 ns for two transmitters, since legacy 802.11a/g receivers may not be able to cope with larger CD values.

Legacy 802.11a/g devices are able to decode the 802.11n preamble up to legacy signal field (L-SIG). This guarantees proper deference behavior when 802.11n frames are detected. The new high throughput signal field (HT-SIG) follows the L-SIG field. There are 48 bits in this field comprising a 16-bit length subfield, 7-bit MCS subfield, and bits to indicate options like LDPC, and the 8-bit CRC. To detect the HT-SIG field, the BPSK constellation is rotated by 90°. The HT short training field (HT-STF) follows the HT-SIG field. This can be used to retrain automatic gain control (AGC). One or more HT long training fields (HT-LTF) follow the HT-STF field. These fields provide channel estimation. The number of HT-LTF symbols is equal to the number of spatial streams. For two spatial streams, the second HT-LTF of the first spatial stream is inverted to create an orthogonal space–time pattern. The receiver can obtain channel estimates for both spatial streams by adding and subtracting the first and second HT-LTF respectively. Channel estimates can then be used to process the MIMO-OFDM data symbols that follow the HT-LTF field.

Recall that it is mandatory to transmit or receive two spatial streams in 802.11n. Optional modes are defined for three and four spatial streams. The optional mixed-mode preamble for four spatial streams is shown in Figure 2.35. In this case, four HT-LTF symbols are encoded with an orthogonal pattern so that the receiver is able to obtain the channel estimates for all four spatial streams (in Figure 2.34, 2 HT-LTF symbols are required for two spatial streams). Together with the short GI option and the use of a 40 MHz channel, four spatial streams can achieve the highest possible raw data rate of 600 Mbit/s.

2.7.5   Greenfield Preamble

As the name implies, the greenfield preamble is applicable to 802.11n deployments that do not include legacy 802.11 a/g stations. In this case, lower training overheads

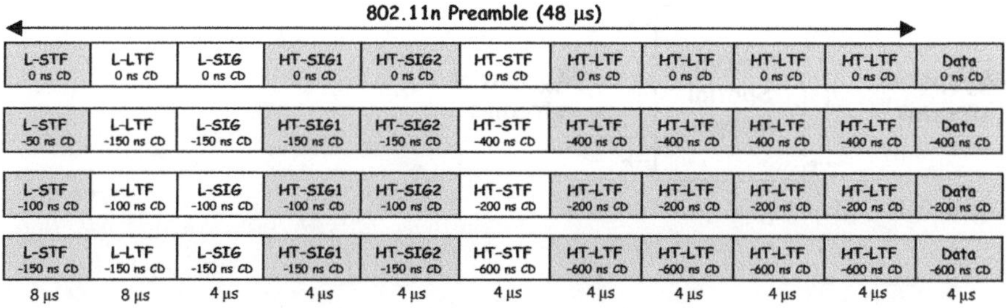

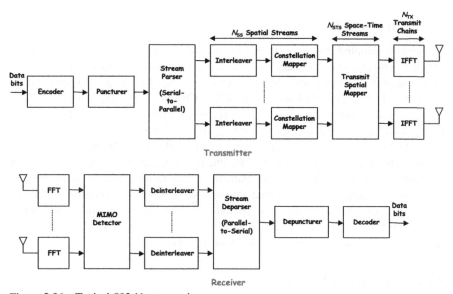

802.11n Preamble (48 μs)
L-STF 0 ns CD | L-LTF 0 ns CD | L-SIG 0 ns CD | HT-SIG1 0 ns CD | HT-SIG2 0 ns CD | HT-STF 0 ns CD | HT-LTF 0 ns CD | HT-LTF 0 ns CD | HT-LTF 0 ns CD | HT-LTF 0 ns CD | Data 0 ns CD
L-STF -50 ns CD | L-LTF -150 ns CD | L-SIG -150 ns CD | HT-SIG1 -150 ns CD | HT-SIG2 -150 ns CD | HT-STF -400 ns CD | HT-LTF -400 ns CD | HT-LTF -400 ns CD | HT-LTF -400 ns CD | HT-LTF -400 ns CD | Data -400 ns CD
L-STF -100 ns CD | L-LTF -100 ns CD | L-SIG -100 ns CD | HT-SIG1 -100 ns CD | HT-SIG2 -100 ns CD | HT-STF -200 ns CD | HT-LTF -200 ns CD | HT-LTF -200 ns CD | HT-LTF -200 ns CD | HT-LTF -200 ns CD | Data -200 ns CD
L-STF -150 ns CD | L-LTF -150 ns CD | L-SIG -150 ns CD | HT-SIG1 -150 ns CD | HT-SIG2 -150 ns CD | HT-STF -600 ns CD | HT-LTF -600 ns CD | HT-LTF -600 ns CD | HT-LTF -600 ns CD | HT-LTF -600 ns CD | Data -600 ns CD
8 μs | 8 μs | 4 μs | 4 μs | 4 μs | 4 μs | 4 μs | 4 μs | 4 μs | 4 μs | 4 μs

Figure 2.35 802.11n mixed mode preamble for four spatial streams.

N_{SS} Spatial Streams
N_{STS} Space-Time Streams
N_{TX} Transmit Chains
Data bits
Encoder
Puncturer
Stream Parser (Serial-to-Parallel)
Interleaver
Constellation Mapper
Transmit Spatial Mapper
IFFT
Transmitter
FFT
MIMO Detector
Deinterleaver
Stream Deparser (Parallel-to-Serial)
Depuncturer
Decoder
Data bits
Receiver

Figure 2.36 Typical 802.11n transceiver.

(8 μs shorter than mixed-mode preamble) are possible, leading to higher net throughput. For example, for 3 spatial streams, the greenfield preamble is 36 μs, whereas the mixed mode preamble is 44 μs. Reservation can be made via the RTS/CTS virtual sensing mechanism. When the medium is successfully reserved, a burst of data frames can be sent using a RIFS of 2 μs.

2.7.6 Transceiver Design

A typical 802.11n transceiver is shown in Figure 2.36. For simplicity, the GI insertion and the RF modules are not shown. Bit interleaving is performed across both spatial streams and subcarriers. This provides spatial and frequency diversity that improve link performance. After interleaving, the scrambled bits are used to generate complex-valued modulation symbols and mapped to QAM carriers. The spatial

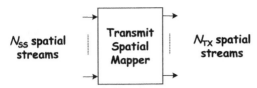

Figure 2.37   Mapping of spatial streams.

stream dependent CD is applied in the frequency domain. Cyclic shift diversity (CSD) insertion prevents unintentional beamforming.

2.7.7   Antenna Selection

The spatial mapping matrix is applied to each subcarrier, as shown in Figure 2.37. This converts N_{SS} spatial stream inputs into N_{TX} transmitter outputs (corresponding to the transmit antenna ports). Mapping can be chosen based on the instantaneous or averaged channel state information (CSI). Channel sounding PPDUs are sent consecutively within a single TXOP. The preamble training fields are used to measure the channel for sounding purposes. Training information is exchanged using the HT Control field. If the transmitter and receiver have ASEL capabilities, training of the transmit and receive antennas can be done one after another. ASEL allows the station to identify the antennas it prefers the transmitting AP to use. 802.11n supports up to eight antennas and up to four spatial streams. If the number of transmitters equals the number of spatial streams, the spatial mapping matrix becomes an identity matrix. To transmit legacy 802.11a/g rates that have only one spatial stream, the spatial mapping matrix reduces to a column of ones.

2.7.8   Subcarrier Mapping

The 20 MHz mode uses 56 subcarriers (data + pilot), whereas 52 subcarriers are used by 802.11a/g. The extra subcarriers increase the throughput at the cost of some extra transmitter complexity to keep the transmitted spectrum within the spectral mask. The legacy part of the mixed-mode preamble and HT-SIG use the same sub-carriers as 802.11a/g (shown as grey blocks in Figure 2.38) [7]. It employs four pilots just like 802.11a/g. The 40 MHz mode uses six pilots. The difference with 802.11a/g is that 802.11n uses space–time mapping for the pilots when transmitting using multiple antennas.

2.7.9   Space–Time Block Coding

STBC is an optional feature that provides extra diversity gain when the number of available transmitters is larger than the number of spatial streams. STBC modes with 1, 2, and 3 spatial streams are specified in 802.11n. It operates on groups of two symbols and maps N_{SS} spatial stream inputs onto N_{STS} space–time streams. Suppose one spatial stream is mapped into two space–time transmit streams. Let $\{d_{ke}, d_{ko}\}$ be

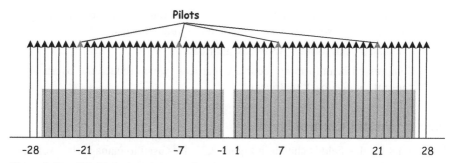

Figure 2.38   802.11n subcarrier mapping.

the data symbols for even and odd symbols for subcarrier k, respectively. STBC maps a single spatial stream input onto two space–time stream outputs $\{d_{ke}, d_{ko}\}$ and $\{-d_{ke}^*, d_{ko}^*\}$. The even symbol contains d_{ke} on the first space–time stream and d_{ko}^* on the second space–time stream. The odd symbol contains d_{ko} on first space–time stream and $-d_{ke}^*$ on the second space–time stream.

2.7.10   Antenna Beamforming

The 802.11n beamformer (at the transmitter) may utilize knowledge of the MIMO channel to generate a steering matrix that improves reception in the beamformee (at the receiver). This optional feature can be implicit or explicit. Beamforming provides extra performance gain when the number of transmit antennas is larger than the number of spatial streams and the transmitter has knowledge about channel. Essentially, it multiplies N_{STS} spatial stream inputs for every subcarrier by an N_{TX} by N_{STS} beamforming matrix. The beamformer replaces the input data vector transmitted in subcarrier k (x_k), which has $N_{STS} \leq N_{TX}$ elements, with $Q_k x_k$, where beamforming matrix Q_k has N_{TX} rows and N_{STS} columns.

$$Y_k = H_k Q_k x_k + n, \tag{2.5}$$

where H_k = channel matrix, Y_k = received vector, and n = noise vector.

The beamforming steering matrix that is computed (or updated) from new channel measurements replaces the existing Q_k for the next beamformed data transmission.

In implicit beamforming, overhead is minimized because the beamformer assumes a reciprocal channel and uses received preambles from the beamformee to calculate the beamforming weights. Transmission direction can be unidirectional or bidirectional. This requires stringent calibration to reduce differences between transmit and receive chains. Correction matrices are computed to ensure that observed channel matrices in both directions of the link are transposes of each other, thus rendering the resultant channel reciprocal.

In explicit beamforming, the beamformer uses the feedback response received from the beamformee to calculate the beamforming feedback matrix for transmit beamforming. There are three types of feedback response:

- CSI containing MIMO channel coefficients derived from the received channel statistics for all subcarriers, receivers, and spatial streams.
- Calculated beamforming feedback matrices.
- Compressed beamforming feedback matrices (which reduce the overhead to transmit the feedback matrices).

2.7.11 MIMO Control Field

The format for this field is shown in Figure 2.39. It is used to manage the exchange of MIMO CSI or transmit beamforming feedback information. It is used in the CSI, noncompressed beamforming, and compressed beamforming frames. The beamformer can discard the feedback response if the timing synchronization function (TSF) time interval minus the sounding timestamp value is greater than coherence time of the propagation medium.

2.7.12 HT Capabilities Element

This is used to advertise the optional HT capabilities of an 802.11n HT station. It is present in management frames, such as Beacon, Association Request/Response, Reassociation Request/Response, Probe Request/Response. The format is shown in Figure 2.40. The Transmit STBC field indicates whether STBC is supported. The

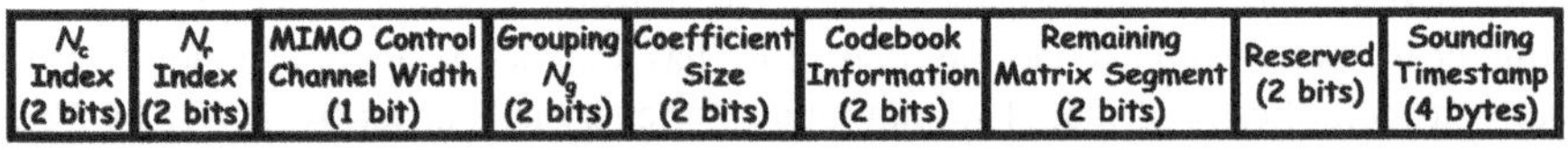

N_c Index (2 bits)	N_r Index (2 bits)	MIMO Control Channel Width (1 bit)	Grouping N_g (2 bits)	Coefficient Size (2 bits)	Codebook Information (2 bits)	Remaining Matrix Segment (2 bits)	Reserved (2 bits)	Sounding Timestamp (4 bytes)

Figure 2.39 802.11n MIMO control field. N_c, number of columns in matrix −1; N_r, number of rows in matrix −1; MIMO control channel width, indicates whether 20 MHz or 40 MHz channel is used; N_g, number of carriers grouped into one (can be 0, 1, or 2); coefficient size, indicates number of bits in representation of real and imaginary parts of each element in matrix (caters for CSI and noncompressed beamforming feedback); codebook information, indicates size of codebook entries; remaining matrix segment, contains remaining segment number for associated measurement report; sounding timestamp, contains lower 4 bytes of TSF timer value.

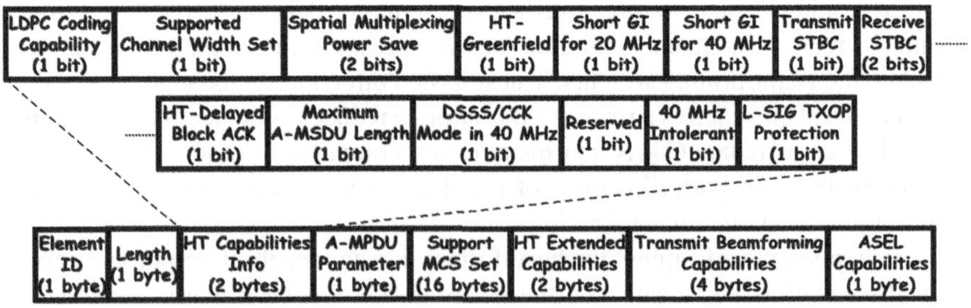

LDPC Coding Capability (1 bit)	Supported Channel Width Set (1 bit)	Spatial Multiplexing Power Save (2 bits)	HT-Greenfield (1 bit)	Short GI for 20 MHz (1 bit)	Short GI for 40 MHz (1 bit)	Transmit STBC (1 bit)	Receive STBC (2 bits)

HT-Delayed Block ACK (1 bit)	Maximum A-MSDU Length (1 bit)	DSSS/CCK Mode in 40 MHz (1 bit)	Reserved (1 bit)	40 MHz Intolerant (1 bit)	L-SIG TXOP Protection (1 bit)

Element ID (1 byte)	Length (1 byte)	HT Capabilities Info (2 bytes)	A-MPDU Parameter (1 byte)	Support MCS Set (16 bytes)	HT Extended Capabilities (2 bytes)	Transmit Beamforming Capabilities (4 bytes)	ASEL Capabilities (1 byte)

Figure 2.40 HT capabilities element.

Receive STBC indicates whether STBC is supported, and if so, whether 1, 2, or 3 spatial streams are supported. The maximum A-MSDU length can be set to 3839 or 7935 bytes. The 8-bit A-MPDU parameter gives the maximum length of the A-MPDU (2 bits) and the minimum MPDU spacing (3 bits). The other 3 bits are reserved. The length of the A-MPDU is given as $2^{(13+MaxExponent)} - 1$ bytes, where MaxExponent takes values from 0 to 3. Clearly, the minimum A-MPDU length is $2^{13} - 1$ or 8191 bytes. This is followed by the minimum time between the start of adjacent MPDUs within an A-MPDU that the station can receive. There are eight possible values: 0 for no restriction, 1 for 1/4 μs, 2 for 1/2 μs, 3 for 1 μs, 4 for 2 μs, 5 for 4 μs, 6 for 8 μs, and 7 for 16 μs.

2.7.13 MAC Enhancements

802.11n provides frame aggregation to minimize the impact of the preamble overhead. The maximum length Ethernet frame (1500 bytes) takes only 40 μs to transmit at 300 Mbit/s. This is the same duration as the mixed-mode preamble for 2 spatial streams. Thus, the net throughput is reduced by 50% with the preamble overhead. There are two aggregation types: MSDU and MPDU frame aggregation. Several MSDUs can be aggregated into one A-MSDU of up to 7935 bytes. Similarly, it is possible to aggregate MPDUs into one A-MPDU of up to 65,535 bytes. All MSDUs within an A-MSDU are targeted to the same destination address. The destination and source addresses of the MSDU are included in the MPDU header. However, an A-MPDU may comprise MPDUs that are targeted to different destination addresses, although the source address remains the same as the MSDU. Additional mechanisms to support frame aggregation include RTS/CTS bandwidth reservation and the use of BAs. The aggregation procedure is shown in Figure 2.41. For a 300 Mbit/s data rate using a 40 MHz channel and two spatial streams (PHY overhead of 40 μs), the total time to transmit a maximum length A-MPDU becomes $(65,535 \times 8)/(300 \times 10^6)$ or 1747.6 μs (ignoring the tail and pad bits, and zero MPDU time spacing). This gives a high data efficiency of almost 98%.

2.7.14 MPDU Header

The 802.11n MPDU header is shown in Figure 2.42. The Frame Control, Duration/ID, Address 1 fields are present in all data frames. Other fields are present only in certain frame types. The QoS Control and HT Control fields are absent in 802.11a/g. The Sequence Control, QoS Control, and HT Control fields are optional fields. For example, they are not present in control frames. The 16-bit Sequence Control contains a 4-bit fragment number and a 12-bit sequence number of a MPDU or management MPDU (MMPDU). The 16-bit QoS Control field is introduced by 802.11e and is present in all data frames when the Subtype field is set to 1. It contains information on the traffic identifier or TID (which identifies the traffic category or stream corresponding to the MSDU and the requested TXOP), TXOP limit imposed by the AP (32–8160 μs), TXOP duration requested by the station (in units of 32 μs), queued data frames at the station, and power-save buffered frames at the AP. The

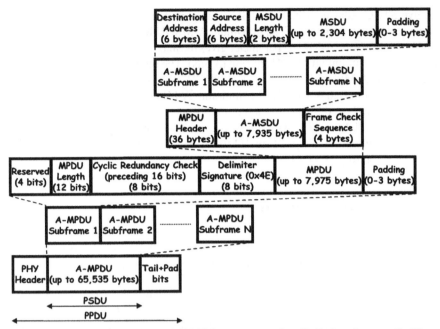

Figure 2.41 802.11n MSDU and MPDU frame aggregation. Delimiter signature 0x4E corresponds to ASCII character "N." MAC padding makes each MPDU/MSDU a multiple of 4 bytes in length. Length field allows receiver to support variable-size MAC payload and locate padding or FCS bytes. More length bits for MSDU than MPDU ($2^{16} = 65,536$ and $2^{12} = 4096$). PHY header is fixed duration in μs.

acknowledgment policy can be normal ACK, no ACK (e.g., for multicast and broadcast operation), no explicit ACK (used in polling messages), or BA.

The 16-bit HT control field is introduced by 802.11n. 802.11n provides a field for link adaptation, allowing the receiver to send MCS feedback (MFB) to the sender. The receiver may recommend the MCS based on received channel estimates and the received signal-to-noise ratio (SNR). The actual algorithm to perform link adaption is vendor-specific. This is an important function that determines spatial multiplexing gain. For example, lower-order modulations, such QPSK and 16-QAM, tend to be more robust against interference and noise, and hence, can typically operate at longer distances but at lower data rates than higher-order modulations. Full spatial multiplexing gains are normally obtained for such modulations. However, a higher-order modulation, such as 64-QAM, may not provide the expected multiplexing gains because the shorter operating range reduces the probability of rich multipath scattering.

2.7.15 Frame Types and MAC Addresses

Just like legacy 802.11 a/g amendments, 802.11n uses three frame types, namely data frames, control frames (e.g., RTS, CTS, ACK, power-save polling messages),

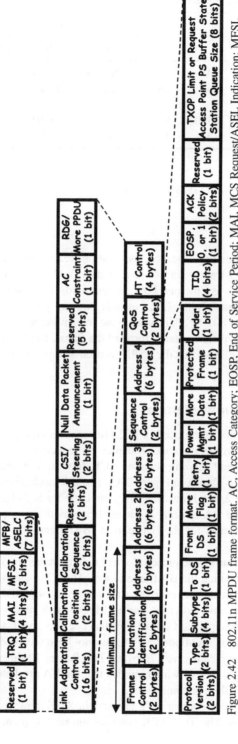

Figure 2.42 802.11n MPDU frame format. AC, Access Category; EOSP, End of Service Period; MAI, MCS Request/ASEL Indication; MFSI, MCS Feedback Sequence Identifier; MFB/ASELC, MCS Feedback and ASEL Command/data; PS, Power Save; RDG, Reverse Direction Grant; TID, Traffic Identifier; TRQ, Training Request.

TABLE 2.6 Address Fields in 802.11n

Function	To DS	From DS	Address 1	Address 2	Address 3 MSDU	Address 3 A-MSDU	Address 4 MSDU	Address 4 A-MSDU
IBSS	0	0	DA[b]	SA[b]	BSSID	BSSID	–	–
From AP	0	1	DA[b]	BSSID	SA[b]	BSSID	–	–
To AP	1	0	BSSID	SA[b]	DA[b]	BSSID	–	–
DS	1	1	RA[a]	TA[a]	DA[b]	BSSID	SA[b]	BSSID

[a]Single-hop (local) significance.

[b]End-to-end significance.

RA, Receive Address; DA, Destination Address; TA, Transmit Address; SA, Source Address.

and management frames (e.g., beacon, association, reassociation, and authentication). Only the control frames do not have a data field. The length of the MAC address field is the same as Ethernet. Four address fields are present in 802.11n (and legacy 802.11a/g) frames (Table 2.6), but only two are present in Ethernet or IP headers. The additional 802.11 addresses are the transmitter and receiver addresses, which denote the stations that actually forward the frame, and this may change with each hop in the network. Note that the BSSID is used when transmitting an A-MSDU over the DS.

2.7.16 Block Acknowledgment

The BA mechanism (first introduced in 802.11e) enables multiple frame transmission per station. It improves network efficiency by reducing the need to recontend for channel resources and sending separate ACK frames. The setup and teardown of the BA mechanism is shown in Figure 2.43. The BA mechanism can be immediate or delayed (Figure 2.44). Immediate BA is suitable for high-bandwidth and low latency traffic. Delayed BA is suitable for applications that can tolerate moderate latency. The frame formats for the BA request (BAR) and BA frames are shown in Figure 2.45.

The 16-bit BA Starting Sequence Control field contains the sequence number of the first MSDU for which the BA frame is sent. The 128-byte BA bitmap is used by the BA frame to indicate the receiving status of up to 64 MSDUs. If bit n of the BA bitmap is set to 1, this indicates that an MPDU with sequence control value equal to [BA Starting Sequence Control + n] has been received. Conversely, if bit n is set to 0, then the aforementioned MPDU has not been received. The sum of the BA request and BA overheads give 176 bytes. This is over 12 times higher than the regular 802.11 ACK (14 bytes), which implies that the BA mechanism is beneficial only if at least 12 MPDUs are transmitted.

In TCP, the sequence number is 32 bits and the ACK number is 32 bits, giving a total of only 8 bytes. The TCP ACK contains the next sequence number of the

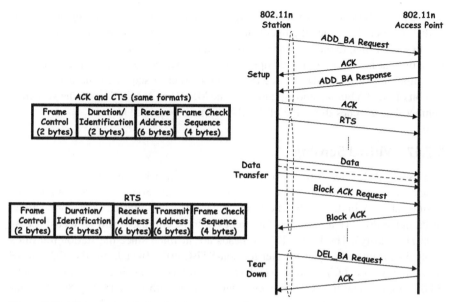

Figure 2.43   BA setup and tear down.

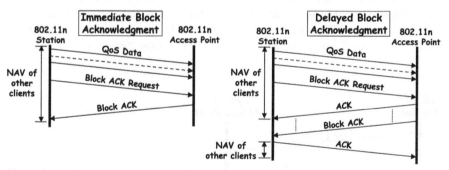

Figure 2.44   Immediate and delayed BA.

Block Acknowledgment Request (BAR) Frame (24 bytes)

Frame Control (2 bytes)	Duration/ ID (2 bytes)	Receive Address (6 bytes)	Transmit Address (6 bytes)	BA Control (2 bytes)	BA Starting Sequence Control (2 bytes)	Frame Check Sequence (4 bytes)

Block Acknowledgment (BA) Frame (152 bytes)

Frame Control (2 bytes)	Duration/ ID (2 bytes)	Receive Address (6 bytes)	Transmit Address (6 bytes)	BA Control (2 bytes)	BA Starting Sequence Control (2 bytes)	BA Bitmap[a] (128 bytes)	Frame Check Sequence (4 bytes)

Figure 2.45   BA request and BA frame. [a]An 8-byte compress BA bitmap can also be used.

data segment the sender is expecting to receive, and acknowledges all segments prior to that number. To reduce the overhead, 802.11n provides a compressed BA bitmap of 8 bytes (64 bits) to indicate the received status of up to 64 MSDUs and A-MSDUs. Each bit that is set to 1 in the compressed BA bitmap acknowledges the successful reception of a single MSDU or A-MSDU in the order of sequence number, with the first bit of the BA bitmap corresponding to the MSDU or A-MSDU with the sequence number that matches the value of the BA Starting Sequence Control.

2.7.17   Virtual Sensing

The use of RTS/CTS virtual sensing provides efficient bandwidth reservation and added reliability in multiframe data transfer by a single 802.11n station or AP. This is illustrated in Figure 2.46. The term "reliability" is preferred over "protection" (used in the 802.11n amendment) in order not to be confused with protected access (802.11 security). The RTS/CTS overheads are justified since the station transmits many data frames. In order to cater for non-STBC 802.11n stations, the 802.11n AP may employ a dual CTS mode, which can be initiated by an 802.11n STBC or non-STBC station. The first CTS prevents interference with other nearby 802.11n stations, whereas the second CTS prevents interference with legacy 802.11a/g systems. The AP issues two CTS and two CF-End frames. The CF-End frame signifies end of contention-free (CF) period. A key distinction here is the removal of carrier sensing before sending the 802.11n RTS frame (i.e., unlike 802.11a/g, SIFS is used instead of DIFS). Thus, all legacy 802.11a/g stations have to defer to the 802.11n RTS (due to carrier sensing), and hence, 802.11n stations have priority over legacy 802.11a/g stations in data transmission. However, this may result in a higher probability of RTS collision, especially after the end of a long transmission period.

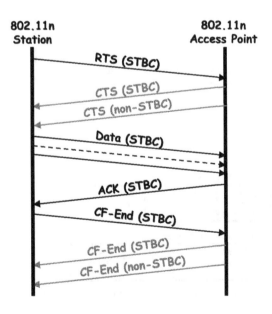

Figure 2.46   Bandwidth reservation in 802.11n using virtual sensing.

2.7.18 Use of 40 MHz Channels

In the 5 GHz band, the use of a 40 MHz mixed-mode preamble or the use of a duplicate 802.11a RTS/CTS in both primary and secondary 20 MHz channels ensures all legacy 802.11a stations can correctly defer to each 40 MHz transmission. At the same time, the 40 MHz 802.11n devices must be able to correctly defer to legacy stations. This requires the 40 MHz station to perform carrier sensing in both the primary and secondary channels. In the 2.4 GHz band, using a 40 MHz channel is more restrictive than the 5 GHz band. Recall that fewer channels are available: only three nonoverlapping 20 MHz channels, two overlapping 40 MHz channels, or one nonoverlapping 40 MHz channel. Thus, due to the greater interference potential, 40 MHz operation should be avoided in the 2.4 GHz band and is turned off by default. Nevertheless, some hardware vendors claim to implement a 40 MHz channel at 2.4 GHz, which offers more than double the available data rate than a 20 MHz channel (compare columns 5 and 7 in Table 2.4). In some cases, a longer GI is used to provide greater resistance to interference at 2.4 GHz since the extra overheads associated with a longer GI will not reduce the data rate significantly (compare columns 7 and 8 in Table 2.4). The overall benefits are a longer operating range than 5 GHz channels and a higher rate than 20 MHz channels using the 2.4 GHz band. However, the WFA mandates that certified 802.11n products must switch from a 40 MHz channel to a 20 MHz channel if other users are detected. Since the 2.4 GHz band is usually congested, the 40 MHz mode is most likely turned off, otherwise frequent switching to a 20 MHz channel may lead to poorer performance. Unlike a 20 MHz channel grid in the 5 GHz band, the 2.4 GHz channels are partially overlapped as center frequencies are specified on a 5 MHz grid. Hence, it is not possible to transmit a mixed-mode preamble that can be correctly received by all legacy devices due to interference from overlapping channels.

2.8 NEW IEEE 802.11 MULTIGIGABIT TASK GROUPS

Two 802.11 multigigabit TGs evolved from the Very High Throughput (VHT) study group formed in 2007. Two frequency bands were considered, namely 5 and 60 GHz. This mirrors the development of the 802.11b (2.4 GHz) and 802.11a (5 GHz) amendments in the late 1990s. The 802.11ac TG was formed in September 2008 and focuses on operating frequencies in the 5 GHz band. The 802.11ad TG was formed in December 2008. It is developing a millimeter-wave amendment in the 60 GHz band. Draft versions of the amendments are available, and these technical specifications are expected to be finalized by end-2012 with ratification in late-2013. 802.11ac is likely to become mainstream for high-rate wireless network deployment, requiring only dual-band devices for backward compatibility to legacy devices. 802.11ad will demand the more expensive option of tri-band devices. WFA certification of draft-802.11ac products will start in early 2013. A bit rate that exceeds 1 Gbit/s is ideal for in-home wireless applications, where several 1080p HD videos can be streamed at a time. HD videos have become pervasive given the intense competition between satellite, cable, and telcos to offer the highest number of HD channels.

2.9 IEEE 802.11ac AMENDMENT

The aim of the TG is to achieve aggregate throughputs beyond 1 Gbit/s in the 5 GHz band. This is the first time an 802.11 amendment attempts to improve the total network throughput (containing concurrent links) rather than improving the throughput of a single link. It is also the first time the 2.4 GHz band is completely omitted. The throughput is increased by several mechanisms: MU-MIMO, wider channel bandwidths (80 and 160 MHz), and 256-QAM. The use of wider bandwidths and 256-QAM only improve the single link throughput. In contrast, MU-MIMO achieves higher network throughput by aggregating multiple links. Unlike dual band 802.11n devices, 802.11ac uses only the 5 GHz band, which reduces cost. Relatively simple stations with 1 or 2 antennas can be used while the AP may have up to eight antennas. Table 2.7 shows a comparison of the mandatory and optional OFDM metrics in 802.11ac and 802.11n.

2.9.1 Multiuser MIMO

Many wireless networks allow multiple active stations to share the available bandwidth. If this sharing is done asynchronously (i.e., using DCF or EDCA), then the overall throughput can only be increased by increasing the link rate for all stations. However, many stations cannot transmit at rates higher than 802.11n because they only have one or two antennas. In addition, the AP can only serve one station at a

TABLE 2.7 Mandatory and Optional OFDM Metrics in 802.11ac and 802.11n

	Mandatory	Optional
802.11n (2.4 and 5 GHz)		
Channel bandwidth	20 MHz	40 MHz
IFFT size	64	128
Data subcarriers	52	108
Subcarrier spacing	312.5 KHz	
OFDM symbol duration	4 μs (800 ns GI)	3.6 μs (400 ns GI)
Modulation	BPSK, QPSK, 16-QAM, 64-QAM	
FEC	BCC	LDPC
802.11ac (5 GHz only)		
Channel bandwidth	20, 40, 80 MHz	160, 80 + 80 MHz
IFFT size	64, 128, 256	512
Data subcarriers	52, 108, 234	468
Subcarrier spacing	312.5 KHz	
OFDM symbol duration	4 μs (800 ns GI)	3.6 μs (400 ns GI)
Modulation	BPSK, QPSK, 16-QAM, 64-QAM	256-QAM
FEC	BCC	LDPC

time. Multiple stations will have to wait for their turn to transmit or receive. MU-MIMO relieves the network bottleneck at the AP by allowing concurrent transmission of different spatial streams to different stations (as opposed to the same station in 802.11n).

In 802.11ac, the MU-MIMO mode is defined with up to eight spatial streams divided across up to four different stations. MU-MIMO is implemented on the AP and not on the station. Thus, it is only relevant for transmissions originating from the AP and directed at the stations. With an 80 MHz channel, it is possible for the AP to send data to four stations simultaneously at 866.7 Mbit/s per station, if each station can demultiplex two spatial streams. The total data rate becomes 3.46 Gbit/s, four times higher than without MU-MIMO. The data rate per station is also higher because MU-MIMO frames can be transmitted at the maximum data rate per station. Without MU-MIMO, each station can only be transmitted to about a quarter of the time such that the effective per-station throughput is a quarter of its maximum. In practice, the throughput of MU-MIMO is reduced slightly since the ACKs are still sent sequentially in time [8]. Depending on the SNRs for all stations, it may not be possible to maintain the maximum data rates because a MU-MIMO link will lose some SNR relative to a single-station link. The use of MU-MIMO will remove the need for virtual SSIDs. In a typical home, an independent spatial stream may serve a room with a user device, eliminating contention in shared channel access and hidden collisions. More importantly, MU-MIMO can separate low and high rate user devices, solving the throughput degradation problem for high rate devices when operated with low rate devices over the same channel.

There are several challenges in implementing MU-MIMO. Link adaptation (i.e., changing the MCS) is needed, but the SNR experienced by the stations may vary. In addition, there are potential switches between single-station and multistation frames. Finally, accurate channel knowledge is needed to minimize interstation interference. The draft amendment provides mechanisms for stations to recommend a certain MCS. It is up to vendors to choose an appropriate mechanism to perform link adaptation. The draft amendment also specifies a single method of explicit compressed feedback to enable both MU-MIMO and single-station beamforming, which will further improve network performance. The single method encourages broad use and ensures interoperability, which is in contrast to 802.11n, where the presence of multiple feedback options prevented the widespread adoption of beamforming. 802.11ac defines an optional beamforming protocol, which consists of the AP sending a null data frame for channel sounding and then receiving feedback from the station, followed by the exchange of beamformed data frames. Currently, the WFA has developed a test for 802.11ac devices that implement the optional beamforming feature.

2.9.2 Use of 256-QAM

The highest order constellation in 802.11 has been 64-QAM since the adoption of 802.11a in September 1999. The use 256-QAM can further boost the data rates of 802.11ac. The maximum code rate of 5/6 remains the same as in 802.11n. With 256-QAM, the maximum single station rate for an 80 MHz device with two spatial streams becomes 866.7 Mbit/s versus 650 Mbit/s for 64-QAM. However, higher SNR is required to operate in this mode.

2.9.3   Available Bandwidth

In 802.11n, channel bandwidths of 20 and 40 MHz are defined. In 802.11ac, two more bandwidths of 80 and 160 MHz are offered. The wider channels not only offer higher data rates, but also greater frequency diversity (the coherence bandwidth becomes much smaller compared with the channel bandwidth), thus allowing high-rate stations to operate robustly using only one antenna. This advantage, coupled with the high data rate for a single stream (up to 866.7 Mbit/s), implies that a single transmitter may be sufficient for the station. For handheld devices, such as tablets and smartphones, the use of a single transmitter allows more cost-effective and energy-efficient RF hardware design than devices equipped with multiantenna transmitters. At the AP, the transmitters may operate in the MU-SISO mode, which simplifies hardware design at the expense of less efficient use of channel bandwidth than MU-MIMO. Each set of OFDM subcarriers allotted to a station in MU-SISO is served by an independent antenna at the AP.

In the 80 MHz mode, two adjacent 40 MHz channels are used (carrier aggregation) with some extra subcarriers to fill the unused subcarriers between two adjacent 40 MHz channels. In the 160 MHz mode, two separate 80 MHz channels can be used without any subcarrier filling in the middle of these two channels. Thus, the two 80 MHz channels do not have to be adjacent. This increases the probability of finding a 160 MHz channel at the cost of additional hardware to send and receive on two non-adjacent 80 MHz channels. Alternatively, a contiguous 160 MHz channel (single carrier) can be used. With a 160 MHz channel, a maximum data rate of 866.7 Mbit/s can be achieved for a single spatial stream using 256-QAM with a 5/6 code rate, and a short GI. However, there are only three 160 MHz channels in the 5 GHz band, which restricts frequency reuse in a manner similar to the three non-overlapping channels in the 2.4 GHz band. Like 802.11n, since the subcarrier spacing is fixed, the data rate is increased by employing more data subcarriers (Table 2.8). Note that the number of data subcarriers for 160 MHz is exactly twice the number of data subcarriers for 80 MHz to allow non-adjacent 80 + 80 MHz channel operation. First generation 802.11ac products are unlikely to support 160 MHz channels.

2.9.4   Modulation and Coding Schemes

Tables 2.9–2.12 show the MCSs for 802.11ac. With 8 spatial streams, a maximum data rate of 6.933 Gbit/s is achievable. MCSs 0 to 7 for 1 spatial stream, 800 ns GI, and channel bandwidths of 20, 40, and 80 MHz are mandatory. Some MCSs are not defined because they result in a fractional number of data bits when all subcarriers are simultaneously transmitted within a symbol period. For example, a code rate of 5/6 for 256-QAM using a 20 MHz channel results in $5/6 \times 8 \times 52$ or 346.7 bits per symbol period. However, there are exactly 720 bits per symbol period for a 40 MHz channel. Hence, the MCS is defined for this channel bandwidth. The unavailability

TABLE 2.8   OFDM Metrics in 802.11ac

Channel bandwidth (MHz)	Subcarrier spacing (KHz)	Total subcarriers (IFFT Size)	Data subcarriers	Pilot subcarriers	Data efficiency (%)
20	312.5	64	52	4	93
40	312.5	128	108	6	95
80	312.5	256	234	8	97
160	312.5	512	468	16	97

of some MCSs leads to a nonlinear variation in the maximum 802.11ac data rates (Figure 2.47). However, with the exception of seven spatial streams at 160 MHz, the rates generally increase linearly as the number of spatial streams increases. In addition, like 802.11n, a higher rate increase is more evident for wider channels. Several draft-ac home routers and a limited number of draft-ac USB adapters are available. Currently, the most popular rate is 1.3 Gbit/s, which can be achieved using an 80 MHz channel, 256-QAM, and three spatial streams.

2.9.5   Interoperability

802.11ac devices have to be fully interoperable with 5 GHz 802.11n and 802.11a devices. 802.11ac only applies to the 5 GHz band since there is no room in the 2.4 GHz band for 80 and 160 MHz channels. Coexistence with existing 802.11a devices is ensured by having a backward compatible preamble. The first part of the 802.11ac preamble is identical to 802.11a up to and including the 802.11a Signal field. The Signal field contains a length field and a 6 Mbit/s rate indication, to ensure that any existing 11a device will defer for the correct frame duration.

2.10   IEEE 802.11AD AMENDMENT

Early 60 GHz standardization efforts were initiated by the 802.15.3c TG for wireless personal area networks (WPANs), which was formed in March 2005 (http://www.ieee802.org/15/pub/TG3c.html). Unlike the WirelessHD specification (issued on October 2007), which adopts OFDM as the only PHY specification, the 802.15.3c amendment (issued on Oct 2009) supports both single-carrier (SC) and OFDM PHYs. 802.15.3c contains a low power option to achieve a minimum throughput of 1 Gbit/s. Longer range (e.g., 10 m) can be achieved with antenna beamforming. Subsequently, the Wireless Gigabit (WiGig) Alliance (http://wirelessgigabitalliance.org), a consortium of CE and chip vendors, was formed in May 2009 to develop and promote 60 GHz wireless communications. The WiGig specification (issued on May 2010) forms the basis of the 802.11ad draft amendment (issued on September 2010). It offers a maximum data rate of 7 Gbit/s, which is higher than the 5.28 Gbit/s rate in 802.15.3c.

TABLE 2.9 MCSs and PHY Rates in 802.11ac (1 and 2 Spatial Streams)

MCS index	Code rate	Modulation	Number of spatial streams	Data rate (Mbit/s) 20 MHz 800 ns GI	20 MHz 400 ns GI	40 MHz 800 ns GI	40 MHz 400 ns GI	80 MHz 800 ns GI	80 MHz 400 ns GI	160 MHz 800 ns GI	160 MHz 400 ns GI
0	1/2	BPSK	1	6.5	7.2	13.5	15	29.3	32.5	58.5	65
1	1/2	QPSK	1	13	14.4	27	30	58.5	65	117	130
2	3/4	QPSK	1	19.5	21.7	40.5	45	87.8	97.5	175.5	195
3	1/2	16-QAM	1	26	28.9	54	60	117	130	234	260
4	3/4	16-QAM	1	39	43.3	81	90	175.5	195	351	390
5	2/3	64-QAM	1	52	57.8	108	120	234	260	468	520
6	3/4	64-QAM	1	58.5	65	121.5	135	263.3	292.5	526.5	585
7	5/6	64-QAM	1	65	72.2	135	150	292.5	325	585	650
8	3/4	256-QAM	1	78	86.7	162	180	351	390	702	780
9	5/6	256-QAM	1	–	–	180	200	390	433.3	780	866.7
0	1/2	BPSK	2	13	14.4	27	30	58.5	65	117	130
1	1/2	QPSK	2	26	28.9	54	60	117	130	234	260
2	3/4	QPSK	2	39	43.3	81	90	175.5	195	351	390
3	1/2	16-QAM	2	52	57.8	108	120	234	260	468	520
4	3/4	16-QAM	2	78	86.7	162	180	351	390	702	780
5	2/3	64-QAM	2	104	115.6	216	240	468	520	936	1040
6	3/4	64-QAM	2	117	130	243	270	526.5	585	1053	1170
7	5/6	64-QAM	2	130	144.4	270	300	585	650	1170	1300
8	3/4	256-QAM	2	156	173.3	324	360	702	780	1404	1560
9	5/6	256-QAM	2	–	–	360	400	780	866.7	1560	1733.3

TABLE 2.10 MCSs and PHY Rates in 802.11ac (3 and 4 Spatial Streams)

MCS index	Code rate	Modulation	Number of spatial streams	20 MHz 800 ns GI	20 MHz 400 ns GI	40 MHz 800 ns GI	40 MHz 400 ns GI	80 MHz 800 ns GI	80 MHz 400 ns GI	160 MHz 800 ns GI	160 MHz 400 ns GI
0	1/2	BPSK	3	19.5	21.7	40.5	45	87.8	97.5	175.5	195
1	1/2	QPSK	3	39	43.3	81	90	175.5	195	351	390
2	3/4	QPSK	3	58.5	65	121.5	135	263.3	292.5	526.5	585
3	1/2	16-QAM	3	78	86.7	162	180	351	390	702	780
4	3/4	16-QAM	3	117	130	243	270	526.5	585	1053	1170
5	2/3	64-QAM	3	156	173.3	324	360	702	780	1404	1560
6	3/4	64-QAM	3	175.5	195	364.5	405	–	–	1579.5	1755
7	5/6	64-QAM	3	195	216.7	405	450	877.5	975	1755	1950
8	3/4	256-QAM	3	234	260	486	540	1053	1170	2106	2340
9	5/6	256-QAM	3	260	288.9	540	600	1170	1300	–	–
0	1/2	BPSK	4	26	28.9	54	60	117	130	234	260
1	1/2	QPSK	4	52	57.8	108	120	234	260	468	520
2	3/4	QPSK	4	78	86.7	162	180	351	390	702	780
3	1/2	16-QAM	4	104	115.6	216	240	468	520	936	1040
4	3/4	16-QAM	4	156	173.3	324	360	702	780	1404	1560
5	2/3	64-QAM	4	208	231.1	432	480	936	1040	1872	2080
6	3/4	64-QAM	4	234	260	486	540	1053	1170	2106	2340
7	5/6	64-QAM	4	260	288.9	540	600	1170	1300	–	–
8	3/4	256-QAM	4	312	346.7	648	720	1404	1560	2808	3120
9	5/6	256-QAM	4	–	–	720	800	1560	1733.3	3120	3466.7

TABLE 2.11 MCSs and PHY Rates in 802.11ac (5 and 6 Spatial Streams)

MCS index	Code rate	Modulation	Number of spatial streams	Data rate (Mbit/s) 20 MHz 800 ns GI	20 MHz 400 ns GI	40 MHz 800 ns GI	40 MHz 400 ns GI	80 MHz 800 ns GI	80 MHz 400 ns GI	160 MHz 800 ns GI	160 MHz 400 ns GI
0	1/2	BPSK	5	32.5	36.1	67.5	75	146.3	162.5	292.5	325
1	1/2	QPSK	5	65	72.2	135	150	292.5	325	585	650
2	3/4	QPSK	5	97.5	108.3	202.5	225	438.8	487.5	877.5	975
3	1/2	16-QAM	5	130	144.4	270	300	585	650	1170	1300
4	3/4	16-QAM	5	195	216.7	405	450	877.5	975	1755	1950
5	2/3	64-QAM	5	260	288.9	540	600	1170	1300	2340	2600
6	3/4	64-QAM	5	292.5	325	607.5	675	1316.3	1462.5	2632.5	2925
7	5/6	64-QAM	5	325	361.1	675	750	1462.5	1625	2925	3250
8	3/4	256-QAM	5	390	433.3	810	900	1755	1950	–	–
9	5/6	256-QAM	5	–	–	900	1000	1950	2166.7	3900	4333.3
0	1/2	BPSK	6	39	43.3	81	90	175.5	195	351	390
1	1/2	QPSK	6	78	86.7	162	180	351	390	702	780
2	3/4	QPSK	6	117	130	243	270	526.5	585	1053	1170
3	1/2	16-QAM	6	156	173.3	324	360	702	780	1404	1560
4	3/4	16-QAM	6	234	260	486	540	1053	1170	2106	2340
5	2/3	64-QAM	6	312	346.7	648	720	1404	1560	2808	3120
6	3/4	64-QAM	6	351	390	729	810	1579.5	1755	3159	3510
7	5/6	64-QAM	6	390	433.3	810	900	1755	1950	–	–
8	3/4	256-QAM	6	468	520	972	1080	2106	2340	4212	4680
9	5/6	256-QAM	6	520	577.8	1080	1200	–	–	4680	5200

TABLE 2.12 MCSs and PHY Rates in 802.11ac (7 and 8 Spatial Streams)

MCS index	Code rate	Modulation	Number of spatial streams	Data rate (Mbit/s) 20 MHz 800 ns GI	20 MHz 400 ns GI	40 MHz 800 ns GI	40 MHz 400 ns GI	80 MHz 800 ns GI	80 MHz 400 ns GI	160 MHz 800 ns GI	160 MHz 400 ns GI
0	1/2	BPSK	7	45.5	50.6	94.5	105	204.8	227.5	409.5	455
1	1/2	QPSK	7	91	101.1	189	210	409.5	455	819	910
2	3/4	QPSK	7	136.5	151.7	283.5	315	–	–	1228.5	1365
3	1/2	16-QAM	7	182	202.2	378	420	819	910	1638	1820
4	3/4	16-QAM	7	273	303.3	567	630	1228.5	1365	–	–
5	2/3	64-QAM	7	364	404.4	756	840	1638	1820	3276	3640
6	3/4	64-QAM	7	409.5	455	850.5	945	–	–	3685.5	4095
7	5/6	64-QAM	7	455	505.6	945	1050	–	–	–	–
8	3/4	256-QAM	7	546	606.7	1134	1260	–	–	–	–
9	5/6	256-QAM	7	–	–	1260	1400	2730	3033.3	–	–
0	1/2	BPSK	8	52	57.8	108	120	234	260	468	520
1	1/2	QPSK	8	104	115.6	216	240	468	520	936	1040
2	3/4	QPSK	8	156	173.3	324	360	702	780	1404	1560
3	1/2	16-QAM	8	208	231.1	432	480	936	1040	1872	2080
4	3/4	16-QAM	8	312	346.7	648	720	1404	1560	2808	3120
5	2/3	64-QAM	8	416	462.2	864	960	1872	2080	–	–
6	3/4	64-QAM	8	468	520	972	1080	2106	2340	4212	4680
7	5/6	64-QAM	8	520	577.8	1080	1200	–	–	4680	5200
8	3/4	256-QAM	8	624	693.3	1296	1440	2808	3120	–	–
9	5/6	256-QAM	8	–	–	1440	1600	3120	3466.7	6240	6933.3

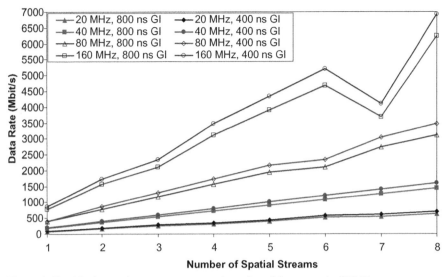

Figure 2.47   Maximum data rate versus number of spatial streams in 802.11ac.

WiGig has developed the PHY/MAC and several protocol adaptation layers (PALs) to enable interoperable devices that take advantage of these high data rates [5]. Each PAL layer sits directly over the MAC layer and is designed to act as an interface between the 60 GHz PHY/MAC and a bus or display interface. The PAL interface model has a simpler and more efficient design that is suitable for cable replacement applications. The IP model is more suited to network access. WiGig supports either or both models of operation. Initially, WiGig has been developing PALs to support audio/visual and high-speed bus interfaces. The WiGig Display Extension (WDE) PAL will support both HDTV through the High-Definition Multimedia Interface (HDMI) and high-quality computer displays through DisplayPort. It has features to support video compression, as well as High-bandwidth Digital Content Protection (HDCP). Two high-speed bus interface PALs are in development and more are planned. The WiGig Bus Extension (WBE) PAL supports the extension of high-speed PCIe over the 60 GHz link. The WiGig Serial Extension (WSE) PAL does the same for USB. It is expected that in the near future, PAL specifications will be created to cover other common high-speed bus interfaces. Among the key applications include compressed and uncompressed/decompressed 720p and 1080p HD video streaming using a wireless HDMI interface.

2.10.1   PHY Specifications

The 802.11ad draft amendment employs two PHY specifications: a mandatory SC PHY and an optional OFDM PHY. At 60 GHz, the delay spread due to multipath is much lower than the 2.4 or 5 GHz bands. For example, the maximum delay spread for a 2 m range is in the order of 5 ns. Hence, multicarrier OFDM transmission is not critical at 60 GHz. In addition, the SC PHY may potentially offer low power,

TABLE 2.13   SC MCSs and PHY Rates in 802.11ad

MCS INDEX	Modulation	Code rate	Number of bits per symbol	Repetition	Data rate (Mbit/s)
0	Differential BPSK	1/2	NA	NA	27.5
1	BPSK	1/2	1	2	385
2	BPSK	1/2	1	1	770
3	BPSK	5/8	1	1	962.5
4	BPSK	3/4	1	1	1155
5	BPSK	13/16	1	1	1251.25
6	QPSK	1/2	2	1	1540
7	QPSK	5/8	2	1	1925
8	QPSK	3/4	2	1	2310
9	QPSK	13/16	2	1	2502.5
10	16-QAM	1/2	4	1	3080
11	16-QAM	5/8	4	1	3850
12	16-QAM	3/4	4	1	4620

low complexity, and hence cost-effective implementation. For instance, data rates of between 385 Mbit/s to 4.62 Gbit/s (Table 2.13) can be achieved using low-power modulation (BPSK to 16-QAM) and simple equalizers can be constructed due to the low delay spread. The overall PHY/MAC of the low-power SC mode is targeted to consume an average power of less than 150 mW. The SC PHY employs a chip rate of 1.76 GHz with 448 chips per symbol, each preceded with a 48-chip GI. The OFDM PHY improves performance in non-LOS use. It offers a 2.64 GHz sampling rate, a 48.4 ns GI (128 samples), 512-sample FFT size (336 for data, 16 for pilot), 5.15625 MHz subcarrier spacing, 242 ns symbol duration, and data rates ranging from 693 Mbit/s to 6.757 Gbit/s (QPSK to 64-QAM), as shown in Table 2.14. The sampling rate for the OFDM PHY is exactly 1.5 times greater than the chip rate for the SC PHY. It is expected that many devices will employ the SC mode, especially personal handsets. Both SC and OFDM PHYs share a common preamble to achieve interoperability. The MCSs are shown in Tables 2.13 and 2.14. The phase of the modulations is typically rotated or staggered to improve the efficiency of power amplification. 802.11ad employs LDPC with a common codeword length of 672 bits and code rates of 1/2, 5/8, 3/4, and 13/16 for both SC and OFDM modes. Simpler Reed Solomon codes can be used for the low power SC mode. The wide range of code rates allows transceivers to choose the best rate for a specific fading channel they encounter. The codes are designed for implementation and include properties to support both layer decoding and fully parallel belief propagation decoding [6].

2.10.2   MAC Specifications

Like the DFIR PHY in the 802.11-1997 standard, the 802.11ad TG defines a new PHY/MAC architecture (Figure 2.48) that is scalable across different usage models,

TABLE 2.14   OFDM MCSs and PHY Rates in 802.11ad

MCS index	Modulation	Code rate	Number of bits per subcarrier	Number of coded bits per symbol	Number of data bits per symbol	Data rate (Mbit/s)
13	Staggered QPSK	1/2	1	336	168	693
14	Staggered QPSK	5/8	1	336	210	866.25
15	QPSK	1/2	2	672	336	1386
16	QPSK	5/8	2	672	420	1732
17	QPSK	3/4	2	672	504	2079
18	16-QAM	1/2	4	1344	672	2772
19	16-QAM	5/8	4	1344	840	3465
20	16-QAM	3/4	4	1344	1008	4158
21	16-QAM	13/16	4	1344	1092	4504.5
22	64-QAM	5/8	6	2016	1260	5197.5
23	64-QAM	3/4	6	2016	1512	6237
24	64-QAM	13/16	6	2016	1638	6756.75

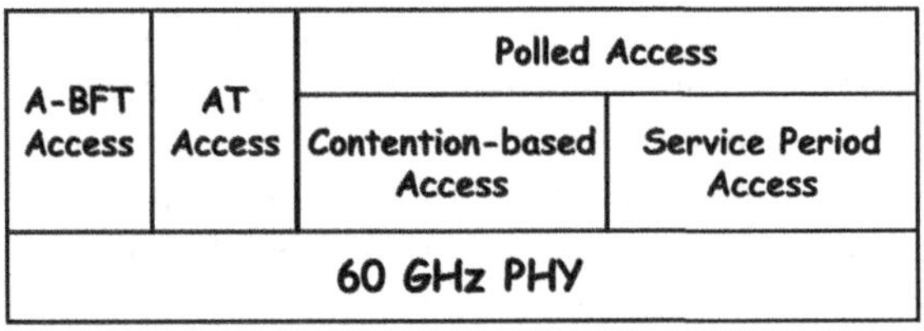

A-BFT Access	AT Access	Polled Access	
		Contention-based Access	Service Period Access
60 GHz PHY			

Figure 2.48   802.11ad protocol architecture.

devices, and platforms. This will accommodate the unique characteristics of the 60 GHz PHY, which requires directional channel access using "pencil" antenna beams. However, rapid transfer of connections between 60 GHz and 2.4/5 GHz bands is supported to maintain seamless network connectivity, in case a mobile device moves out of the range of a 60 GHz connection. The device can quickly switch channels and fall back to a lower rate connection in one of the other unlicensed bands. This process is known as fast session transfer (FST). The handoff between 60 and 2.4/5 GHz connections may require separate radio transceivers, which increases cost.

A personal BSS (PBSS) is defined to enable quick connections to TVs/projectors and fast sync-and-go file transfers. Although a PBSS is an ad-hoc network similar to an IBSS, a key difference is that the PBSS station can further assume the role of a PBSS central point (PCP), and only the PCP transmits beacon frames. The beamforming direction is established before association and data transfer. This is achieved using the beacon time (to discover new stations), the association beam-forming training (A-BFT), and the announcement time (AT). These additional steps are absent in other 802.11 amendments when establishing a connection. Data transfer occurs on one or more service periods (SPs) and/or contention-based periods (CBPs). The SP is negotiated between the AP/PCP and the station or dynamically allocated where only prescribed stations can access the channel. The reliability of the SP can be enhanced using the RTS/CTS virtual sensing mechanism. 60 GHz transmissions tend to be more vulnerable to hidden collisions than 2.4/5 GHz transmissions, as multiple beams can be directed at a common receiver simultaneously. The CBP is time-scheduled by the AP/PCP and is based on EDCA. Other features to support efficient high-speed data transfer include aggregated MPDUs, BA, and advanced security using the AES-Galois counter mode (AES-GCM) that requires fewer computations per bit than earlier AES modes in 802.11. For example, GCM offers a 50% reduction in AES operations than CCM. Unlike GCM, which can efficiently provide authenticated encryption at speeds of 10 Gbit/s or more, CCM is not suited for high-speed implementations because it cannot be pipelined or parallelized. The GCM protocol header in 802.11ad is the same as the CCMP header, but the MIC has been extended from 8 to 16 bytes.

2.10.3 Beamforming Protocol

The protocol comprises two independent phases: sector level sweep (SLS) and beam refinement protocol (BRP). The SLS enables communication at the control PHY rate (i.e., MCS 0) and provides transmit training. The BRP enables receiver training and estimates the best adaptive weight vector (AWV). Beam tracking is supported during data communications. The receiver sector sweep (RSS) allows the device with a simple antenna (e.g., a smartphone) that may not have sufficient transmit gain to reach a receiver using an omnidirectional antenna (e.g., an 802.11ad TV or laptop) with multiple beamforming arrays. Thus, RSS allows the device to connect at a longer range. The transmit sector sweep (TSS) embeds each frame with a countdown indicator, a sector ID, and an antenna ID. The best antenna sector and its ID at the responder are fed back to the initiator using the sector sweep feedback and the sector sweep ACK, as illustrated in Figure 2.49.

2.10.4 60 GHz Implementation

A key challenge for 60 GHz operation is in overcoming the high signal attenuation and absorption from transmitter to receiver. This implies short-range communications for low-power devices, typically confined within a small room. The Friis equation can be used to compute this effect:

$$P_r = \frac{P_t G_t G_r \lambda^2}{(4\pi R)^2},$$

$$(2.6)$$

where P_r is the received power, P_t is the transmitted power, G_t is the transmitter antenna gain, G_r is the receiver antenna gain, λ is the wavelength, and R is the

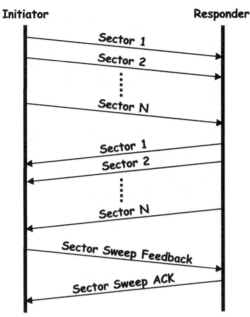

Figure 2.49   Antenna sector selection.

distance from transmitter to receiver. Typically, 60 GHz systems will operate at a received power that is 10 dB higher than 2.4 or 5 GHz systems because the total noise power from the wider bandwidth is much higher. Furthermore, there is a loss of about 21–28 dB relative to the 2.4/5 GHz bands due to the shorter wavelength at 60 GHz. Some of the loss can be offset by reducing the maximum operating range. The remaining loss must be compensated for by increasing the antenna gain. Fortunately, high antenna gains with small antenna sizes are feasible at 60 GHz, because, for a given antenna aperture, gain scales inversely with the square of the wavelength. For a perfectly efficient antenna system:

$$\text{Antenna Gain} = \frac{4\pi A_e}{\lambda^2},$$

(2.7)

where A_e is the effective aperture area. An antenna gain of at least 15 dBi may be required for an acceptable range with 10 mW of transmit power, so some directionality in the transmission will be needed to extend the range. This is very important in making 60 GHz operating range similar to microwave operation. The short wavelength (~ 5 mm) means that a 16-element array with 1/2 wavelength spacing will occupy a space of about 20×20 mm. Thus, multiple antenna arrays can be integrated into a single chip but multiple RF transceiver chains may be needed. The integrated antennas also eliminate the need for wires to carry the signal to and from the chip. They help reduce packaging cost and reduce transceiver vulnerability to electrostatic discharge during fabrication/assembly, removing the need for antistatic devices that add capacitance and degrade performance. The 1/2 wavelength spacing comes about because that is the maximum antenna element spacing in a planar array to prevent grating lobes (i.e. unwanted beams). The assumption is that the antenna design should have the maximum gain for a given number of antenna elements. A 1/4 wavelength spacing may offer some advantages, such as independent fading, but it will have a higher complexity. For 60 GHz systems, there is plenty of bandwidth to provide frequency diversity to help overcome fading. However, large-scale diversity (e.g., two or more antenna arrays spaced far apart) will be beneficial to cope with shadowing from people and other objects.

There are other challenges associated with 60 GHz operation. The data rates fall rapidly with distance, resulting in a pronounced tradeoff in data rate/range. Thus, the peak rate may only be relevant for the first few feet. Objects or obstructions may also block and absorb the RF signals in a manner similar to optical signals, which implies the network may only work well in an environment that is fairly uncluttered. Environmental factors (e.g., rain and snow) lead to more signal loss in outdoor deployment. The migration from 5 to 60 GHz may take some time (compare the gradual migration from 2.4 to 5 GHz, which took more than a decade). A limited number of vendors are developing 60 GHz chipsets, which may ultimately impact cost. Currently, 60 GHz technology may be more costly than free-space optics using DFIR light sources, which also offer a significant amount of bandwidth. Although 60 GHz technology is driven by the feasibility of silicon CMOS RF integrated circuits, which are a cheaper alternative to power-hungry gallium-arsenide and silicon-germanium transceivers, some 60 GHz chipsets still consume over 1 W of power to operate wideband channels and antenna arrays, making them unsuitable for battery-operated handheld devices.

REFERENCES

[1] Wi-Fi Alliance, http://www.wi-fi.org.
[2] IEEE 802.11 Working Group, http://ieee802.org/11.
[3] 802.11 Standard Download, http://standards.ieee.org/about/get/802/802.11.html.
[4] B. Bing, *All in a Wi-Fi Network: A Comprehensive Workbook on Wireless LAN Technologies*, 2005.
[5] C. Hansen, "WiGig: Multigigabit Wireless Communications in the 60 GHz Band," *IEEE Wireless Communications Industry Perspectives*, December 2011.
[6] M. Fossorier, M. Mihaljevic, and H. Imai, "Reduced Complexity Iterative Decoding of Low-Density Parity Check Codes Based on Belief Propagation," *IEEE Transactions on Communications*, Vol. 47, No. 5, pp. 673–680, May 1999.
[7] B. Bing (ed.), *Emerging Technologies in Wireless LANs: Theory, Design, and Deployment*, Cambridge University Press, 2007.
[8] R. Van Nee, "Breaking the Gigabit Barrier," *IEEE Wireless Communications Industry Perspectives*, April 2011.

HOMEWORK PROBLEMS

2.1. Legacy wired 802.3 LANs specify a single data rate and a maximum operating range. For example, 10Base5 (Thick Ethernet) refers to a fixed data rate of 10 Mbit/s operating at a maximum distance of 500 m, whereas 10Base2 (Thin Ethernet or Cheapernet) refers to a fixed data rate of 10 Mbit/s at a maximum distance of 200 m. Both standards employ the coaxial cable as the transmission medium with different connectors (DB-15 AUI connector for 10Base5 and BNC T-connectors for 10Base2). In the case of 10BaseT, the data rate is 10 Mbit/s using unshielded twisted pair (UTP) wiring with RJ-45 connectors for connections to a central hub. The UTP length limitation is 100 m. Explain why a maximum operating range is not defined in 802.11. Hint: The received signal power is the same for the same span of cable in a wired network. For 10Base2, the two ends of each wire segment are connected to BNC terminators to prevent signal reflections. In addition, legacy 802.3 LANs employ baseband (unmodulated) transmissions with Manchester coding to provide periodic voltage transitions for clock recovery and synchronization, thereby improving reception reliability.

2.2. Explain how duplicate frames can occur in wireless LANs. In this case, the same data frame is received more than once. Suppose an 802.11 frame is received with a length error. Will this also result in a CRC error?

2.3. Explain whether a station can establish an infrastructure connection with an AP and an independent (ad-hoc) connection with another station simultaneously.

2.4. How would you modify the CSMA mechanism illustrated in Figure 2.14 such that first come, first serve service (FCFS) is achieved? In other words, a data frame that arrives first is always transmitted earlier than other late arriving frames from other stations. If each station has two or more data frames to transmit, can your modified mechanism still achieve FCFS service?

2.5. Many network devices are limited not by the number of bits they can transmit per second but by the number of frames they can process per second. 802.11 employs variable frame lengths. This minimizes the total number of frames sent, thereby increasing usable throughput. A simple fragmentation/reassembly mechanism is incorporated into the 802.11 MAC layer. Each fragment of a data frame contains a sequence number for reassembly purposes. A fragmentation threshold determines the maximum length of a frame above which the frame will be fragmented. Explain whether frame fragmentation is needed in 802.11n. Note that the maximum Ethernet frame length of 1518 bytes is optimized for wired LANs, not 802.11 wireless LANs. Explain whether 802.11 fragmentation can be applied to broadcast or multicast frames. Assuming the same amount of data bits to be transmitted, compare the overheads for sending short frames versus sending fragments. In a wired network, the fragmentation of long data frames allows short voice frames to be interleaved among the data fragments, reducing delay. In a wireless LAN, can frame fragmentation serve the same purpose?

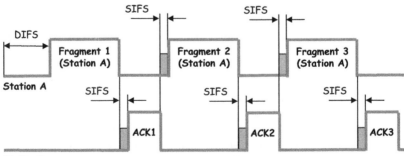

2.6. In 802.11n, why is the PHY header specified in seconds and not bytes (like the MAC header and payload)? Why is the MPDU spacing within an A-MPDU specified in µs? Note that the MSDU spacing within an A-MSDU is demarcated by the address and length fields (specified in bytes).

2.7. Unlike polling protocols, contention protocols, such as DCF, do not require the actual identities (addresses) of stations with data to transmit to be known in advance. Why is this feature important for wireless LANs?

2.8. The maximum spectral efficiency for 802.11n is 15 bit/s/Hz (600 Mbit/s over a 40 MHz channel bandwidth). Compute the maximum efficiency for 802.11ac and ad.

2.9. Compare the advantages and disadvantages of using the RTS frame versus using the optional polling mechanism for multiframe data transmission. Provide a numerical example to illustrate any gains in network throughput for one method versus the other. Assume that if the RTS frame is corrupted, then the backoff mechanism will be activated to resolve the collision. Will the RTS mechanism perform better in the 5 GHz band than the 2.4 GHz band? Note that the polling frame (issued by the AP) contains the NAV reservation information.

2.10. The dual CTS mechanism provides added reliability in data transmission. Since all stations can understand the legacy CTS (and the legacy CF-End), the AP need not issue the 802.11n CTS and 802.11n CF-End, thereby reducing overheads. Explain whether this reasoning is valid. Since the NAV is specified in the CTS, is CF-End necessary? Will all other stations within the wireless subnet receive the RTS frame if it gets to the AP successfully? If so, will this mean that the transmission of the legacy 802.11 CTS (hence the dual CTS) is unnecessary? Why is the transmit address included in RTS, BA, and ACK, but omitted in CTS?

2.11. Peer-to-peer (p2p) traffic requires high bandwidth usage and can strain the inbound and outbound wireless links for an extended period of time. The figure below illustrates bandwidth usage at a hotel where a BitTorrent p2p session started around 00:00, rapidly transferring file segments on the outbound (consuming over 1 Mbit/s), and the session ended about 12 hours later. Explain how 802.11n can provide improved network performance over legacy 802.11a/g networks when supporting p2p applications. Justify your reasoning in terms of the number of simultaneous stations supported, the fairness in network access, and the achievable throughput. Bonus: Why is the inbound bit rate lower (in some instances, considerably lower) than the outbound bit rate?

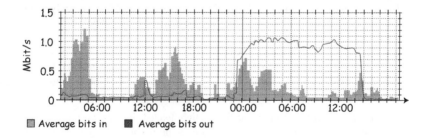

2.12. Explain the use of the BSSID when transmitting an 802.11n A-MSDU over the DS. Why is it that the MSDU address has end-to-end significance, whereas the A-MSDU address only has local significance (within the same wireless subnet)? If a station roams from one wireless network with ESSID 1 to another wireless network with ESSID 2, what will happen to the network connection?

2.13. Derive the IFS values for PIFS and DIFS in the 2.4 and 5 GHz bands. How many slot times are needed to exceed DIFS and SIFS in these bands? Explain whether the propagation delay can be absorbed within the carrier sensing interval. If so, compute the carrier sensing intervals for the 2.4 and 5 GHz bands. Why does the legacy 10 Mbit/s wired Ethernet standard employ a longer slot time of 51.5 μs? Why is carrier sensing used in the PCF mode? Discuss the impact on the performance of the DCF when carrier sensing is not performed continuously within the DIFS and slot time. Since the slot time is shorter than the SIFS in the 5 GHz band, evaluate the implications.

2.14. In Reference [2], it was shown that the performance of DCF is affected by the ratio of the slot time over frame transmission time. The ratio must be reasonably small (e.g., lower than 0.01) for DCF to operate efficiently. Thus, at high wireless data rates (e.g., 54 and 135 Mbit/s), DCF may not operate efficiently as shown in the table below (for a 9 μs slot time and 1 spatial stream). Should the DCF be deactivated for high rate 802.11 systems? Is this a good reason to send the RTS without carrier sensing, as in 802.11n? If the ratio is greater than 1, what does this imply? How would you compute this ratio for 2 or more spatial streams?

Ratio of Slot Time over Frame Duration under Different Data Rates

Frame length	2 Mbit/s	11 Mbit/s	54 Mbit/s	135 Mbit/s
1518 bytes	0.0015	0.0082	0.0402	0.1000
512 bytes	0.0044	0.0242	0.1187	0.2966
64 bytes	0.0352	0.1936	0.9492	2.273

2.15. An 802.11n station transmits several 1500-byte Ethernet frames using two spatial streams via an 802.11n AP. Taking into account the limits imposed by the MPDU length, compute the maximum number of MSDUs that can be aggregated into a single A-MSDU for each spatial stream. Verify that the length of this A-MSDU is longer than the minimum A-MPDU length. If the 802.11n link rate is chosen to be 300 Mbit/s (green-field mode), and there are no other contending stations or transmission errors, compute the minimum time to transmit the two A-MSDUs (one for each spatial stream), including all 802.11n control frames and frame overheads, but excluding overheads due to BA set up and tear down. The dual CTS mechanism is disabled (i.e., only one CTS is issued). From this time interval, derive the maximum net throughput of the transmission using the Ethernet payload. Repeat the procedure for the case when the 802.11n station sends the frames to a legacy 54 Mbit/s 802.11a/g AP. Suppose the received BA map indicates that one of the MSDUs is corrupted during the transmission. Compute the minimum time needed to retransmit this MSDU. Assume that the BA mode has been terminated after the earlier transmission of the A-MSDUs.

2.16. DCF is very effective when the medium is light loaded since the protocol allows nodes to transmit with minimum delay. Due to a finite propagation delay along the transmission medium, there is a probability of two or more nodes simultaneously sensing the medium as being free and transmitting at the same time, thereby causing a collision. The ratio between propagation delay and frame transmission time is another important parameter that determines the performance of DCF. DCF degrades rapidly when the transmission time for the data frame becomes small compared with the propagation delay. Consider a typical wireless LAN that spans 30 m with an rms delay spread of 50 ns. The link propagation speed is 3×10^8 m/s (meters per second). The propagation delay becomes 100 ns, giving a maximum data rate of 1/(50 ns) bit/s or 20 Mbit/s (conservative estimate). If the frame length is 10,000 bits, then the frame transmission time becomes 0.5 ms. The ratio of the propagation delay to average frame transmission time becomes 0.0002 (much less than 1), a value small enough for efficient DCF operation. If the ratio is greater than 1, what does this imply? How would you compute this ratio for two or more spatial streams? In the figure below, the entire 802.11 frame transmission for station C is corrupted due to the overlapping (colliding) transmission from station A. Suggest ways to shorten this collision interval.

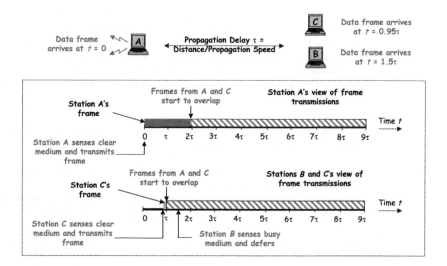

2.17. Consider two types of stations in the wireless subnet: (1) 802.11n stations that employ RTS (with no carrier sensing) before data transmission, and (2) legacy 802.11 a/g stations that perform carrier sensing before data transmission. Each wireless subnet comprises only one transmitting station. In addition, consider infrastructure and independent network architectures. Can the backoff mechanism be used to resolve hidden collisions? Since the PCF can eliminate all hidden collisions, why did the 802.11 standard persist with the use of DCF and virtual sensing (RTS/CTS)? In a PBSS, since a PCP can be elected, why is RTS/CTS still needed?

2.18. Devise a more efficient BA bitmap than the 128-byte 802.11 BA bitmap without using data compression techniques. Take advantage of the fact that successive MSDU sequence numbers increment by 1 (802.11 MSDUs are assigned a single modulo-4096 counter, starting from 0 and incrementing by 1 for each MSDU). If a BA bitmap indicates the total number of MSDUs received (up to 64) and the address(es) of any missing MSDU(s), will this lead to a bitmap of less than 128 bytes? How robust is such a bitmap in recovering from errors? Repeat for the case when the sequence number of any missing MSDU(s) is used instead.

2.19. Some of the first attempts to centralize wireless LANs employ a virtual private network (VPN), which uses IP security (IPSec) and nonstandard TCP ports. IPSec standardizes the authentication and encryption procedures. It may employ password-or certificate-based authentication. Supplied as either server software or an appliance, the VPN separates the wireless LAN from the wired LAN. To maintain a TCP connection when moving from wired Ethernet to wireless 802.11, client software from the same vendor is normally installed on the station. Because VPNs operate at layer 3, they can work with a variety of layer 2 APs, including cheap APs with no security features. Compare the VPN approach with the use of switches that control a variety of APs using layer 2 802.1X.

2.20. Eavesdropping is a less tractable problem than detecting rogue APs because eavesdroppers are normally passive. Soft APs can be set up using a laptop, and they appear legitimate because they employ a similar name as the legitimate AP. These mobile APs can be located physically closer to the user than the legitimate AP, potentially drawing more victims. Thus, this type of attack is much easier than phishing, which involves setting up a fraudulent website to lure users. Such APs can also be hard to trace because they can be turned off anytime. Free 802.11 hotspots and small-office/home-office networks normally employ isolated APs that are particularly vulnerable to these attacks. Corporate users can protect themselves by using VPNs. However, some product vendors feel that WPA or VPN security is an overkill for small 802.11 networks and that easing security setups may help consumers. They advocate changing the SSID or turning off the SSID broadcast for consumers. Discuss the pros and cons of such a solution. The ultimate solution, however, may be to force even listeners to transmit from time to time. Discuss the implications of such a solution.

2.21. Consider a wired network that serves a group of smaller wireless subnets. Is there a need to change IP addresses or reauthenticate stations that roam among the wireless subnets? Is it possible to build a wireless network that serves a group of smaller wired subnets? Note that a wireless subnet need not correspond to a wired subnet.

2.22. Consider the following method to locate an 802.11 station by timing how long it takes for frames to travel to and from an AP. This may prevent stations located outside a house or office from accessing an indoor 802.11 network. Identify the potential flaws when using this method to locate a station. What will be the primary factor that determines the accuracy? Will this be more effective than employing encryption?

2.23. Ultra-wideband (UWB) employs short pulses of energy and spreads them over a wide range of frequencies (typically 500 MHz chunks in the 3.1 GHz to 10.6 GHz frequency range) using OFDM or DSSS. As a result, UWB devices operate at very low power (typically in the region of microwatts for short-range communications as mandated by FCC). Explain whether UWB devices should employ carrier sensing methods to detect ongoing UWB transmissions. Give two compelling reasons to support your thinking. High-rate 60 GHz systems can handle HD video streams between audio/visual equipment without the latency associated with compression. In early 2009, companies such as LG and Panasonic demonstrated 60 GHz wireless HDTV prototypes that can stream up to 4 Gbit/s at up to 30 feet. Compare 60 GHz technology with WiMedia's short-range high-speed wireless USB specification, which is based on UWB technology. List the key differences between 60 GHz and UWB technologies. Which of these two technologies is more suited for short-range wire replacement? Wouldn't it be more convenient to use a wire to connect devices located at short distances? Explain whether a MAC protocol is needed for 60 GHz communications.

2.24. 802.11n employs the RTS/CTS virtual sensing mechanism to reserve the wireless medium for multi-frame data transmission. Consider the following reservation protocol. The station senses the medium and transmits a data frame. Upon successful transmission, the medium is automatically reserved for the same station to transmit more data frames. Note that in this case, frame aggregation overheads for A-MSDU and A-MPDU are unnecessary. The protocol is therefore applicable to 802.11n, as well as legacy 802.11a/g systems, requiring only 1 bit (e.g., using the More Flag in the Frame Control field) to indicate the presence or absence of more data frames. Assuming there are no errors in transmission, will this new reservation protocol perform better than the RTS/CTS mechanism? Justify your answer with a numerical example.

2.25. Although rate fallback is possible, a common maximum rate is defined for 802.11 (e.g., 11 Mbit/s for 802.11b, 54 Mbit/s for 802.11g, 300 Mbit/s using two 20 MHz channels for 802.11n). Why are uplink/downlink rates not specified in 802.11, unlike other wireless access standards (e.g., 802.16, LTE)? Why is rate capping important for public 802.11 networks, and how can this be enforced?

2.26. Consider an 802.11 home gateway with a patch antenna that allow the user to focus signals toward a specific direction (similar in concept to beamforming antennas except that here, the user can control the direction of the transmission), eliminating nosy neighbor snooping on 802.11 home wireless networks. Discuss the pros and cons of doing this.

2.27. CSMA (and DCF) has a vulnerable period corresponding to the maximum propagation delay (T) between the stations because collisions will only occur if frame arrivals are spaced less than T seconds apart. Will the vulnerable period change in the presence of hidden collisions? Explain how CSMA performance will be affected by hidden stations when (1) the transmit range of each station is different and (2) the sensing range is different. Suppose a multichannel CSMA protocol performs carrier sensing sequentially over one or more channels in order to detect whether each channel is busy or idle. Consider the following situation where station A operates using channel 1 and station B uses channels 1, 2, 3, and 4 simultaneously. In this case, station A will only carrier sense on one channel, whereas station B will need to carrier sense on all four channels sequentially (i.e., channel 1, followed by channel 2, then channel 3, and finally channel 4). Explain whether the vulnerable period will be affected in multichannel CSMA operation.

2.28. The figure below shows the operation of the DCF. Stations B and C only wish to transmit one data frame. Complete the figure, showing all frame transmissions. What can you conclude about the fairness of the DCF?

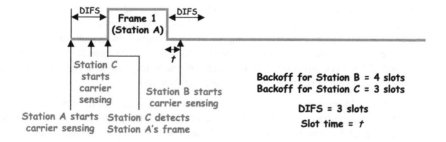

2.29. The figure below shows station C transmits to station D, and at the same time, station A transmits to station B. Explain whether a hidden collision will occur. Suppose virtual sensing (i.e., RTS/CTS) is used by station A. What will happen to station C's transmission? Station D is sometimes known as the exposed station.

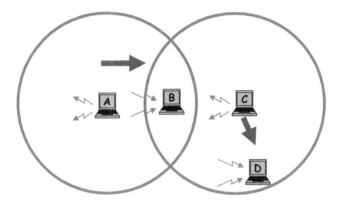

2.30. Consider the following protocol that attempts to minimize the probability of RTS collisions (including hidden RTS collisions) using the beacon frame that is issued by the 802.11 AP. The beacon frame announces a set of contention slots in each cycle, each slot corresponding to the duration of an RTS frame. Each active station chooses a slot with equal probability $1/N$, where N is the number of slots. To reduce the overheads, RIFS is used in place of SIFS before the transmission of an RTS frame. RTS frames that are correctly received by the AP are acknowledged, and an appropriate NAV is specified for the following data transmission period. The number of slots in each cycle is variable. A greater number of slots can be allocated after a long data transmission period. For simplicity, assume that there are three slots for each contention cycle serving three active stations with data to transmit. Compare the efficiency of this protocol versus the use of native RTS with backoff to resolve collisions.

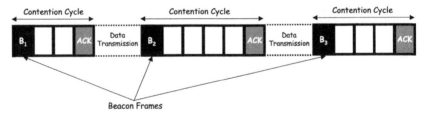

2.31. Consider a point-to-point 802.11 technology that can send data from a city to remote areas up to 60 miles away with a peak throughput of 6.5 Mbit/s using directional antennas. Point-to-point transmission implies no user access between the transmitting and receiving antennas. Explain how time slots (that remove the need for carrier sensing) and ACKs (to confirm correct frame reception) can be implemented. Hint: Need to think about avoiding unnecessary timeouts and retransmissions.

2.32. Anti-802.11 paint is a special paint that blocks wireless signals from passing through walls. Aluminum-iron oxide particles are mixed with the paint. These metal particles resonate at the same frequency as 802.11 and other radio waves, so signals cannot pass through the thin layer of pigment. The paint was developed by the University of Tokyo and is said to be the first that can block microwave frequencies where 802.11 and other higher-bandwidth communications occur rather than just low-frequency wireless like FM radio. The Tokyo paint can reportedly block frequencies all the way up to 100 GHz, with a 200 GHz-blocking paint now in the works. The wireless-blocking paint can also block lower-frequency signals from cellphones. Movie theaters have long been interested in finding a legal way to keep cellphones silent during screenings. Electronic jammers that actively block wireless signals are illegal, but passive materials that prevent wireless signals from getting through are not. Compare this method of confining radio signals with free-space optics (FSO) in terms of practical network deployment considerations (e.g., security, throughput, hidden collisions). In FSO, any opaque material will block the optical signals. Some people envisage a future where the light-emitting diodes (LED) can serve the dual purpose of providing a light source and broadband connectivity at the same time. The technology may be applied within a home, an office, or even in a football stadium where limited RF bandwidth restricts the number of simultaneous cellular connections. Evaluate the merits of such a system, taking into account bidirectional data transmission and optical interference.

2.33. There are many advantages enjoyed by infrared light over radio as a medium for wireless transmission. For instance, infrared light possesses an abundance of unregulated bandwidth (typical data rates greater than 1 Gbit/s), is immune to radio interference, does not require regulatory license, provides highly secure links, achieves reasonable range (greater than 1 km with directional beams), and infrared components are small, cheap, and consume little power. Infrared communications use part of the electromagnetic spectrum just below visible light as the transmission medium. Being near in wavelength, infrared light possesses essentially all the physical properties of visible light. Like visible light, infrared light operate at very high frequencies. This implies it travels in straight lines and cannot penetrate opaque objects and physical obstructions (e.g., walls, partitions, and ceilings), and will be significantly attenuated passing through windows. This characteristic helps to confine optical infrared energy within a single room, virtually eliminating interference problems and unauthorized eavesdropping. However, infrared light will pass through open doorways, reflect off walls and bounce around corners just like sunlight and office lighting. Since infrared light have a longer wavelength than visible light, it is invisible to the naked eye under most lighting conditions. Will hidden collisions become more pronounced with infrared communications? Justify your answer by comparing with 5 and 60 GHz RF systems and by considering the impact of wall partitions.

2.34. In 802.11b, the symbol rates range from 1 to 1.375 Msymbol/s, as shown in Table A. However, the symbol rate for 802.11a is 250 Ksymbol/s, which corresponds to a symbol duration of 4 μs. The modulation-dependent parameters are shown in Table B. Why is the 802.11a symbol rate significantly lower than 802.11b?

2.35. When two wireless devices establish a secure connection, they normally swap cryptographic keys, which are unique codes used to encrypt their transmissions. In a MITM attack, the attacker attempts to broadcast his own key at the exact moment that the key swap takes place and simultaneously mask out the signal from the legitimate sender. If he is successful, one or both of the devices will mistake him for the other, and he will be able to intercept their transmissions. Password protection can thwart MITM attacks, assuming the attacker does not know the password. However, it is not always a valid assumption. For example, at a hotel or airport that offers 802.11 service, all users are generally given the same password, which means that

TABLE A DSSS Modulation Parameters (All Rates are Mandatory)

Data rate (Mbit/s)	Symbol rate (Msymbol/s)	Bits per symbol	Code length	Modulation
1	1	1	11-bit Barker Code	DBPSK[a]
2	1	2	11-bit Barker Code	DQPSK[b]
5.5	1.375	4	8-bit CCK	DQPSK
11	1.375	8	8-bit CCK	DQPSK

[a]Differential binary phase shift keying.

[b]Differential quadrature phase shift keying.

TABLE B OFDM Modulation Parameters (Mandatory Rates in Bold)

Modulation	Code rate	Coded bits per subcarrier	Coded bits per OFDM symbol	Data bits per OFDM symbol	Data rate in Mbit/s (20 MHz channel)
BPSK	**1/2**	**1**	**48**	**24**	**6**
BPSK	3/4	1	48	36	9
QPSK	**1/2**	**2**	**96**	**48**	**12**
QPSK	3/4	2	96	72	18
16-QAM	**1/2**	**4**	**192**	**96**	**24**
16-QAM	3/4	4	192	144	36
64-QAM	2/3	6	288	192	48
64-QAM	3/4	6	288	216	54

any one of them may launch an MITM attack against the others. Similarly, someone who has just quit a company will know the password of its network and may launch MITM attacks. In 802.11 home networks, users either do not bother to protect them or they select simple passwords. Consider the following solution (developed by researchers from the Massachusetts Institute of Technology) to detect an MITM attack without using a password. After transmitting its encryption key, the legitimate sender transmits a second string of numbers related to the key by a known mathematical operation. Whereas the key is converted into a wireless signal in the ordinary way, the second sequence of numbers is encoded as alternating bursts of radiation and silences. If an attacker tries to substitute his key for the legitimate sender's, he will have to send the corresponding sequence of bursts and silences. Because that sequence will differ from the legitimate one, the overlapping sequences will look to the receiver like a new sequence, which will not match up with the transmitted key, indicating an MITM attack. Evaluate the pros and cons of using this method compared with the password method. List any issues in practical implementation. If the attacker transmits a long sequence of bursts and silences, it may eventually outlast the legitimate sender's sequence. The receiver will still think that the sequence is valid. Suggest a solution to overcome this possibility.

2.36. There are several basic steps in maintaining wireless security. The first step is to detect all wireless devices in the environment, including wireless APs that are purchased out of security policy. These APs can inadvertently expose the wired network inside the firewall, making it vulnerable to attack. Locating the positions of the rogue APs and stations once they have been detected represents the next step, and this could involve handheld devices or advanced antenna technology. A key aspect of network security is the need for end users to adhere to recommended security practices, such as keeping up-to-date virus software and intrusion detection software on their laptops. There are conflicting requirements of security and convenience. End users desire a simple, quick logon using stored passwords on user devices. However, for stronger authentication, particularly in enterprise networks, two separate credentials from the user, a password and a time-changing code are typically required. This is similar to the credentials required to withdraw cash from an ATM: you must present both a password (something you know) and the appropriate ATM card (something you have). Compare the security framework of 802.11 with the ATM analogy. What level of privacy and authentication would you recommend for enterprise applications, public hotspots, and residential networks? Can these methods prevent the user from being diverted to a malicious network involving a soft AP, such as a laptop?

2.37. The 802.11p amendment provides wireless access in vehicular environments (WAVE). Devices operate in the 5.9 GHz (5.850 to 5.925 GHz) dedicated short-range communications (DSRC) band for intelligent transportation systems (ITS). Wireless communications over LOS distances (of less than 1 km) between vehicles and vehicles on roadsides can be achieved. Units with wireless interfaces may reside on high-speed vehicles (on-board units) or on roadsides (e.g., streetlamps). Roadside APs may provide efficient traffic offloading for connected vehicles. The roadside units can also share information (e.g., hazards, congestion) with passing vehicles and with safety, highway, and traffic control authorities. To ensure privacy, the information from the cars will not identify the vehicle and IP/MAC addresses can be changed periodically. To mask the point of origin, cars will transmit only limited data until they have traveled a certain distance from the starting points. 802.11p enables new applications such as road safety and emergency services. Drivers may receive information about road conditions, red lights, and hazards 300 m to 500 m ahead on highways and 100 m ahead in cities. By allowing vehicles to stop less abruptly, gas consumption can be minimized. Emergency vehicles may be able to warn vehicles ahead of them to pass and to control traffic lights to give them right of way. Thus, 802.11p can alleviate congestion and improve safety and energy efficiency of automotive transportation. However, intrusive uses such as tracking cars that speed or ignore stop signs may discourage car owners from adopting the system. Reliability and low latencies are critical but note that the information exchange may require only a low bandwidth link. Current 802.11 association process may exceed 100 ms. To provide priority to public safety communications, 802.11p uses a different MAC strategy than standard 802.11. It also uses smaller channel bandwidths of 5 and 10 MHz due to the limited bandwidth in the 5.9 GHz band. Typical data rates range from 6 to 27 Mbit/s using OFDM. Explain how you would modify the 802.11 standard to achieve low association latencies and priority for public safety communications. Do you think roadside units should be involved in forwarding frames to the destination vehicle, or should vehicles be allowed to communicate directly with each other? Why are short frame lengths specified for higher vehicular speeds (see table below)?

Speed	Frame error rate (FER)
140 km/h	<10% (1000-byte payload)
200 km/h	<10% (64-byte payload)
283 km/h	<10% (64-byte payload)

2.38. Power over Ethernet (PoE) delivers DC power over twisted pair Ethernet cables. It allows wireless APs to be powered directly by network cable only (i.e., one cable instead of two) and removes the need for AC-to-DC converters and power cords. This shortens the installation cycle and enables remote reboot or reset of APs. PoE was originally invented and intended for IP phones (PoE provides the dial-tone during power outages, just like a landline PBX). It was ratified as IEEE 802.3af standard (June 2003) and has a limit of 48 V, 15.4 W, and 400 mA. Even at these limited levels, they are sufficient to power up APs. The updated IEEE 802.3at-2009 PoE standard (also known as PoE+) provides up to 25.5 W of power. PoE typically works by inserting DC power on top of the data signal. Once the DC power signal reaches a compatible device, the signal is separated and used to provide electricity. It eliminates extra power cables and outlets, as well as expensive universal power supplies (UPSs). For example, in a typical wireless LAN, the administrator may employ a switch, a set of APs, standard CAT 5 cabling, and UPS on the main core. Only devices that present an authenticated PoE signature will receive power, thus preventing damage to noncompliant equipment. Explain whether regular 120 V AC can be carried over CAT 5 or CAT 6 twisted pair cabling. How is PoE different from broadband power lines that carry both AC electric power and data signals?

2.39. Location management using 802.11 can lead to many applications. Some retail stores are using 802.11 to provide relevant service information to shoppers depending on their location in a store. Major airports provide flight and restaurant information at different locations to passengers with 802.11-enabled devices. Location service can be an invaluable aid in rapidly finding/fixing hardware problems and security threats. The geographic location of users is a critical consideration in the application of secure protocols for wired networks, and this can be extended to wireless networks. A wireless location service can also prevent loss or theft of valuable mobile assets (i.e., physical security with location service versus information security with 802.11i). Using voice over Wi-Fi, the location of a local distressed caller can be quickly identified in emergency situations. Location management is used in workflow automation and inventory management and tracking. It can potentially pinpoint the locations of malicious devices and APs. APs provide the simplest way to locate mobile users. The accuracy improves with the density of AP (e.g., an AP in every office). Several methods can be used. The triangulation method compares the signal strength of several APs. RF fingerprinting (wireless location signatures) compares signals with an existing radio reference model. This method is especially effective when determining a position relative to fixed physical barriers, such as walls or immovable objects. A third method involves RF identification (RFID) tags or readers, which can be placed around windows or doors to track movement into a room/building. Evaluate the pros and cons of each method. Explain whether these methods can achieve an accuracy that surpasses GPS (typically up to 5 m in accuracy). Note that GPS satellites broadcast at two frequencies, 1.57542 GHz (L1 signal) and 1.2276 GHz (L2 signal). The primary GPS information is obtained from the L1 signal, which occupies roughly 20 MHz of RF bandwidth. A second augmentation signal, located at 1.17645 GHz, is used by authorized parties to improve the accuracy to 1 m. The satellite network uses a CDMA spread-spectrum technique where the low bit rate message is encoded with a high-rate pseudo-random code that is different for each satellite. The receiver must be aware of the codes for each satellite to reconstruct the actual message. GPS may perform poorly in areas with tall buildings, which block the LOS signal to the satellite. In addition, reflected GPS signals due to multipath may make it difficult for the receiver to determine the precise location. Explain whether GPS receivers on mobile devices can be used to locate trapped individuals after a devastating earthquake.

2.40. The 802.11y amendment focuses on PHY operation in the 3.65–3.7 GHz frequency band. The band was opened for public use by FCC in May 2005. The band does not allow unlicensed operation but allows nonexclusive licensed operation (i.e., "light" license). In this case, it is possible for multiple wireless networks, designed according to different standards or specifications, to operate in the same location. High transmit powers are allowed in this band. For example, APs can operate with a peak power of 25 W over a 25 MHz channel. Mobile devices can operate with a peak power of 1 W over a 25 MHz channel. This facilitates deployment of long-range 802.11 infrastructure and mesh networks. The operation in this band requires the use of a contention-based protocol, such as DCF. The sensitivity of energy detection (ED) is also increased. This improves sharing of radio spectrum with other wireless networks. ED is used in carrier sensing to determine if the medium is currently busy. It prevents interference with fixed satellite service and radiolocation services that are primary users of 3.65 GHz band. Explain whether the use of ED will degrade the performance of the DCF. Why is a contention-free protocol, such as PCF, not used in this band?

2.41. 802.11 RF channels should never be configured by the user because known interfering sources (e.g., other 802.11 users) can be transient, whereas unknown interfering sources (e.g., microwave ovens) can only be detected by the radio hardware. The 802.11k amendment allows stations to perform RRM and send detailed layer 1 and 2 channel measurement reports to the AP. It is designed to be implemented in software. This will eliminate the need for manual frequency planning, allowing the AP to become a plug-and-play device. If a station reports that it is at the edge of the wireless coverage area, and the serving AP redirects the station to an adjacent AP, will this help eliminate hidden collisions? If 802.11k is implemented, is it necessary to allow the user the option of selecting a specific channel? Clearly, a user will not have access to the channel measurement reports.

2.42. The 802.11r amendment allows fast BSS transitions for secure roaming of 802.11 stations. Dual-mode phones are spurring enterprise interest in voice over Wi-Fi, which can replace cellular voice in in-building networks and hotspots, or provide voice service to 802.11 devices that do not have a cellular radio. Thus, 802.11r will meet a significant need. It enables the station to establish a state at the new AP before making a transition. The faster handoff solution addresses the needs of security, low latency, and QoS resource reservation. For example, it reduces roaming delays associated with 802.1X authentication and shortens the time to reauthenticate and reestablish a connection after a user moves from one AP to another. This is achieved by distributing the encryption keys throughout the infrastructure. Although 802.11i provides optional mechanisms, such as pairwise master key caching and preauthentication to minimize roaming times, they have not been broadly implemented by vendors. In 802.11i, when a station roams to a new AP, it must exchange association messages with the AP. After a user's login credentials have been authenticated, a master session key is derived. 802.11r ensures that the authentication processes and encryption keys are established before a handoff takes place. Authentication occurs only once when a station first enters the mobility domain. Subsequent handoffs within the domain do not require reauthentication. To achieve this objective, 802.11r introduces a new key-management hierarchy. The highest level key holder (usually the wireless LAN administrator) is responsible for deriving the keys for the lower-level key holders (i.e., the APs). 802.11r uses a one-way hash function to ensure that a lower-level key, if compromised, cannot be used to derive the original master key. 802.11r also allows a station to request for QoS resources (e.g., for IP telephony services) on the target AP before making a connection. To do this, the station may need to change to the channel of the new AP or stay on its current channel and use its current AP to communicate with other candidate APs. The 802.11v amendment defines layer 2 management of 802.11 stations, such as monitoring, configuring, and updating channel usage and collocated interference, in either a centralized or distributed manner. The procedures allow the wireless infrastructure to control parameters on the wireless stations, such as identifying which network or AP to connect and define mechanisms for BSS transition management, multicast diagnostic and event reporting, and efficient beacon mechanisms. Identify the role of 802.11r and 802.11v in the following figure and justify your answer. In this figure, thin APs (with minimum 802.11 functionality) connect to 802.11 switches, which are in turn connected to the enterprise wired network and the external cellular network via unlicensed mobile access (UMA). With a UMA mobile backhaul, a session may be initiated on one network, such as an 802.11 or cellular network, and gets handed off to another network (802.11 or cellular), and then returned to the network on which the session was initiated.

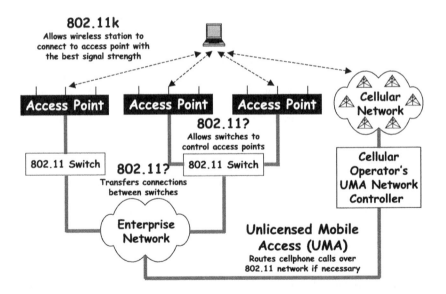

2.43. 802.1X EAP can be used to provide user authentication and WEP key distribution. The periodic rekeying option of 802.1X generates a new pair of keys for the AP at a rate that is faster than the keys can be cracked. The AP in turn uses EAP-TLS authentication to send these keys to the stations. If static keys are used for WEP encryption, will 802.1X play a direct role? Note that frequent user reauthentication (not network card re-authentication) can be used to enhance the security of static keys. Since enterprises require both wired and wireless security, is relogin necessary when changing from one connection to another?

2.44. An 802.11 temperature sensor sends low rate information (e.g., 200 bytes every 5 minutes) to an AP but requires the connection to be secured. Suggest methods to achieve this aim without incurring excessive overheads. Suppose several malfunctioning sensors start sending frames to the AP using the wrong encryption key every 5 minutes. Will this cause the AP to interpret the frames from the sensors as frames sent by hackers to create a DoS attack? Describe a possible solution to this problem.

2.45. From information theory, the capacity of a bandwidth-limited, additive white Gaussian noise (AWGN) channel is:

$$C = W \log_2 \left(1 + \frac{P}{N_0 W} \right),$$

where C is the capacity in bits/second, W is the bandwidth in Hz, P is the signal power in watts, and N_0 is the noise density in watts/Hz. A typical 802.11n system may operate with a bandwidth of 40 MHz and a SNR ratio, corresponding to $P/N_0 W$, of 25 dB. This yields a channel capacity of 332 Mbit/s. For a 60 GHz system, the bandwidth will increase by a factor of 50 to 2 GHz. By keeping other factors constant, compute the increase in channel capacity. How will this capacity be affected by a delay spread of 5 ns (for a typical network range of 2 m)?

2.46. Limiting permissible MAC addresses is a simple access control method to enforce wireless LAN security when the secrecy of the information to be transmitted is not important (e.g., information generated by temperature sensors). The AP, for example, can be configured to allow only certain MAC addresses on the network. Describe the weaknesses of this method. How effective is this method in preventing identity theft (e.g., MAC address/SSID spoofing)? Compare this method with another access control method where the AP requires the user to employ a unique identifier (available on the AP) before any connection can be established. Thus, a network connection cannot be hijacked by a third party unless the attacker knows the identifier (which is equivalent to a secret key). Explain whether user authentication is needed in this case. Is this equivalent to the case when the AP does not broadcast its SSID?

2.47. Explain whether user authentication without encryption is sufficient to ensure strong security. In other words, is access control more important than privacy (contrast WEP, which provides privacy but no access control)? Justify your reasoning for a corporate and a public hotspot environment. A corporation typically does not require charging. Justify the use of WEP for devices such as printers, which may be located in remote areas but still need a connection to the office LAN.

2.48. The 802.1X layer 2 authentication protocol may stop DoS attacks (e.g., jamming, disconnections, and rerouting) by blocking or limiting wireless network access on a user-by-user or port-by-port (i.e., AP-by-AP) basis. This means that network repairs can be better managed, and selected sections of the network can be shut out. 802.1X also provides built-in mechanisms for distributing wireless encryption keys. Is 802.1X more effective than a layer 3 VPN or the transport layer security protocol (RFC 5246), previously known as secure socket layer (SSL)? Can each of these methods operate independently to protect the security of wireless LANs? How can these methods maintain a single wireless infrastructure where trusted users are given access inside the corporate firewall, and guests and visitors are placed outside? Hint: Browser-based security provides the greatest compatibility and ease of use for guest users and visitors as many wireless devices have a browser. Which of these methods is best equipped to handle the low-level security of Internet websites (usually not authenticated, not controlled, and not trusted)? Which of these methods is the most bandwidth efficient when supporting encrypted VoIP calls (normally requiring low data rates)?

2.49. The SSID is issued by the AP when it receives a probe request from a station. The probe request is a management frame requesting for the SSID, supported data rates, and other information, including vendor specific information. The probe response (issued by the AP) is a management frame containing information requested by the probe request plus timestamp, beacon interval, MAC parameters, and other network-related information. How is an 802.11 station able to receive SSIDs issued by multiple APs at the same time? This will happen when a station refreshes the wireless network list or when it is set up for the first time. A possible length for the SSID is 0 byte. Explain whether such a SSID can be useful.

2.50. Evaluate the implications of using a 40 MHz channel in the 2.4 GHz band. How will this impact the probability of hidden collisions (initiated by stations and APs), the possibility of frequency reuse, and the performance of CSMA? Note that in this case, carrier sensing will have to be performed on a broader range of frequencies than a 20 MHz channel. Explain whether it is possible to employ a 40 MHz channel using channel 165 in the 5 GHz band.

2.51. Verify the 802.11n data rates in Tables 2.4 and 2.5, taking into account the number of spatial streams, and the different channel bandwidths and GIs. Suppose compressed video traffic is transmitted using 576-byte chunks at a constant rate of 30 chunks per second. Select the appropriate 802.11n rate to support the video stream. Verify that that the maximum data rate for a single 802.11ac stream using a 160 MHz channel, 256-QAM, 5/6 code rate, and short GI is 866.7 Mbit/s. Verify that all fractional bits per symbol have been omitted in Tables 2.9–2.12. In Tables 2.4, 2.5, and 2.9–2.12, why are the data rates for the 40 MHz channel not exactly twice the rate of the 20 MHz channel (assuming the same GI)?

2.52. The SSID allows multiple APs to be recognized as a single logical LAN. A station can only be associated with a single SSID. Without modifying the 802.11 frame format, explain how a single AP can accommodate multiple "virtual" SSIDs that share the same frequency channel and bandwidth. How can a new station associate with a "virtual" SSID belonging to the same 802.11 network card? For each virtual SSID, explain whether it is possible to establish a different authentication and encryption scheme. In addition, describe how the range of a wireless network can be extended by configuring the AP as a wireless bridge using virtual SSIDs.

2.53. Dual band 802.11n systems employ two radios, each operating in the 2.4 or 5 GHz bands, to achieve full duplex operation. In this case, when the 2.4 GHz radio is transmitting, the 5 GHz radio is ready to receive. Explain whether it is possible to achieve full duplex transmission using only one radio (i.e., either 2.4 GHz or 5 GHz) and multiple channels. For example, two independent 20 MHz channels in the 5 GHz band can be employed, one for transmitting, the other for receiving.

2.54. Suppose a 2.4 GHz wireless LAN comprises a mixture of 802.11g and 802.11n stations. If the 802.11g stations are only sending low rate data intermittently (e.g., temperature sensor information) and the 802.11n stations are running bandwidth-demanding video applications, will all stations experience degraded performance? Repeat for the case when both 802.11g and 802.11n stations are running high rate video applications. Justify your reasoning with the following statistics. Assume there are five 802.11g and five 802.11n stations continuously transmitting 500-byte data frames alternately using the same 2.4 GHz channel. The 802.11g stations transmit at a data rate of 54 Mbit/s, whereas the 802.11n stations transmit at 130 Mbit/s. Ignoring overheads due to ACKs and others, compute the individual rates of each station type. Repeat for the cases when there are one 802.11g station and nine 802.11n stations, and when there are nine 802.11g stations and one 802.11n station. Generalize to m 802.11g stations and n 802.11n stations. Will frame aggregation (applicable only to 802.11n stations), virtual sensing or the PCF help alleviate any performance bottlenecks for the 802.11n stations? Clearly, the full migration of 802.11n to the 5 GHz band will eliminate any interference from the 2.4 GHz band, including interference from 802.11g radios.

2.55. 802.11 can be used to enable automated utility meter readings. For example, in Burbank, California, 802.11 meters are employed to collect gas, water, and electric readings, which are then transferred to an 802.11 municipal network. Alternatively, drive-by meter reading collection can be enabled. In addition, the meter can act as a smart gateway to the customer's residence, providing additional services, such as Internet services, energy conservation, monitoring, and control (e.g., changing thermostat readings), and fire, emergency, outage, billing, and customer satisfaction information acquisition. Describe how you would allocate the 2.4 and 5 GHz channels to operate such a network. How can drive-by meter reading be enabled quickly without the need to reconnect to every meter?

2.56. Suppose a three-stream 802.11n home gateway serves a number of two-stream user devices. Will the gateway be able to transmit at the maximum available rate of 450 Mbit/s or is it limited to the maximum 2-stream rate of 300 Mbit/s? Hint: How can the gateway detect the HT capabilities of the user device? Since three-stream user devices are rare, USB adapters with three-stream 802.11n capability have become available. Since USB 2.0 operates at a maximum data rate of 480 Mbit/s, explain whether there will be a performance bottleneck when the user device is transmitting or receiving using three spatial streams. Bonus: Explain why the HTTP performance of three streams may be comparable with two streams.

2.57. Explain how MIMO can be useful for environmental monitoring in shallow water communications, where signals can be affected by waves and reflections off the ocean's top and bottom surfaces. Dolphins and whales use acoustic waves to communicate when they are thousands of miles apart. Can the same waves be used by 802.11 to transmit information wirelessly in underwater deployments?

2.58. Explain how you would optimize the selection of the 5 GHz channels to form a 160 MHz channel for 802.11ac operation, based on the number of radio transceivers and the overall transmit power. ∎

CHAPTER 3

IEEE 802.16 STANDARD

The 802.16 Working Group (WG) was formed in August 1998. The overarching goal of the WG was to specify a wireless metropolitan area network (WMAN) air interface for fixed and mobile broadband wireless access. Three physical layers (PHYs) were considered, namely:

- Single carrier (SC)
- Orthogonal frequency division multiplexing (OFDM)
- Orthogonal frequency division multiple access (OFDMA).

802.16 defines the standard interface between the core medium access control (MAC) sublayer and any convergence sublayer (CS). The interface allows multiple encapsulated payloads, such as IP, Ethernet, and even legacy ATM. Like many wireless standards, the data rates are scalable and can be lowered if a longer operating range is desired. Like 802.11, the 802.16 standard is available for free download [1]. Unlike 802.11, where data packets are transmitted asynchronously in the forward and reverse directions, 802.16 may allow two-way simultaneous (full-duplex) communication. The IEEE 802.16 standard originally adopted a reservation-based time division multiple access (TDMA) mechanism, which precludes the occurrence of collisions when a reservation request is successfully received by the base station (BS). Subsequently, an OFDMA interface with subchannelization was specified for channel access. This channel access method allows the possibility of a frequency reuse of 1, which leads to more flexible deployment since it removes the need to perform frequency planning. In this chapter, we will discuss the organization of the 802.16 standard, including physical layer transmission, adaptive modulation and coding, MAC frame formats, service flows and scheduling types, and mobility support.

3.1 OVERVIEW OF IEEE 802.16

The initial focus of the WG was a fixed wireless access network where multipath is negligible. Short wavelength line-of-sight (LOS) licensed frequency bands (10 to 66 GHz) were selected for this purpose. In these bands, channel bandwidths of 25 or 28 MHz are typical. The first version was completed on October 2001 (802.16-2001)

Broadband Wireless Multimedia Networks, First Edition. Benny Bing.
© 2013 John Wiley & Sons, Inc. Published 2013 by John Wiley & Sons, Inc.

and published on April 8, 2002. This standard employs SC modulation due to the LOS connectivity requirement. It supports many licensed frequencies (e.g., 10.5, 25, 26, 31, 38, and 39 GHz). The subsequent interest on non-LOS bands (2–11 GHz) led to the ratification of the 802.16a and 802.16d amendments to 802.16-2001. In these bands, multipath may be significant. The ability to support near-LOS and non-LOS (NLOS) scenarios may require additional PHY functionality, such as power management, interference mitigation, and multiple antennas. The 802.16d amendment is essentially an uplink (UL) enhancement to 802.16a. The 802.16-2004 standard, encompassing the earlier amendments, was ratified on June 2004 to support fixed and nomadic deployments. The 802.16e-2005 amendment to 802.16-2004 was subsequently added in October 2005 to support mobility, including vehicular mobility. This provided the basis for Worldwide Interoperability for Microwave Access Industry (WiMAX) System Release 1. An enhanced version that includes FDD support was provided in 802.16e-2009 [2]. We shall collectively refer the 802.16e-2005 and 802.16e-2009 amendments as 802.11e. A consolidated standard, 802.16-2009, was issued in May 2009. Newer amendments include multihop relay (802.16j), improved coexistence mechanisms for license-exempt operation (802.16h), and an advanced air interface with data rates of 100 Mbit/s mobile and 1 Gbit/s fixed (802.16m). 802.16m provides the basis for WiMAX System Release 2 and is backwards compatible to System Release 1. Higher reliability networks (P802.16n) and enhancements to support machine-to-machine applications (P802.16p) are two ongoing projects. The 802.16-2004 standard and 802.16e amendments are fairly well deployed. As of end-2009, WiMAX service providers covered more than 600 million people with over 500 deployments in over 140 markets worldwide [3].

The 802.16-2004 standard supports one MAC layer and three PHY layers, namely SC, 256-carrier OFDM, and 2048-carrier OFDMA. The European Telecommunications Standards Institute (ETSI) High Performance Metropolitan Area Network (HiperMAN) is identical to 802.16-2004 but has only 1 PHY layer, the OFDM PHY layer. The 256-OFDM comprises 200 active subcarriers: 192 data, 56 guard, and 8 pilot. The WiMAX Forum focuses on 256-OFDM for fixed access. The cyclic prefix (CP) is configurable with 8, 16, 32, and 64 additional samples. The guard interval (GI) to symbol interval ratios are 1/4, 1/8, 1/16, and 1/32. The use of Turbo codes is optional. Known preambles provide synchronization and channel estimation. They comprise two binary phase shift keying (BPSK) OFDM symbols on the downlink (DL) and one BPSK OFDM symbol on the UL. The frame duration is variable: 2, 2.5, or 5 ms. The short duration allows the possibility of omitting adaptive channel estimates for fixed wireless deployments, thereby reducing bit overheads.

802.11e supports mobile operation limited to licensed bands below 6 GHz and was accepted by the International Telecommunication Union (ITU) as a 3G International Mobile Telecommunications 2000 (IMT-2000) standard in 2007. 802.16e provides mobility management, intercell handoff (handover), user-device power conservation, and promises to support speeds of up to 80 miles/h. In addition, enhanced performance can be expected even in fixed and nomadic environments. The asymmetrical link structure caters for handheld devices, such as smartphones. The amendment offers 1.25, 3.5, and 5 MHz channel bandwidths. It supports 256-carrier

OFDM and scalable OFDMA (S-OFDMA), including 128/512/1024/2048-carrier OFDMA. There is no 256-OFDMA to avoid confusion with 256-OFDM. The larger FFT size in 2048-OFDMA allows a smaller CP. The WiMAX Forum focuses on OFDMA for mobility. The use of FEC schemes, such as low-density parity coding (LDPC) and convolutional Turbo coding (CTC) are optional. CTC employs iterative and parallel decoding of two or more concatenated convolutional codes. CTC can be used to support hybrid ARQ (HARQ). Multiple input multiple output (MIMO) is defined for the OFDMA mode only and supports up to four transmit antennas and 1, 2, and 4 receive antennas. Data security is supported using Advanced Encryption Standard (AES) in the cipher block chaining (CBC) mode and the Extensible Authentication Protocol (EAP). Table 3.1 illustrates the 802.16 amendments up until 802.16e.

3.2 BASIC IEEE 802.16 OPERATION

The basic point-to-multipoint operation of 802.16 is shown in Figure 3.1. The DL transmission from the BS is followed by UL transmission from a fixed subscriber station (SS) or a mobile station (MS). We shall refer the SS and MS collectively as stations, unless there are 802.16 features designed for a specific station type. The DL transmission must occur first before any UL transmission can occur. This is to allow the stations that are located at different distances from the BS to be synchronized via ranging operations. Figure 3.1 also shows a time division duplex (TDD) frame format, where DL and UL transmissions are subdivided into frames that comprise discrete time slots. DL transmissions are based on time division multiplexing (TDM), since there is only one BS transmitting. TDMA is used on the UL and individual subscribers are allocated time slots serially by the BS. A time gap allows the BS and stations to switch from transmit to receive and vice versa in order to prevent overlapping transmission from different users.

3.2.1 Reference Model

The 802.16 reference model defines several layers as shown in Figure 3.2. In this figure, only the data and control planes are shown and the management planes are excluded. The service access point (SAP) defines the interface between two adjacent layers. Services of lower layers are available to higher layers. The model defines two fundamental data units. A MAC layer service data unit (SDU) is exchanged between protocol layers of the same device. A protocol data unit (PDU) is exchanged between peer entities (i.e., between transmitting and receiving devices) of the same protocol layer.

The CS provides packet classification for the higher network layers. It determines which MAC connection a specific packet shall be carried and can be based on IP classifiers. Optional payload header suppression (PHS) is also available to suppress unnecessary Ethernet/IP header information for improved bandwidth utilization. The common part sublayer (CPS) provides fragmentation of long SDUs or concatenation of several short SDUs. UL fragmentation allows segmentation of

TABLE 3.1   Evolution of 802.16 Amendments

	802.16 (October 2001)	802.16a (January 2003)	802.16d (June 2004)	802.16e (October 2005)
Description	Based on local multipoint distribution service (LMDS)	Based on multichannel multipoint distribution service (MMDS) and HiperMAN	Uplink enhancement to 802.16a	Adds handover, power save to 802.16-2004
Frequency	10–66 GHz	2–11 GHz	2–11 GHz	2–6 GHz
Propagation conditions	LOS	Non-LOS	Non-LOS	Non-LOS
Downlink bit rate	32–134 Mbit/s 134 Mbit/s with 28 MHz channel	Up to 75 Mbit/s 75 Mbit/s with 20 MHz channel	Up to 75 Mbit/s 75 Mbit/s with 20 MHz channel	Variable 15 Mbit/s with 5 MHz channel
Channel bandwidth	20, 25 MHz (United States), 28 MHz (Europe)	Scalable multiples of 1.25, 1.5, 1.75 MHz, up to 20 MHz	Scalable multiples of 1.25, 1.5, 1.75 MHz, up to 20 MHz	Similar to 802.16-2004 but with subchannelization
Modulation	Single carrier, BPSK, QPSK, 16-QAM, 64-QAM	256-OFDM, BPSK, QPSK, 16-QAM, 64-QAM	256-OFDM, 2048-OFDMA, BPSK, QPSK, 16-QAM, 64-QAM	256-OFDM, scalable 128/512/1024/2048-OFDMA
MAC protocol	TDMA	TDMA	TDMA using 256-OFDM, 2048-OFDMA	Scalable OFDMA
Mobility	Fixed	Fixed	Fixed and nomadic	Fixed and mobile
Network topology	Point to point and point to multipoint	Point to point and point to multipoint	Point to point, point to multipoint, mesh	Point to point, point to multipoint, mesh
Typical cell radius	1–3 mi	3–30 mi	3–30 mi	1–3 mi

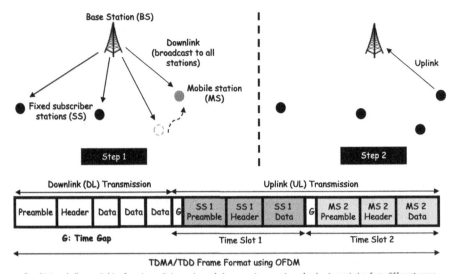

Figure 3.1   Basic 802.16 operation (point-to-multipoint broadcast).

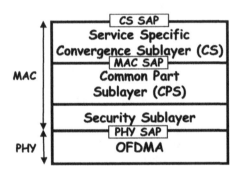

Figure 3.2   Data and control planes of the IEEE 802.16 reference model.

long packets, which reduces timing jitter for real-time services, such as voice traffic. It also allows accurate bandwidth allocation and faster packet retransmission. However, fragmentation incurs more overheads due to the additional header information. Conversely, UL concatenation allows bundling of multiple short packets, thereby reducing UL transmission overheads by transmitting several packets at once.

The security sublayer ensures secure user identification and authentication. It employs X.509 digital certificates and key exchange based on privacy key management (PKM). Key encryption is based on CBC mode of the Triple Data Encryption Standard (3DES) or the AES. Encryption support for multicast operation using the Internet group management protocol (IGMP) is also available. Two-way EAP authentication support is introduced in 802.16e to allow the BS to be authenticated by the user. Only the MAC PDU (MPDU) payload is encrypted. The generic MAC header is always sent in plaintext. It contains two encrypted key sequence (EKS) bits of the traffic encryption key (TEK) sequence number that is used to encrypt the PDU. Several procedures are defined for key management. The station sends an authorization request to the BS. The BS verifies the station's service authorization. If the station is approved for authorization, the BS will send an authorization reply. Two TEKs with overlapping lifetimes are maintained to prevent data transfer disruption due to key expiration.

3.2.2 Frequency Bands

802.16 supports both license-exempt (unlicensed) and licensed band operation. Licensed bands appear to be more popular due to less restrictions on the transmit power limits. The 2.3, 2.5, 3.5 GHz are the primary licensed bands. The 5.725–5.850 GHz band is the primary license-exempt band. The license-exempt section of the standard is called WirelessHUMAN. It supports dynamic frequency selection (DFS), allows frequency detection, and avoids interference (e.g., radar systems operating in 5 GHz frequency band). WirelessHUMAN supports only the TDD mode, whereas FDD operation is for licensed band operation only. This is a reasonable restriction since FDD support requires the availability of a pair of channels with frequency separation for simultaneous transmission and reception. It is difficult to find the availability of such pairs of channels in an unlicensed band.

802.16 specifies several coexistence mechanisms, which are policies and MAC enhancements that enable coexistence among license-exempt systems, as well as between such systems and primary (licensed) users. The coexistence of 802.16 devices with other wireless devices can be a challenge since it employs a TDMA MAC protocol, which cannot sense ongoing transmissions. Coexistence methods include coordinated and uncoordinated mechanisms. Coordinated mechanisms support scheduling transmission for nearby cells. Uncoordinated mechanisms are aimed at improving coexistence of 802.16 with other network types (e.g., 802.11). The uncoordinated coexistence protocol (UCP) comprises three features:

- Extended quiet period (EQP)
- Adaptive extended quiet period (aEQP)
- Listen before talk (LBT).

In addition to DFS, 802.16 supports dynamic channel selection (DCS) for automatic selection of the operating channel. In DFS, the system must move to another channel if the primary system is detected. In DCS, channel can be selected based on other criteria (e.g., performance of the 802.16 network).

3.3 IEEE 802.16-2004 STANDARD

802.16-2004 adopts a number of UL/DL modulation and coding schemes (MCSs) as shown in Table 3.2. Adaptive modulation is used to balance the data rates and channel quality when stations are positioned at different distances from the BS. It is employed during the initial ranging operation and as MSs roam to different

TABLE 3.2   802.16-2004 UL/DL MCSs

MCS ID	Modulation	Code rate	CINR (dB)	Receiver SNR (dB)	Bits per symbol	Peak data rate for 5 MHz bandwidth (Mbit/s)
0	BPSK	1/2	13.9	3.0	0.5	1.89
1	QPSK	1/2	16.9	6.0	1	3.95
2	QPSK	3/4	18.65	8.5	1.5	6.00
3	16-QAM	1/2	23.7	11.5	2	8.06
4	16-QAM	3/4	25.45	15.0	3	12.18
5	64-QAM	2/3	29.7	19.0	4	16.30
6	64-QAM	3/4	31.45	21.0	4.5	18.36

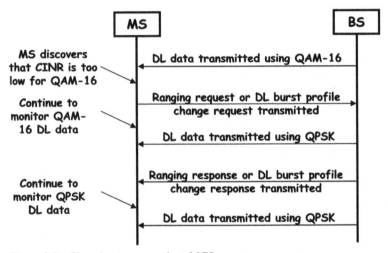

Figure 3.3   Changing to more robust MCS.

locations. The channel quality is evaluated using power measurements, which is contained in the burst profile. 802.16-2004 employs a fixed bit error rate (BER) threshold of 10^{-6} corresponding to a specified carrier to interference plus noise ratio (CINR). This affects data rate, range, robustness, and types of supported applications. An outer Reed–Solomon block code is concatenated with an inner, mandatory convolutional code (1/2 rate with constraint length of 7). This works well with random bit errors. Bit-interleaving can be introduced to reduce the effect of burst errors. Optional support for Turbo codes is provided. Optional LDPC is added in 802.16e. The peak data rate of 18.36 Mbit/s corresponds to a spectral efficiency of 3.67 bit/s/Hz. Figure 3.3 shows the steps for changing to a more robust MCS. Figure 3.4 shows the steps for changing to a less robust MCS.

3.3.1   Frame Format

The 802.16-2004 frame format is shown in Figure 3.5 (some time gaps are not shown). The UL contention slots are transmitted using quadrature phase shift keying

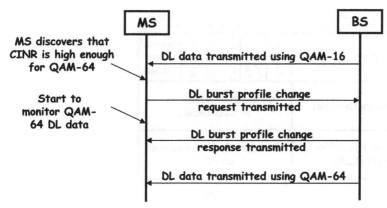

Figure 3.4   Changing to less robust MCS.

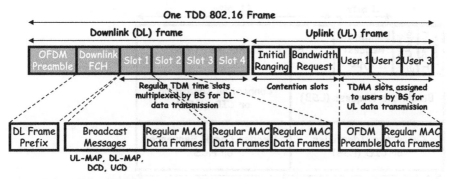

Figure 3.5   802.16-2004 frame format. *Note*: Some guard intervals not shown. FCH, frame control header; MAP, bandwidth allocation message; DCD, downlink channel descriptor; UCD, uplink channel descriptor.

(QPSK). The bandwidth grants issued by the BS are valid for a limited duration. For example, user requests successfully transmitted on the UL bandwidth allocation message (MAP) may be granted bandwidth by the BS for the UL frame on the current or next frame.

The TDD frame duration can be 0.5, 1, or 2 ms. The frame control header (FCH) is transmitted after the preamble, and this specifies the length of one or more DL data bursts immediately following the FCH. Each data burst comprises a number of MPDUs and a burst cyclic redundancy check (CRC). The location and profile of the first DL burst is specified in the DL frame prefix (DLFP), which is part of the FCH. The UL and DL MAPs specify the station receiving or transmitting in each burst. The DL MAP refers to current frame irrespective of the duplexing mode. In the TDD mode, the UL MAP is broadcast one frame ahead. In FDD, the UL MAP takes into account the round-trip signal propagation delay and the MAP processing time. The UL MAPs are transmitted at about 250 times per second, and specify the station receiving or transmitting in each burst, as well as the UL OFDMA subchannels that each station is transmitting. The UL or DL burst profile refers to the MCS,

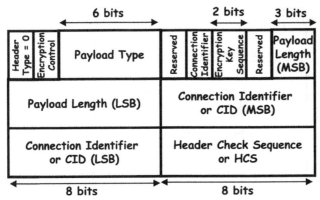

Figure 3.6   Generic MAC header.

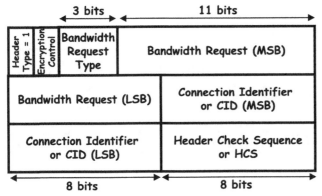

Figure 3.7   Bandwidth request MAC header.

preamble type, and guard time that are used in a data burst. This is specified by the UL channel descriptor (UCD) and the DL channel descriptor (DCD) messages transmitted by the BS at periodic intervals. Different portions of the DL frame can be divided into zones. Each zone can be mapped to a predefined OFDMA subchannel or antenna diversity zone (e.g., AAS, MIMO). The generic and bandwidth request headers are shown in Figures 3.6 and 3.7, respectively. All services are connection-oriented, and connectionless services are mapped to a connection with a connection ID (CID). Since the CID is 16-bit, this implies a total of up to 65,535 connections are possible. The transport connections carry user data. The management connections are used during the initial ranging process.

3.3.2   Multiple Antenna Transmission

802.16 supports two types of optional spatial antenna diversity for frequencies below 11 GHz: Adaptive Antenna System (AAS) and MIMO system. MIMO is able to minimize MAC protocol overheads through broadcast transmission. It performs well

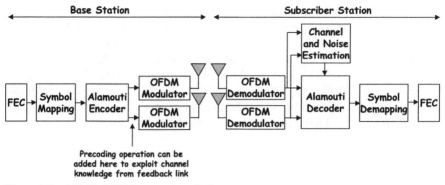

Figure 3.8   MIMO-OFDM transceiver design.

Figure 3.9   TDD frame with AAS.

in a rich scattering environment. When combined with OFDM, MIMO achieves spatial, frequency, and temporal diversity. However, MIMO tends to be preferred for indoor environments, where rich multipath reflections from walls and structures enhance the performance of the technology. In outdoor environments, the BS is normally located in an area where there is clear LOS. Figure 3.8 shows a typical 2×2 MIMO-OFDM transceiver design based on the Alamouti space–time coding (STC). This design is applicable to the DL transmission from the BS to the station. Although each transmit antenna has its own OFDM chain, they employ the same local oscillator for synchronization purposes. Receive diversity is an implementer's decision.

3.3.3   Adaptive Antenna System

The AAS employs directional antenna beams from multiple antenna arrays that radiate only in specific directions. AAS provides angular segregation of wanted signal from interferers. It attempts to logically partition the wireless link into non-interfering sets called zones (Figure 3.9). This helps enlarge the signal coverage area while reducing interference and transmit power. Identical information is transmitted simultaneously from all antennas arrays at BS when performing antenna selection. AAS requires uncorrelated channels.

The AAS BS supports both AAS and non-AAS SSs. Broadcasting can be a problem for AAS SSs that lie outside the antenna beams from the BS (e.g., new SSs). A single broadcast cannot be used to communicate with all AAS SSs. Thus, there is a need to partition an 802.16 TDMA frame into AAS and non-AAS zones.

The AAS zone is explicitly specified in the broadcast MAP from the BS. This comprises the frame control header (AAS-FCH) or DL frame prefix (AAS-DLFP). The UL MAP allows concurrent transmissions from multiple AAS SSs. This can be exploited by the BS to achieve spatial diversity and increase UL capacity. AAS is supported by the OFDMA PHY for FFT sizes larger than 128. This is achieved by creating an AAS diversity MAP zone (DMZ). The DMZ contains the AAS DL preamble and the AAS-DLFP. It can be used in conjunction with superframes to allow the BS to spend more time in the AAS mode.

3.4 IEEE 802.16E AMENDMENT

Figure 3.10 shows the 802.16e frame format. The two-dimensional (time–frequency) UL and DL data bursts are organized as "tiles." This provides greater flexibility, granularity, and scalability in bandwidth allocation by the BS. OFDMA is the main PHY mode in 802.16e. The OFDMA symbol slots are smaller than the TDMA time slots. The duration and bandwidth of the data bursts can be varied via subchannelization. The UL bursts follow in sequence. Subchannelization is mandatory for the OFDMA PHY (both UL and DL), and only the point-to-multipoint network configuration is supported. Just as in 802.16-2004, the DL data is transmitted first, but unlike 802.16-2004, no TDMA MAC is required.

3.4.1 Subcarrier Allocation

The subcarrier allocation depends on the time–frequency mapping and the FFT-size. The mapping of data to subcarriers occurs in two steps. First, the data are mapped into one or more data slots on one or more subchannels. Each data slot and subchannel are then mapped to the physical subcarrier. In full usage of subchannels (FUSC), which only applies to DL transmission, the training is common to all subchannels,

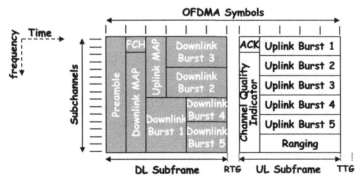

Figure 3.10   802.16e frame format. RTG, receive–transmit transition gap; TTG, transmit–receive transition gap.

TABLE 3.3   Subcarrier Allocation for FFT with Different Sizes

Method	128-FFT	512-FFT	1024-FFT	2048-FFT
DL PUSC	3	15	30	60
UL PUSC	4	17	34	70
DL FUSC	2	8	16	32
AMC	2	8	16	32

across the entire bandwidth. Subcarriers in the subchannels are allocated in a pseudorandom manner, enabling frequency hopping that provides interference mitigation. In partial usage of subchannels (PUSC), subcarriers are organized into "tiles" comprising 14 subcarriers by two symbols on the DL and four subcarriers by three symbols on the UL. Training subcarriers are contained in each tile, and tiles are pseudorandomly allocated across the bandwidth. The DL tiles are shared by several subchannels. The third subcarrier allocation method, adaptive modulation and coding (AMC), is only applicable to the AAS. Tiles are composed of nine subcarriers, and adjacent allocations are allowed. Tile Usage of Subchannels (TUSC) is also applicable to AAS only. It is similar to PUSC, but each subchannel has its own tiles. Constant 4×3-sized tiles are used on the DL and UL. The symmetric allocation facilitates adaptive antennas. The subcarrier allocation for FFT with different sizes is shown in Table 3.3.

3.4.2   Control Mechanisms

The control mechanisms vary depending on whether OFDM or OFDMA is chosen for PHY transmission. For OFDM, ranging is an important step. Initial ranging allows registration of the station with the BS. Maintenance ranging periodically adjusts the PHY parameters of the station. Bandwidth requests can be sent to the BS via polling, piggybacking over packets, or can be contention based. UL power control allows the BS to compute the effects of changing modulation, bandwidth allocation, and subchannels to maximize bandwidth utilization. For OFDMA, two fast feedback methods are needed for time-critical UL PHY parameters (e.g., SNR, MIMO configuration parameters). Because OFDMA deals with multiple users, it employs CDMA ranging based on a contention-based method. Pseudorandom codes for initial ranging, maintenance ranging, handover ranging, and transmission of bandwidth requests. These codes allow multiple ranging transmissions to be resolved simultaneously. Unlike traditional CDMA, codes are modulated in frequency domain with six to eight ranging channels.

3.4.3   Closed-Loop Power Control

Closed-loop power control is mandatory in 802.16e. It allows the BS to control the transmit power of the MS and improves the responsiveness of AMC. Closed-loop power control can be employed for other objectives, such as ranging and adjustment of OFDM/OFDMA parameters. During initialization, the MS reports to the BS its current transmit power. The BS can use both maintenance ranging and fast power control MAC messages to inform the MS to adjust its power levels.

3.4.4 OFDM/OFDMA Implementation

GIs or CPs are necessary to protect the FFT demodulation region against corruption by multipath. The burst profile and symbol description contains the GI information. It is computed based on link conditions (e.g., changes in the multipath delay spread). The duration should be greater than the delay spread so that the receiver can begin FFT sampling within the CP. The GI computation for various FFT sizes is shown in Table 3.4. The governing relations are as follows:

Subcarrier spacing (S_s) = Oversampling ratio × bandwidth $(B)/N_{FFT}$

Oversampling ratio = 8/7 (fixed ratio for 802.16)

Symbol time $T_{FFT} = 1/S_s$, $T_{GI} = GI_{ratio} \times T_{FFT}$

If $B = 3.5$ MHz, $N_{FFT} = 256$, $GI_{ratio} = 1/16$, then $S_s = 15.625$ KHz, $T_{FFT} = 64$ μs, $T_{GI} = 4$ μs.

The 802.16 standard defines PAPR reduction sequences and PAPR levels for single-antenna and multiple-antenna systems on both UL and DL. It employs a pseudorandom code for synchronization to the pilot subcarriers. Codes are modulated on a separate subcarrier before the IFFT operation. The pilot subcarrier patterns can be modified in OFDMA 802.16e systems. 256 144-bit length codes can be created using a predefined 16-bit polynomial generator.

3.4.5 Transmit Diversity

The optional MIMO mode is defined for the OFDMA PHY only. 802.16e adds closed-loop transmit diversity to open-loop schemes. The BS sends a channel sounding command to obtain the channel state information (CSI) from the station. The station reports MIMO weights for the BS to use for best DL reception. Closed loop diversity can be a challenge under high-mobility situations because the CSI may

TABLE 3.4 Guard Interval for FFT with Different Sizes

B (MHz)	1.25	2.5	5	10	20
N_{FFT}	128	256	512	1024	2048
GI_{ratio}	1/8				
S_s (KHz)	11.16				
T_{FFT} (μs)	89.6				
T_{GI} (μs)	11.2				

quickly become outdated. Long-term channel statistics can be employed to improve the channel estimate. MIMO support is defined only for the FUSC and PUSC subcarrier allocation methods. A transmitter with four antennas can be accommodated. Besides STC, spatial multiplexing is allowed. This increases the capacity at the expense of multiple receiver chains and no diversity gain. Frequency hopped diversity coding (FHDC) is also allowed. In this case, data encoding is performed over two frequency channels, instead of two OFDM symbols. Unlike the OFDM PHY, STC and spatial multiplexing are defined for the UL and only for PUSC. PUSC also allows virtual MIMO, where antennas belonging to different stations can transmit simultaneously, achieving multiuser diversity when channels from different stations are mutually independent (uncorrelated).

3.5   IEEE 802.16 MEDIUM ACCESS CONTROL

The DL and UL transmissions are scheduled independently by the BS and stations do not hear each other. The scheduling takes into account the signal propagation delays. Fast scheduling is based on the channel quality information (CQI). The MAC layer supports QoS mechanisms that are mostly extracted from the Data Over Cable Service Interface Specification (DOCSIS) cable standard. Examples include QoS, security, and the reservation MAC protocol (802.16-2004). The 802.16 MAC protocol also supports fragmentation, packing, and efficiency enhancement. Automatic retransmission request (ARQ) support is also provided to mitigate errors in unreliable channels, such as mobile and unlicensed channels.

3.5.1   Duplexing

Both time and frequency division duplexing (TDD and FDD) methods are supported, including FDD with half-duplex stations, a cost-effective FDD implementation. In this case, the station may not request to transmit and receive at the same time. However, the transmission must be activated before reception, otherwise the station may lose synchronization. It is mandatory that at least one duplexing mode be supported. Only one duplexing type is enabled at any one time.

3.5.2   Uplink Transmission

UL transmission is arbitrated via TDMA or OFDMA, which is more suited for licensed band operation (compare CSMA, which is more suited for unlicensed operation). Individual subscribers are allocated time slots serially by the BS. Automatic channel acquisition and ranging are supported, as well as multiple PHY specifications customized for the frequency band of use. The TDMA MAC supports circuit-switched voice connections, but does not specify details of traffic management even though QoS mechanisms are provided. Device and user authentication capabilities are added in 802.16e.

3.5.3 Downlink Transmission

DL transmission is via TDM since there is no contention on the DL, and only one BS is transmitting. Transmission time by each station is limited by the duration of the time slot. The physical slot (PS) is defined to be QPSK (4-QAM) symbols in the SC PHY and 4/(sampling frequency) for the OFDM and OFDMA PHYs [5]. Reservation minislots are defined for the SC PHY only and equals 2^nPS, where $n = 0$, 1, 2, . . . , 7. These minislots allow reservation requests to be transmitted by the stations to reserve for longer data slots.

3.5.4 Polling Mechanisms

There are three polling types, namely unicast, multicast, and broadcast polling. Unicast polling gives the station a contention-free opportunity to request for extra bandwidth. It minimizes delay and timing jitter among all polling types. Multicast polling provides contention-based bandwidth requests for a group of stations. Broadcast polling provides contention-based bandwidth requests for all stations. It minimizes overheads associated with unicast polling, but is useful for stations whose services are idle for some time.

3.5.5 Hybrid Automatic Repeat Request

HARQ is an optional mechanism in 802.16 and may only be used for the OFDMA PHY. HARQ is useful for improving the UL, where the signal can be weak. As the name implies, it is implemented in conjunction with the MAC and PHY layers. HARQ combines error correction and ARQ, and adds redundancy only when needed. The ARQ block is part of the SDU, and the block length is negotiated at connection setup. HARQ employs the block error rate (BLER) metric, which is measured at the transport layer, and CINR to assess the channel quality. The appropriate MCS can then be chosen based on the channel quality. HARQ uses the stop-and-wait ARQ protocol. Positive and negative acknowledgements (ACKs/NACKs) are used to indicate whether each data packet has been received correctly or not. For example, the UL ACKs/NACKs provide feedback for packets transmitted using the DL HARQ. A dedicated channel is used for transmitting the ACKs/NACKs. CTC is used to support HARQ. There are two HARQ types: incremental redundancy and chase combining. Both types will transmit a new data packet in response to an ACK. However, HARQ will transmit the next subpacket (for incremental redundancy) or retransmit the old packet (for chase combining) in response to a NACK. In incremental redundancy, the receiver tries to decode using the first and second subpackets. The process continues until success or up to the fourth subpacket. The ability to perform error correction can be adjusted according to the degree of channel fading and interference experienced by the transmitted packets. Frequent or highly accurate channel quality measurements are not needed.

3.5.6   Bandwidth Allocation

QoS can be provisioned with UL service flows, scheduling types, and dynamic service establishment. Grants are sent through the UL MAP. The BS decision is based on the requested bandwidth and the QoS requirements versus available resources. The scheduling algorithm is not specified to allow vendor differentiation. The UL service flow scheduling types include:

- Best effort service (e.g., email and Web browsing), which is the only scheduling type that omits the UL grant scheduling type parameter, hence no throughput or delay guarantees
- Unsolicited grant service
- Real-time polling service
- Non–real-time polling service
- Extended real-time variable rate service.

3.5.7   Service Flows

The service flows are higher layer flows that characterize the service profiles. Each service flow is mapped to a connection with extra connections for management and control. The QoS parameters for service flows are listed in Table 3.5. Multiple service flows per station are allowed. Thus, different traffic types (e.g., data, voice, and video) from the same station can be separately identified. However, there is only one transport connection per service flow. The service flow ID (SFID) is assigned

TABLE 3.5   Service Flow QoS Parameters

Traffic priority	Eight levels are defined for prioritizing service flow and should not override the minimum reserved rate of a connection of lower priority.
Request/transmission policy	This specifies the attributes for the UL service flows, broadcast bandwidth requests, UL piggyback requests, and PDU formation, such as packing, fragmentation, payload suppression, and CRC inclusion.
Maximum sustained traffic rate	This defines the peak rate of service flow and relates to the SDUs at the input to the CS (i.e., MAC headers and CRC are excluded).
Maximum latency	This is defined as the maximum delay between the ingress of the packet to the CS and the forwarding of the SDU to the air interface.
Minimum reserved traffic rate	This specifies the minimum rate reserved for the service flow and prevents starvation of data from a connection when averaged over time. It is only honored when sufficient data is available for scheduling.
Tolerated jitter	This specifies the delay jitter for the service flow.

by the BS and exists for all service flows (active or inactive). It provides the BS and stations a primary reference of a service flow. The CID maps the SFID to the internal MAC layer. The BS creates, modifies, and deletes service flows. Service flows are activated using dynamic service addition (DSA). A two-phase activation mode is supported. This mode allows resources to be assigned to the admitted flow, but only activated after the end-to-end negotiation is completed.

3.5.8   Unsolicited Grant Service

Unsolicited grant service (UGS) is a TDM section that follows immediately after the FCH in the DL frame. It is designed to support constant bit rate (CBR) real-time service flows that generate fixed-size packets on a periodic basis (e.g., VoIP). It provides fixed-size grants periodically so as to guarantee bandwidth to the station without the recurring overhead and latency of station requests. There are several mandatory QoS parameters, which are listed below:

- Maximum sustained traffic rate
- Maximum latency
- Tolerated jitter
- UL grant scheduling type
- Request/transmission policy.

3.5.9   Real-Time Polling Service

Real-time polling service (RTPS) is designed to support variable bit rate (VBR) real-time service flows that generate variable length data packets on a periodic basis (e.g., streaming MPEG video). It offers real-time, periodic unicast polls, which are bandwidth request opportunities. The service meets the flow's real-time needs and allows the station to specify the length of the desired grant. It requires more request overhead than UGS and supports variable grant lengths for optimum data transport efficiency. An extended real-time polling service (on/off UGS) for silence-suppressed VoIP applications is also available. Several mandatory QoS parameters are listed below:

- Minimum reserved traffic rate
- Maximum sustained traffic rate
- Maximum latency
- UL grant scheduling type
- Request/transmission policy.

3.5.10   Non–Real-Time Polling Service

The non–real-time polling service (NRTPS) is designed to support non–real-time service flows that require variable size data grants frequently (e.g., high-bandwidth file transfer). The service offers regular unicast polls to assure that the flow receives request opportunities even during network congestion. The BS typically polls in an interval on the order of one second or less. The station may use contention request opportunities, as well as unicast request opportunities. The mandatory QoS parameters include:

- Minimum reserved traffic rate
- Maximum sustained traffic rate
- Traffic priority
- UL grant scheduling
- Request/transmission policy.

3.5.11   Extended Real-Time Variable Rate Service

Extended real-time variable rate service is a polling service introduced by 802.16e for managing traffic rates, transmission policies, bandwidth, and content delivery. It improves latency and timing jitter. Unlike UGS, it supports real-time applications with variable bit rates and rich IP multimedia applications (e.g., IP-TV) that make use of streaming video. It also supports voice applications that may become inactive for substantial periods of time (e.g., VoIP with silence suppression). It also provides multicast and broadcast services where single frequency network operation can be achieved using OFDMA. The service enables high data rate coverage at the cell edge. Key service information elements include:

- Maximum sustained traffic rate
- Minimum reserved traffic rate
- Maximum latency
- Request/transmission policy.

3.5.12   Multicast Support

Multicast support allows the BS to transmit to a group of stations on the same channel without duplicating transmissions. 802.16 employs the DL MAP to inform subscribers about multicast traffic. Transmission is performed using the lowest PHY mode common to all stations. To ensure proper multicast operation, the CID for the service is the same for all stations on the same channel that participate in the connection. ARQ and HARQ are disabled.

3.5.13   Mobility Support

Handovers, or handoffs, allow the MS to switch from the serving to the target BS. Several handover methods are defined:

- BS initiated handovers
- MS initiated handovers

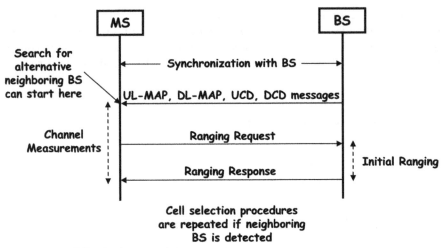

Figure 3.11 Cell selection procedures.

- Macrodiversity or soft handovers (optional)
- Fast BS switching (optional).

Before the initiation of a handover, network topology is acquired with the assistance of the backbone network. This includes network topology advertisement, neighbor BS scanning by the MS, and target BS association. The actual handover process involves cell reselection, handover initiation, handover decision, DL synchronization with the target BS, initial ranging, service termination with the current BS, authorization, and registration. The decision may originate either at the MS, the serving BS, or on the network. The algorithm for handover decision is vendor specific. The standard specifies location update conditions for the MSs. The MS can be associated with multiple BSs. The MSs may obtain addresses using mobile IP or the dynamic host configuration protocol (DHCP). The cell selection procedures are shown in Figure 3.11.

3.5.14 Power Conservation

802.16 stations can operate in the sleep mode. The BS employs a traffic indication map that is transmitted periodically and informs sleeping stations about stored information and synchronizes waking stations. Additional power save modes include three power save classes with a method for computing different sleep and listening windows that support different traffic types.

3.6 IEEE 802.16M AMENDMENT

The 802.16m Task Group (TG) was formed on December 2006 to develop a 1 Gbit/s wireless interface. The 802.16m amendment was ratified in March 2011. In a typical mobile environment, a data rate in the region of 100 Mbit/s on the DL can be expected. However, Samsung recently demonstrated an impressive 802.16m DL rate of 330 Mbit/s at the 2010 CEATAC show in Tokyo. 802.16m is also known as WiMAX Release 2.0, the next-generation WiMAX standard. Like Long Term Evolution (LTE) Advanced, 802.16m has been accepted by the ITU as an official technology for the IMT-Advanced (IMT-A) 4G wireless system. New features include multihop relay functions, self-organization and self-optimization functions, plug-and-play indoor BS for femtocell operation, and interference mitigation. 802.16m introduces a superframe concept that supports improved MIMO modes with emphasis on multiuser MIMO (MU-MIMO), and similar baseband processing requirements in the UL and DL. Up to eight DL and four UL spatial streams can be supported. MU-MIMO with one stream per user is also supported. The 802.16m MS can be connected at mobility speeds of up to 350 km/h (500 km/h in some cases), which are typical speeds of high-speed trains. 802.16m supports the use of femtocells for improved throughput at the cell edge or in areas adversely affected by obstructions.

The increased bandwidth is based on band or channel aggregation involving contiguous or noncontiguous channels to provide an effective bandwidth of up to 100 MHz. The channels may not have the same bandwidth and may reside on different frequency bands. A single MAC protocol controls the adjacent and nonadjacent RF carriers. The primary RF carrier is used by the BS and MS to exchange data or control traffic. The secondary RF carrier is an additional carrier used by the multicarrier-capable station. These carriers are not necessarily located in adjacent bands. If adjacent carriers are used, guard bands that are normally in place to separate the carriers can be used for transferring data. 802.16m is backwards compatible to 802.16e systems. However, the 802.16e station does not recognize channel aggregation and continues to use only one carrier. High-speed zones are introduced in a frame and are only applicable to 802.16m devices.

With interference mitigation, the 802.11m BS may request stations to perform interference measurements at their location. The BS can then use the information gathered from different stations to adjust its power settings within the cell and coordinate with neighboring BSs employing the same frequency.

3.6.1 UL/DL Adaptive Modulation and Coding Schemes

A set of 16 MCSs is defined in 802.16m. 802.16m supports QPSK, 16-QAM, and 64-QAM modulation schemes in the DL and UL. By comparing Table 3.6 with Table 3.2, the key differences from 802.16-2004 include more modulation choices and more granular code rate (rate matching) for 802.16m. The coding methods are convolutional coding and CTC with variable code rate and repetition coding. For coding purposes, groups of symbols are used. The control channel is coded with a mother code of rate 1/5. For QPSK MCS with 31/256 code rate, 31 data bits are converted to 155 coded bits. To reach 256 bits, some of the 155 bits are repeated. Upon reception, likelihoods of repeated transmitted bits are combined, and then the 155 bits are decoded by the rate 1/5 decoder to give 31 data bits. For 64-QAM MCS with 237/256 code rate, the same 1/5 rate tail-biting convolution code is used for the control

TABLE 3.6 802.16m UL/DL MCSs

MCS ID	Modulation	Code rate
0000	QPSK	31/256
0001	QPSK	48/256
0010	QPSK	71/256
0011	QPSK	101/256
0100	QPSK	135/256
0101	QPSK	171/256
0110	16-QAM	102/256
0111	16-QAM	128/256
1000	16-QAM	155/256
1001	16-QAM	184/256
1010	64-QAM	135/256
1011	64-QAM	157/256
1100	64-QAM	181/256
1101	64-QAM	205/256
1110	64-QAM	225/256
1111	64-QAM	237/256

channel. The 237 data bits are encoded into 1185 bits, which are then punctured to give 256 bits. The signal constellation can be rearranged to improve the power efficiency of multilevel modulation. The constellations for successive HARQ retransmissions can be modified to reduce the packet error rates caused by the variation in bit reliabilities when bits are mapped to the signal constellation. Bit reliabilities are averaged over several HARQ retransmissions. The HARQ is implemented with the MAC and PHY layers. The ACK/NACK feedback applies to DL data only.

3.6.2 DL MIMO Enhancement

The BS supports single and MU-MIMO, including spatial multiplexing, beamforming, and a number of transmit diversity schemes. Flexible adaptation between these modes is possible. Vertical encoding (i.e., single codeword) employs one encoder block (or layer). Horizontal encoding (i.e., multiple codewords) employs multiple encoder blocks (or layers). The minimum configuration is 2×2. In single-user MIMO, one user can be scheduled over one resource unit. It employs vertical encoding. Closed-loop TDD/FDD systems employ unitary codebook-based precoding, which can improve the resilience of spatial multiplexing but requires full channel knowledge at the transmitter. The optimal precoding matrix may be selected based on codebook design criteria that minimize the average distortion (e.g., error rate) of a predefined channel model (e.g., Rayleigh or Rician).

In MU-MIMO, different users can be scheduled within one resource unit. Horizontal encoding supports up to two streams with two transmit antennas and up to four streams with four and eight transmit antennas. Only one stream is transmitted to each MS, and both unitary and nonunitary precoding are supported. In unitary precoding, columns of the precoding matrix are orthogonal to each other. Conversely, in nonunitary precoding, columns of the matrix are not orthogonal to each other (beamforming is activated with this precoding method). Stream-to-antenna mapping depends on the MIMO scheme. The CQI and rank feedback are transmitted to assist the BS in rank adaptation, rate adaptation, and mode switching. For spatial multiplexing, the rank is defined as the number of spatial streams to be used by a station. Multi-BS MIMO is also supported. The BSs may collaborate to serve stations at the cell edge. This not only reduces intercell interference but also offers multiplexing rate and diversity gain.

3.6.3   UL MIMO Enhancement

The BS schedules the resource blocks and determines the MCS and MIMO parameters (mode, rank, etc.). It supports 1, 2, or 4 transmit antennas, and two or more receive antennas. Thus, the minimum configuration is one transmit and two receive antennas (i.e., 1×2). Unitary codebook-based precoding is supported for both TDD and FDD. Open-loop single-user MIMO is useful for high-mobility speeds since CSI is not required. Note that 802.16e supports UL/DL space–time block coding (STBC) as the main open-loop MIMO scheme. Open-loop single-user MIMO supports up to four antennas with a rate of 4. The transmit diversity mode supports two and four transmit antennas with a rate 1, STBC, space–time frequency block coding (SFBC) and rank 1 precoding. The spatial multiplexing mode supports two and four transmit antennas with rates 2, 3, or 4. Rate 2 employs two and four transmit antennas with and without precoding. Rate 3 employs four transmit antennas with precoding. Rate 4 employs four transmit antennas. Open-loop and closed-loop multiuser MIMO are supported. A station with one antenna can operate in the open-loop mode.

3.6.4   Frame Format

The 802.16m frame format achieves highly granular bandwidth allocation. The 20 ms 802.16e frame structure is too long for high-speed access. In 802.16m, a 20 ms superframe is used instead, which is subdivided into equally sized 5 ms frames with each frame consisting of eight subframes (Figure 3.12). The superframe

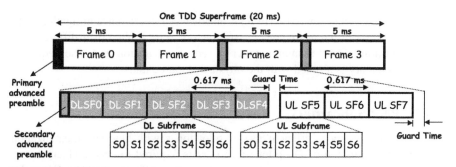

Figure 3.12   802.16m frame format.

header carries short-term and long-term system configuration information. Within each 5 ms frame, the transmission direction can be changed once. The DL/UL time allocations of 7/1, 6/2, 5/3, or 4/4 can be achieved in a frame. By switching the transmission direction every 5 ms, access delay and HARQ retransmission delay are reduced. The three subframe types depend on the size of the CP, namely types 1, 2, and 3, that employ 6, 7, and 5 OFDMA symbols, respectively. The short durations of the subframes allow quick access to the 802.16 network.

3.6.5   Advanced Preambles

DL synchronization is hierarchical since primary and secondary preambles are employed. The primary preamble marks the beginning of a superframe. It is common to a group of sectors or cells and assumes a fixed 5 MHz bandwidth. It can be used to facilitate location-based services. The frequency reuse is 1. The secondary preamble is repeated every frame and spans the entire system bandwidth. It carries the cell ID. Up to 512 distinct cell IDs are carried by the secondary preambles. These preambles employ a frequency reuse of 3 to mitigate intercell interference.

3.6.6   Resource Blocks

The physical resource unit (time–frequency domain) comprises 18 contiguous subcarriers by 6 contiguous OFDMA symbols. The logical resource unit (frequency domain) comprises 18×6 subcarriers that occupy 196.88 KHz of bandwidth. It is used to achieve frequency diversity gain. To form distributed and localized resource units, the subcarriers are partitioned into guard and user subcarriers. The number of pilot and data subcarrier depends on the MIMO antenna mode and number of symbols within a subframe. The UL subframes are partitioned into a number of frequency partitions. Each partition comprises a set of physical resource units and can include localized and/or distributed physical resource units. This is different from a legacy 802.16 system, where either localized or distributed subchannels (each comprising a group of subcarriers) are supported. The smallest unit for constructing a distributed resource unit is a tile. The UL tile size is six contiguous subcarriers by six contiguous OFDMA symbols. The size of the localized resource unit is the same as a physical resource unit. The UL/DL resource partitioning and mapping are similar.

Fractional frequency reuse (FFR) is an advanced interference mitigation technique that allows different frequency reuse factors to be applied over different frequency partitions. The maximum number of frequency partitions is 4. The frequency partition boundary is aligned with the physical resource units.

3.6.7   Pilot Subcarriers

802.16m provides more efficient use of pilot tones with new subchannelization schemes. Transmission of DL pilot subcarriers is required to enable channel estimation, channel quality measurement, and frequency offset estimation. The pilot

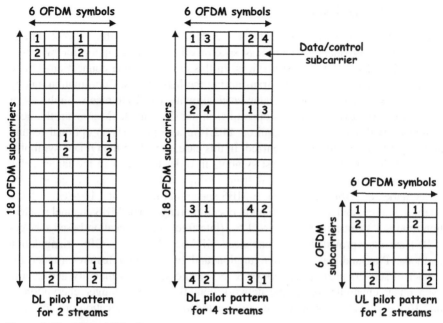

Figure 3.13   UL and DL pilot patterns.

structure is defined for up to eight spatial streams (Figure 3.13). To overcome the effects of pilot interference among the neighboring sectors or BSs, an interlaced pilot structure is utilized by cyclically shifting the base pilot pattern such that the pilots of neighboring cells do not overlap [4]. The UL pilot structure is defined for up to four spatial streams with orthogonal patterns. The 18 × 6 UL RBs use the same pilot patterns as the DL counterpart for up to four spatial streams. The pilot pattern for the 6 × 6 tile structure is different, as shown in Figure 3.13.

3.6.8   MAC Layer

802.16m has an efficient MAC header for small payload. The new header consists of the extended header indicator, flow ID, and payload length fields. It assigns a unidirectional service flow with specific QoS requirements. The service flow is mapped to a transport connection with a flow ID. The parameter set is negotiated between the BS and station. The 802.16m MAC layer supports fast and reliable transmission of MAC management messages. It also supports both network-controlled and MS-assisted handovers. With dynamic power management, sleep and listening windows can be adjusted depending on traffic patterns and HARQ operations. To achieve this, the MS needs to monitor the paging message from the BS. Idle mode efficiency is improved by enabling the MS to periodically monitor for broadcast traffic without registering with a specific 802.16m BS.

3.6.9 Enhanced Services

The enhanced multicast broadcast service (E-MBS) is available in 802.16m to provide greater broadcast and multicast spectral efficiency and support for switching between broadcast and unicast services. In this case, the same connection ID and security association are used for transmitting the MBS flows. Coordination among the BSs may be achieved at the frame level, allowing the BS to receive multicast or broadcast traffic even when it is operating on the sleep mode. Moreover, the need to register with the BS is not required. 802.16m also provides an optional feature allowing BSs to be time synchronized. In this case, the CP of the OFDM symbol is set to a value greater than unicast symbols in order to deal with the larger delay spreads when signals are received from multiple BSs. In terms of performance benchmarks, 802.16m achieves an E-MBS spectral efficiency of 2 bit/s/Hz (1.5 km operating range) and 4 bit/s/Hz (0.5 km operating range). The MBS channel reselection time is 1 s (intrafrequency) and 1.5 s (interfrequency). Location-based services (LBS) are also provided using enhanced GPS-based and non-GPS triangulation methods. Plug-and-play self-configuration and optimization features (under the banner of self-organizing network or SON) ensure service availability and quality even under changing traffic and network conditions. The stronger encryption in 802.16m ensures the confidentiality of user identity and user location privacy.

3.6.10 Summary of 802.16m Features and Performance

The new 802.16m features are listed in Table 3.7 (adapted from [3] and [4]). Note that the station must be capable of processing two spatial streams (on the DL), even though it may transmit using 1 spatial stream (on the UL). The spectral efficiency and cell-edge performance of 802.16m is provided in Table 3.8 for low mobility deployments [3]. The spectral efficiency per user is averaged over 10 stations. The

TABLE 3.7 Summary of Key 802.16m Features

Antenna configuration	DL: 2×2 (baseline), 4×2, 4×4, 8×8
	UL: 1×2 (baseline), 1×4, 2×4, 4×4
Operating bandwidth	5–20 MHz (up to 100 MHz via band aggregation)
Handoff interruption	<30 ms
Link layer/user plane	<10 ms DL or UL
Control plane, idle to active	<100 ms
Location based services	Location determination latency <30 s
	MS-based position accuracy <50 m
	Network-based position accuracy <100 m
Operating frequencies (IMT licensed bands below 6 GHz)	698 to 960 MHz (262 MHz)
	1.71–2.025 GHz (315 MHz)
	2.11–2.2 GHz (90 MHz)
	2.3–2.4 GHz (100 MHz)
	2.496–2.69 GHz (194 MHz)
	3.4–3.6 GHz (200 MHz)

TABLE 3.8   802.16m Performance

Parameter	Antenna configuration	Performance
Peak DL spectral efficiency	2 × 2 MIMO	8.5 bit/s/Hz
	4 × 4 MIMO	17.0 bit/s/Hz
Average DL spectral efficiency	4 × 2 MIMO	3.2 bit/s/Hz
		0.32 bit/s/Hz/user
DL cell-edge user throughput	4 × 2 MIMO	0.09 bit/s/Hz/user
Peak UL spectral efficiency	1 × 2 SIMO	4.6 bit/s/Hz
	2 × 4 MIMO	9.3 bit/s/Hz
Average UL spectral efficiency	2 × 4 MIMO	2.6 bit/s/Hz
		0.26 bit/s/Hz/user
UL cell-edge user throughput	2 × 4 MIMO	0.11 bit/s/Hz/user

TABLE 3.9   802.16m Peak DL Rates (5 : 3 DL/UL Ratio for TDD)

	10 MHz (TDD)	20 MHz (TDD)	2 × 20 MHz (FDD)
2 × 2 MIMO (802.16e-2005)	37 Mbit/s		
2 × 2 MIMO (802.16e-2009)	40 Mbit/s	83 Mbit/s	141 Mbit/s
2 × 2 MIMO (802.16m-2011)		110 Mbit/s	183 Mbit/s
4 × 4 MIMO (802.16m-2011)		219 Mbit/s	365 Mbit/s

peak DL and UL rates for 802.16m in comparison with 802.16e-2005 and 802.16e-2009 are shown in Tables 3.9 and 3.10 respectively. By comparing the DL and UL data rates for 802.16e-2009 and 802.16m (2×2 MIMO, 20 MHz TDD and 2×20 MHz FDD), an improvement of 30–50% can be observed. This is primarily due to the improved link budget of at least 3 dB in 802.16m.

3.7   WiMAX FORUM

The WiMAX Forum is similar in function to the Wi-Fi Alliance (WFA). The non-profit association was formed in June 2001 to define profiles for end-to-end 802.16 system deployments and to develop a testing and certification program called WiMAX Forum Certified. This program ensures compliance and worldwide interoperability of products from different vendors. It promotes the adoption of fixed wireless access standards (802.16-2004) and mobile wireless access (802.16e). The WiMAX Forum focuses on the OFDM PHY for fixed wireless access and OFDMA PHY for mobile wireless access. It currently has over 170 members, includ-

TABLE 3.10 802.16m Peak UL Rates (5 : 3 DL/UL Ratio for TDD)

	10 MHz (TDD)	20 MHz (TDD)	2 × 20 MHz (FDD)
2 × 2 MU-MIMO (802.16e-2005)	17 Mbit/s		
2 × 2 MU-MIMO (802.16e-2009)	23 Mbit/s	46 Mbit/s	138 Mbit/s
2 × 2 MU-MIMO (802.16m-2011)		70 Mbit/s	188 Mbit/s
2 × 4 MU-MIMO (802.16m-2011)		140 Mbit/s	376 Mbit/s

ing leading equipment manufacturers, component suppliers, service providers, and system integrators. Plugfest is carried out at a WiMAX Forum contracted testing site. The week-long event is aimed at validating and verifying equipment interoperability. The product is considered to be interoperable if it is able to send and receive data with two other vendors' equipment for a selected certification profile. The WiMAX Forum also develops higher layers for 802.16. For the station, these layers include IP layer functions and the CS. For the BS, layers include the mobility agent, radio access network, and the CS. The forum initially focused on the TDD operational mode. FDD specifications were added in August 2009. This represents an opportunity for integration between LTE (where FDD is a popular option) and WiMAX.

3.8 WIRELESS ACCESS USING WiMAX

Prior to 802.16, proprietary fixed wireless access products have enjoyed some measure of commercial success. For example, fixed wireless connections between enterprise buildings and backhaul operations have been employed. Using WiMAX, business customers pay less than using a T1 line. Fixed connections are also useful in "green field" deployments where there is no infrastructure (e.g., in rural areas and developing countries). In these deployments, mobility support is less critical and hence a cellular-based system may not be needed.

3.8.1 WiMAX Deployment

Clear has deployed mobile WiMAX extensively in the United States. Merged partner Sprint owns the 2.5 GHz licensed spectrum in markets covering 85% of U.S. population. Roughly 120 MHz of bandwidth is available in most major metropolitan areas, considerably higher available capacity compared to 3G cellular. At \$35/month for mobile Internet service, it was cheaper than 3G/4G cellular and comparable with cable/DSL. Clear's WiMAX service (http://www.clear.com) is currently available in 80 cities and reaches 130 million people. As of the first quarter of 2011, there were a total of 6.15 million subscribers. Clear provides several services:

- VoIP service ($25/month)
- Basic home Internet ($20–$40/month) with rate caps at 0.5 Mbit/s for the UL and 1.5 Mbit/s for the DL
- Mobile Internet ($35–$50/month) with a rate cap at 1 Mbit/s for the UL, but no rate cap for the DL
- Home and mobile Internet ($60/month) with a rate cap at 1 Mbit/s for the UL, but no rate cap for the DL
- Day pass ($10/day).

The same flat rate applies for service access in a different state, which means that one can roam from Atlanta to Las Vegas without paying additional fees. A software platform with open APIs and an SDK are offered to developers. Sprint partners with HTC, a handset manufacturer, to launch a series of 4G WiMAX Android smartphones. A similar approach has been adopted by other carriers. For example, Verizon launched the Android-based Xoom tablet from Motorola that supports LTE. AT&T partners with Apple to launch 3G/4G iPhone and iPad devices that are equipped with an SDK.

The first U.S. WiMAX service was launched in September 2008 in Baltimore, Maryland. There were 182 BSs covering 75% of the city. Each wireless cell is partitioned into three sectors with 3.7–5 Mbit/s on the DL and 1.8–2.6 Mbit/s on the UL. These rates are a significant improvement over 3G cellular rates. Typical 3G rates are 678 Kbit/s on the DL and 520 Kbit/s on the UL. Other U.S. deployments include Portland (January 2009), Atlanta/Las Vegas (June 2009), Philadelphia (October 2009), New York, Los Angeles, San Francisco, and more. In Portland, the network performance exceeded Clear's expectations. In addition, the service attracted subscribers twice as quickly as it did earlier in pre-WiMAX networks. Clear also provides wireless Internet service in the Hawaiian islands Maui and Honolulu, and in 28 other markets. The service is currently portable, not mobile. WiMAX is becoming a key access technology for vertical markets, such as smart-grid utilities, digital signage and video, telematics, industrial personal digital assistants, and surveillance networks. Smart grids improve the efficiency, reliability, resiliency, and overall quality of the power utility system through the integration of communications technologies.

WiMAX can leverage on new powerful wireless technologies that are better than 3G cellular. It possesses virtually all wireless technologies of LTE Advanced with the added benefit of computer chipset support from Intel. Like LTE, WiMAX has the ability to combine residential wireless Internet access with mobile access. WiMAX also provides a cost advantage for backhaul links. Backhaul links connect cell towers to the Internet. Traditional carriers spend billions of dollars a year on a wired backhaul. Wireless links have almost no ongoing cost. These links allow the service provider to break even with fewer than 10% of the people in its coverage area subscribing. LTE and WiMAX networks may interoperate and merge under the IMT-A framework. WiMAX has the advantage of being mature for deployment and has a headstart over LTE. LTE technology deployment has only started in late 2010.

Figure 3.14 Cradlepoint WiMAX/Wi-Fi router.

3.8.2 WiMAX/Wi-Fi Router

Figure 3.14 shows the Cradlepoint WiMAX/Wi-Fi broadband router (http://www.cradlepoint.com) that creates a local Wi-Fi network from a WiMAX signal and allows any existing Wi-Fi device to connect to a WiMAX network. The right module provides WiMAX connectivity, whereas the left Wi-Fi/Ethernet module allows the broadband connection to be shared within the home or office. A similar LTE/Wi-Fi router is also available.

REFERENCES

[1] IEEE 802.16 Standard Download, http://standards.ieee.org/about/get/802/802.16.html.
[2] WiMAX Forum, *Mobile System Profile Specification, Release 1.5 FDD Specific Part*, August 1, 2009.
[3] WiMAX Forum, *WiMAX and the IEEE 802.16m Air Interface Standard*, April 2010.
[4] S. Ahmadi, "An Overview of Next-Generation Mobile WiMAX Technology," *IEEE Communications Magazine*, pp. 84–98, June 2009.
[5] C. Eklund, et al., *WirelessMAN: Inside the IEEE 802.16 Standard for Wireless Metropolitan Area Networks*, IEEE Press, 2006.

HOMEWORK PROBLEMS

3.1. 802.16 employs a broad range of licensed and unlicensed frequencies from 2 to 66 GHz. What are the implications for interoperability between products from different vendors? Is interoperability an important issue for fixed wireless access? If the MAC protocol is changed from TDMA to CSMA, will this solve the interoperability problem to some degree?

3.2. Explain why lower speed caps are needed for Clear's home Internet WiMAX service than for mobile Internet. Hint: Consider office usage for mobile Internet.

3.3. What are the potential performance bottlenecks associated with the Cradlepoint router? Will there be interference between the Wi-Fi and WiMAX antennas (e.g., Wi-Fi antennas transmitting when WiMAX antennas are receiving)? Bonus: How does the router compare with the femtocell approach?

3.4. A reservation MAC protocol is a request and grant mechanism where the station trans-mits a request to the BS that then grants bandwidth resources. An example is the res-ervation TDMA mechanism in 802.16. What do you think is the biggest problem for adopting reservation protocols for wireless access? How can this problem be solved? Hint: The same problem applies to wired networks, such as copper, cable, and fiber access.

3.5. Compare the use of single-carrier operation in 802.16 and 60 GHz 802.11ad. Why is the use of single carrier justified for these cases?

3.6. Why is a voice-related feature like UGS included in 802.16? Do you think 802.11 suffers a significant disadvantage by not having this feature?

3.7. The IP-based Internet network was originally designed to carry data traffic. It has since been extended to carry voice traffic (e.g., VoIP), and, more recently video traffic. 802.11 was designed for indoor networks but has been extended to outdoor and longer-range networks. Discuss the challenges in extending 802.16 (designed for long-range outdoor access) networks for indoor use.

3.8. In the TDMA/TDD MAC protocol used by 802.16, the DL transmission is first broad-cast, followed by the UL transmission. Can this order be reversed?

3.9. Carrier sensing performance degrades as the operating range increases. How is this disadvantage overcome in TDMA (used by 802.16)?

3.10. What are the drawbacks of employing a cellular standard for fixed wireless access?

3.11. The figure below compares the performance of MIMO versus AAS for a 4×4 configu-ration with a SNR of 10 dB. The Rician channel model is used due the likely presence of a LOS path in outdoor deployments. The K factor indicates the ratio of the direct path power over the diffuse (scattered) power. What can you conclude about the per-formance of MIMO versus AAS in 802.16 deployments? How can these systems be optimized?

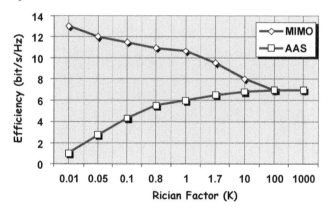

3.12. Is two-way EAP authentication support necessary in 802.16? Hint: Compare this approach with EAP authentication used in 802.11 to detect rogue APs.

3.13. Suppose an OFDM subchannel consists of 24 data subcarriers with 1 data symbol per subcarrier. The UL slot comprises 1 subchannel $\times$ 3 OFDM symbol periods. The DL slot comprises 1 subchannel $\times$ 2 OFDM symbol periods. Thus, there are 72 data symbols per slot for the UL and 48 symbols for the DL. There are more pilot subcarriers (for synchronization) on the UL (8) than the DL (4). The DL can support higher modulation, such as 64-QAM with a 5/6 coding rate compared with the UL, which can reliably support 16-QAM with a 3/4 coding rate. Compute the maximum DL and UL bit rates (excluding coding overheads), assuming that the PUSC tiles are repeated every 5 ms. Comment on the asymmetry of the rates.

3.14. HARQ requires that a copy of previously received packets be retained. Thus, the processing of the next packet depends on the previous one, and the addresses of packets originating from which stations must be known. Explain whether HARQ is appropriate for 802.11 networks operating in unlicensed bands. ∎

CHAPTER **4**

LONG TERM EVOLUTION

The first commercial Long Term Evolution (LTE) networks were deployed on a limited scale in Scandinavia at the end of 2009. Currently, 49 LTE operators have launched commercial services, while 285 operators have committed to commercial LTE network deployments or are engaged in trials, technology testing, or studies. Big operators include AT&T, Verizon Wireless, and China Mobile. In the United States, at the start of 2012, Verizon's LTE network has rolled out 400 markets covering over 200 million points of presence (POPs). Currently, subscribers have 10 LTE devices to choose from, including three smartphones, two tablets, two hotspot devices, and three USB modems. Outside of the operator's LTE footprint, customers automatically connect to the 3G network. AT&T has set up 72 markets with over 100 million POPs. Worldwide LTE deployments are detailed in [1].

Heterogeneous networks are expected to be an integral component of future LTE network deployments. The ability to coordinate and manage interference in networks is key to enabling additional gains in capacity and performance, which is needed to keep up with exponentially growing traffic demands. LTE is currently being deployed in the traditional macrocell network architecture. However, there is tremendous interest from cellular operators and vendors to include small, inexpensive, self-configurable base stations in that architecture. It is expected that a large number of these small cells will be multimode, supporting concurrent operation of LTE/LTE-Advanced, 3G, 2G, 802.11, and sharing backhaul, provisioning, and management functions.

In this chapter, we first provide a quick overview of the High Speed Packet Access (HSPA) and the HSPA+ standards, which are the precursors to LTE and widely considered as 3G and 3.5G cellular technologies. We then proceed to describe the various components of LTE and LTE-Advanced. The final section offers some interesting perspectives on antenna design for 4G wireless handsets.

4.1 HIGH SPEED PACKET ACCESS

The 3GPP wideband CDMA-HSPA is a mobile broadband service that first began with wideband CDMA (WCDMA). The first WCDMA service was offered in Japan

---

Broadband Wireless Multimedia Networks, First Edition. Benny Bing.
© 2013 John Wiley & Sons, Inc. Published 2013 by John Wiley & Sons, Inc.

in October 2001. The first HSPA service was launched in the United States in October 2005. There are over 150 million business users and consumers worldwide today. HSPA offers a DL rate of at least 2 Mbit/s. The maximum rate is 14.4 Mbit/s, but rates of 3.6 or 7.2 Mbit/s are more typical. The UL rate ranges from 2 to 5.8 Mbit/s. The newer HSPA Evolution or HSPA+ (3GPP Release 7) supports a peak DL rate of either 21 Mbit/s or 28 Mbit/s. It employs 16-QAM instead of QPSK to double the UL peak data rate to 11.5 Mbit/s. Connection set-up times are reduced from 1 second (typical with WCDMA) to below 100 ms. Over 1700 HSPA user devices were launched by 190 manufacturers with 283 networks in 119 countries worldwide. Fifty-four operators in 33 countries have committed to HSPA network deployments.

Future 3GPP releases for HSPA+ provide aggregations of up to four carriers and combines 64-QAM with 2×2 MIMO in a 5 MHz carrier to give a peak DL rate of 42 Mbit/s. This supports the high traffic volume caused by a few heavy data applications. Big operator networks now carry 5–10 TB per day.

4.2   LONG TERM EVOLUTION

Unlike 3G cellular networks, LTE allows operators to use new and wider spectrum, which leads to higher user data rates and lower latency. The standard defines a flat Internet Protocol (IP) network architecture that reduces the number of network elements and provides more flexibility. For example, terminals have only two states, idle and active, which helps reduce control plane latency and signaling. The LTE standard was first published in March of 2009 as part of the Third Generation Partnership (3GPP) Release 8 specifications (December 2008). The air interface and the network architecture evolved from 3GPP Release 6 and are therefore known as Evolved UMTS (universal mobile telecommunications system) Terrestrial Radio Access Network (E-UTRAN) and Evolved Packet Core (EPC), respectively. The system architecture evolution (SAE) addresses the evolution of the overall LTE system architecture, including the core network, to support multiple radio access technologies. 3GPP Release 9 (end-2009) achieves a latency below 25 ms and rates of up to 84 Mbit/s. 3GPP Release 10 (mid-2011) improves the rates to up to 168 Mbit/s [2]. LTE allows coexistence with legacy and existing 3GPP/3GPP2 systems, such as GSM/EDGE, CDMA 2000, and HSPA/HSPA+.

LTE provides multicarrier channel-dependent resource scheduling and a fast handover or handoff mechanism to support RT applications involving voice and video. LTE enhances cell-edge performance by employing UL power control and intercell interference coordination (ICIC). The enhanced multimedia broadcast multicast service (E-MBMS) is achieved via connections to the EPC. The E-UTRAN air interface is based on orthogonal frequency division multiplexing (OFDM). The key features are listed in Table 4.1. The peak DL data rate of 300 Mbit/s corresponds to a user handset employing 4×4 MIMO. This rate reduces to 150 Mbit/s when 2×2 MIMO is used.

TABLE 4.1 Key LTE Features

Peak data rates	Shared channel access of up to 300 Mbit/s on DL and 75 Mbit/s on UL (20 MHz bandwidth).
Mobility support	Up to 300 or 500 km/h depending on frequency band.
User plane latency	<5 ms radio network delay for small IP packets in unloaded conditions.
Control plane latency	<100 ms (idle to active states).
Coverage (cell size)	5–30 km.
Capacity	>200 active users per 5 MHz cell.
Duplexing	Supports frequency division duplex (FDD) and time division duplex (TDD) modes of operation.
Channel bandwidth	1.4, 3, 5, 10, 15, and 20 MHz (fixed symbol length for all bandwidths).
Channel access and modulation	Orthogonal frequency division multiple access (OFDMA) on DL, single-carrier frequency division multiple access (SC-FDMA) on UL; QPSK, 16-QAM, and 64-QAM signal constellations.
Number of OFDMA subcarriers	72, 180, 300, 600, 900, and 1200, with fixed 15 KHz subcarrier spacing.
Multiple antenna modes	Supports 2×2 and 4×4 DL MIMO modes, multiuser MIMO (MU-MIMO), and space–time frequency block coding (SFBC).

4.2.1 Evolved Packet Core

The EPC is an all-IP multiaccess core network that reduces deployment cost. The EPC provides IP connections to the IP multimedia subsystem (IMS) and Internet domains while supporting 3GPP and non-3GPP services. It can be connected to other 3GPP and non-3GPP radio access networks (RANs), such as 802.16 and 802.11. The EPC ensures optimized handovers between different RANs. The policy and charging rules function (PCRF) handles QoS management and charging. The home subscriber server (HSS) handles subscriber management and security. Because the network is "flat," only two node types are supported using one control plane and one user plane nodes:

- **CP node**: Mobility management entity (MME)
- **UP node**: Serving gateway (S-GW) and packet-network gateway (P-GW).

The LTE base station is called the evolved NodeB (eNB), and the mobile terminal is known as the user equipment (UE). As shown in Figure 4.1, the eNBs are interconnected using the X2 interface. The eNBs can also connect to the EPC elements, such as the MME and S-GW, using the S1 interface. The radio network controller in the eNB negotiates handovers between eNBs. LTE supports dynamic radio resource assignment via the eNB. For example, the eNB can take advantage of good channel quality to allow more efficient use of bandwidth.

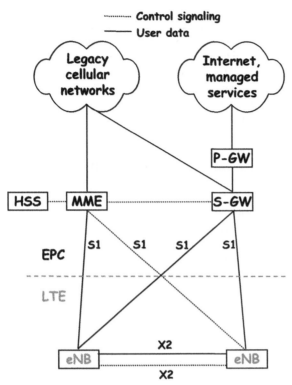

Figure 4.1 Evolved packet core.

The key LTE functional blocks are shown in Figure 4.2. The eNBs provide user plane (PDCP, RLC, MAC, and PHY) and control plane (RRC) terminations for the UE. The S-GW performs "bearer" plane functions by forwarding and routing user data packets. It also manages handovers between eNBs and triggers the paging of an idle UE when new DL data arrives. The MME performs control signaling for LTE. It is responsible for idle mode UE tracking and paging, connection activation/deactivation, and choosing the S-GW for new connections, as well as handovers and user authentication. The P-GW provides connectivity to external packet data networks. The UE may have simultaneous connections to more than one P-GW. The P-GW performs policy enforcement, packet filtering, charging support, lawful interception, and packet screening. It also supports seamless mobility between LTE and non-LTE networks (e.g., 802.16, Evolution-Data Optimized [EVDO]). The S1-flex mechanism enables eNBs to be shared by service providers in order to reduce the cost of owning and operating the LTE network. A UE is connected to the appropriate MME, S-GW, or P-GW entity (each entity separately owned by the service provider) based on the identity of the service provider it is subscribing.

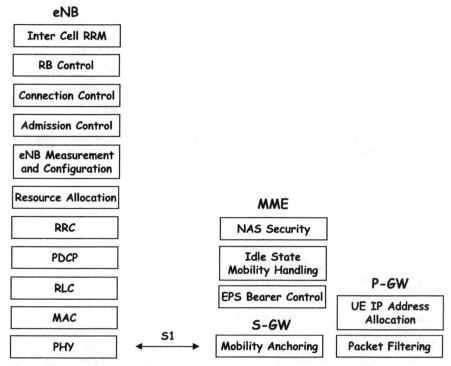

Figure 4.2   LTE functional blocks.

4.2.2   Frequency Bands

LTE employs microwave bands, such as 2.6 GHz, to ensure access to necessary bandwidth for higher data speeds. Lower bands below 1 GHz are also allowed, and they include 700, 800 (digital dividend bands), and 850/900 MHz. These bands are ideal for efficient delivery in a large geographical coverage, especially in rural areas, and they improve in-building penetration of mobile broadband services. Since handovers are minimized, they are also well suited for high-mobility applications, such as Internet connectivity for cars and high-speed trains. Verizon Wireless has developed technical LTE device specifications for its 700 MHz LTE network. AT&T may also deploy a 700 MHz LTE network.

4.2.3   Physical Layer

LTE employs SC-FDMA on the UL to improve the power efficiency of the UE. SC-FDMA employs DFT-precoded OFDM, which takes into account channel quality, resulting in a smaller peak-to-average power ratio (PAPR) than regular OFDM. It also improves cell-edge performance. The precoding process transforms the modulated data symbols into the frequency domain. The data symbols are therefore DFT

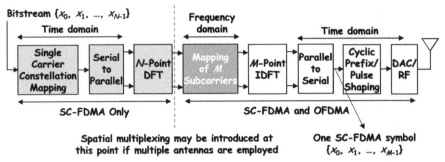

Figure 4.3   SC-FDMA transmitter.

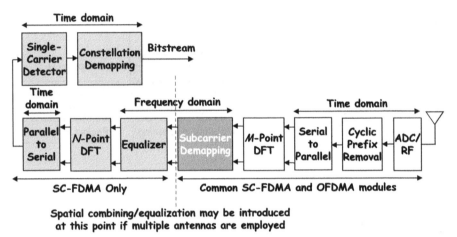

Figure 4.4   SC-FDMA receiver.

spread, and each subcarrier may contain information on multiple data symbols, as opposed to a single symbol in OFDMA and traditional OFDM. However, unlike OFDMA, SC-FDMA requires contiguous frequency bands. LTE employs OFDMA on the DL to simplify receiver baseband processing. Since a channel equalizer is not required, UE cost and power consumption is minimized. Adaptive modulation and coding is provided for the data channel. The UL/DL modulations include QPSK, 16-QAM, and 64-QAM. Convolutional and Turbo coding at 1/3 code rate are supported with an interleaver. The design of a typical SC-FDMA transmitter and receiver is shown in Figures 4.3 and 4.4, respectively. The transmitter employs single carrier modulation, hence the subcarriers are not independently modulated, leading to reduced PAPR. The receiver incorporates a lower complexity frequency-domain equalizer. An equivalent time-domain equalizer requires a large tap size for time-domain filtering when applied to wideband channels. An OFDMA receiver may require separate equalizers and detectors for each subcarrier.

4.2.4   UL Subcarrier Allocation

The UL symbol to subcarrier mapping can be distributed or localized. Localized or grouped subcarrier allocation is employed on the UL. Such channel-dependent scheduling provides frequency diversity gains. However, subcarriers may interact at cell edge (with overlapped coverage areas). When this occurs, channels comprising a group of subcarriers may collide. A distributed method for subcarrier allocation may reduce the impact of such collisions.

4.2.5   MIMO Modes

Multiple antennas at the UE are supported, with two receive and one transmit antennas being mandatory. This means that the UE must be able to spatially demultiplex two spatial streams (that are transmitted by the eNB) even if it is able to transmit just one spatial stream to the eNB. Alternatively, the UE may employ the additional antennas to support DL receive diversity. At the eNB, the use of two transmit antennas is the baseline configuration. Thus, for the LTE DL, 2×2 is the baseline configuration. LTE supports a single layer (single spatial stream) for the UL per UE but up to four layers for the DL per UE. MU-MIMO for UL and DL, and 2×2 and 4×2 closed- and open-loop MIMO modes are supported.

Transmit diversity on the UL improves system performance. This may be employed when channel-dependent scheduling is not possible or incurs high overheads (e.g., for low-rate applications, such as VoIP). DL MIMO modes involve dynamic switching between spatial multiplexing and SFBC. SFBC provides increased robustness using transmit diversity (e.g., using Alamouti STC), and can be complemented by frequency-switched transmit diversity when four antennas are used. Spatial multiplexing with up to 4 antennas at the eNB and UE are supported. Simultaneous data streams (known as layers) can be carried spatially by a single radio interface. These streams are mapped to the antennas by a $N_{\text{antenna}} \times N_{\text{layer}}$ precoding matrix (Figure 4.5). The optimum precoding matrix (which gives the highest capacity) is selected from a codebook known to the eNB and the UE. The selection is

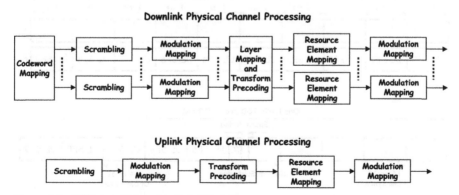

Figure 4.5   LTE downlink and uplink RF chains.

based on UE feedback (i.e., PMI, RI, CQI). When $N_{layer} = 1$, this gives the beamforming vector, also referred to as codebook-based beamforming. Noncodebook-based beamforming is also supported. The UE must estimate the beamformed channel. Although single-user MIMO (SU-MIMO) transmission provides high data rates per user, MU-MIMO provides higher network capacity, supporting more users per cell. Beamforming improves coverage, resulting in fewer cells to cover in a given area. LTE also supports cycle delay diversity (CDD) by adding specific cyclic shifts or delays for different antennas that support spatial multiplexing, effectively introducing multipath at the receiver. On the UL, the UE may employ two transmit antennas using a single RF chain. Alternatively, if MU-MIMO is supported, simultaneous transmissions from two UEs on the same time/frequency resource are allowed, and each UE employs a single antenna. An open-loop spatial multiplexing scheme is supported for high-mobility UEs. In this case, predefined settings for spatial multiplexing and precoding are employed.

4.2.6 Frame Format

As shown in Figure 4.6, TDD, FDD, and half-duplex FDD frame formats are supported in LTE. In FDD, there is a one-to-one relation between UL and DL subframes. For instance, there are 10 subframes for the DL and 10 subframes for the UL. Each subframe comprises two slots and is capable of carrying between 12 and 14 OFDM symbols. Thus, there are 20 slots in a frame. Each slot is 0.5 ms in duration, and comprises a cyclic prefix (CP) and a fixed symbol duration of 66.7 µs. There are two types of CP. The duration of a short CP is 5.2 µs (160 samples) for the first symbol and 4.7 µs (144 samples) for all other symbols, thus giving 14 OFDM symbols per subframe. This implies the CP overhead roughly ranges from 7% to 8% overheads. The extended CP takes up 16.7 µs (512 samples) for all symbols, giving 12 OFDM symbols per subframe. Thus, the extended CP overhead is quite significant, about 25%. The location of the primary synchronization signal is the middle of subframes 0 and 5.

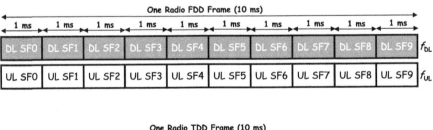

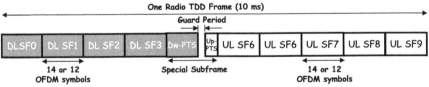

Figure 4.6   FDD and TDD frame formats.

TABLE 4.2   UL and DL Subframe Configurations for LTE TDD

Configuration	DL-to-UL switch periodicity	Subframe number									
		0	1	2	3	4	5	6	7	8	9
0	5 ms	D	S	U	U	U	D	S	U	U	U
1	5 ms	D	S	U	U	D	D	S	U	U	D
2	5 ms	D	S	U	D	D	D	S	U	D	D
3	10 ms	D	S	U	U	U	D	D	D	D	D
4	10 ms	D	S	U	U	D	D	D	D	D	D
5	10 ms	D	S	U	D	D	D	D	D	D	D
6	5 ms	D	S	U	U	U	D	S	U	U	D

D, DL subframe; S, Special subframe; U, UL subframe.

In TDD, seven different UL/DL subframe configurations are supported (Table 4.2). DL/UL periods of 5 or 10 ms and DL/UL ratios from 2 : 3 to 9 : 1 are supported. Split subframes are known as special frames, and they are required because the UL and DL transmission is discontinuous within a single TDD frame. These 1-ms special subframes consist of the DL pilot time slot (DwPTS), guard period, and the UL pilot time slot (UpPTS). The DwPTS is transmitted first and contains 3–12 OFDMA symbols; two of these symbols are used for control signaling. The location of the primary synchronization signal is in the third OFDMA symbol of the DwPTS. The UpPTS comprises one or two OFDM symbols and can be used for the transmission of UL sounding signals (for estimating channel quality) and random access (for timing alignment with the eNB). The guard period ranges from 2 to 10 OFDM symbols (140–667 μs). This is sufficient to handle round-trip propagation delays for cell sizes of up to 100 km. In addition, it is useful for co-existence with time-duplex synchronous CDMA (TD-SCDMA) and other legacy TDD radio access systems.

4.2.7   Physical Resource Blocks

In LTE, physical resource elements are organized as a two-dimensional grid in the time and frequency domains. The smallest unit of bandwidth allocation is called a resource element (RE), which comprises an OFDM symbol on a subcarrier. A resource block (RB) consists of multiple REs. A set of consecutive RBs can be combined into an RB group (RBG). For a 20 MHz channel bandwidth, the RBG comprises four RBs. The minimum unit of scheduling is an RB, which corresponds to a single 1 ms subframe and 12 subcarriers. The UL/DL RBs are controlled by the eNB. It assembles 12 subcarriers with a total bandwidth of 180 KHz over a 0.5-ms slot. The scheduling is not done at subcarrier granularity to limit the control signaling overheads. The subcarrier spacing is fixed at 15 KHz (11.16 KHz for 802.16e). QPSK, 16-QAM, and 64-QAM are the UL/DL modulation schemes. For the UL, 64-QAM is optional for the UE.

The number of RBs can be adjusted for 6 different channel bandwidths (Table 4.3). UE support for the 20 MHz channel is mandatory, which corresponds

TABLE 4.3 Relationship between Channel Bandwidth, Subcarriers, and RBs

Channel bandwidth (MHz)	1.4	3	5	10	15	20
Subcarrier spacing (KHz)	15					
Number of occupied subcarriers	72	180	300	600	900	1200
Number of RBs per slot	6	15	25	50	75	100
FFT/IFFT size	128	256	512	1024	1536	2048
Sampling rate (Msamples/s)	1.92	3.84	7.68	15.36	23.04	30.72

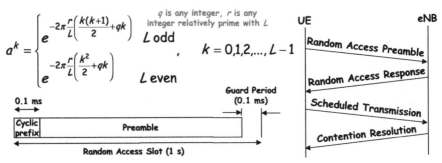

$$a^k = \begin{cases} e^{-2\pi\frac{r}{L}\left(\frac{k(k+1)}{2}+qk\right)} & L \text{ odd} \\ e^{-2\pi\frac{r}{L}\left(\frac{k^2}{2}+qk\right)} & L \text{ even} \end{cases}, \quad k = 0,1,2,\ldots,L-1$$

q is any integer, r is any integer relatively prime with L

Figure 4.7   Resynchronization procedure.

to a bandwidth of 100 RBs (100×12 subcarriers $\times$ 15 KHz $\approx$ 20 MHz). A short CP supports 12 subcarriers $\times$ 7 OFDM symbols. An extended CP supports 12×6 (15 KHz) or 24×3 (7.5 KHz) symbols. Multiple RBs can be assigned to individual users. This is handled by a scheduling function at the eNB. It is based on the radio condition, traffic volume, and QoS requirements. A typical assignment comprises the physical RB and one of 16 modulation and coding schemes (MCSs). The short frame duration allows scheduling of resources based on the current channel quality. It also allows dynamic resource allocation, and fast channel quality and acknowledgment (ACK) feedback. Resources for the DL control signaling can be varied, and they typically occupy the first three OFDM symbols in each subframe.

UL transmissions are aligned with the frame timing of the eNB. When the timing is not aligned or alignment was lost during a period of inactivity, a random access procedure is performed to acquire alignment. The preamble is based on a constant amplitude zero autocorrelation (CAZAC) polyphase sequence [3], which provides constant amplitude, zero circular autocorrelation, flat frequency response, and low circular cross-correlation between two different sequences (Figure 4.7).

4.2.8   Packetization Framework

The LTE packetization framework is shown in Figure 4.8. The MAC layer performs the mapping between the logical channels and transport channels, schedules UL/DL

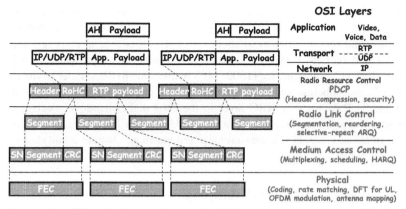

Figure 4.8   LTE packetization framework. FEC, forward error correction; PDCP, packet data convergence protocol; RoHC, robust header compression; AH, application header, SN, sequence number.

resources for the UE, and selects the most appropriate transport format. The logical channels are characterized by the information carried by them. The transport channels are distinguished by the radio characteristics (e.g., MCS) used to transmit the data. LTE supports a two-layered retransmission involving a fast hybrid automatic repeat request (HARQ) protocol at the MAC layer and the selective-repeat ARQ protocol at the radio link control (RLC) layer. The LTE ARQ feedback incurs a low overhead. Soft combining with incremental redundancy reduces latency. Most errors are corrected by HARQ, which uses a single bit error-feedback mechanism. HARQ performs asynchronous retransmissions on the DL and synchronous retransmissions in the UL. Retransmissions of HARQ blocks occur at predefined periodic intervals. Hence, no explicit signaling is required to indicate to the receiver the retransmission schedule. Asynchronous HARQ offers the flexibility of scheduling retransmissions based on link conditions. Selective-repeat ARQ is rarely employed and is used to handle residual errors. There is tight interaction between the MAC and RLC protocols. This reduces protocol overhead and allows dynamic scheduling of shared resources in packet-switched access.

The RLC layer handles upper layer protocol data units (PDUs), ARQ retransmission, concatenation, segmentation, and reassembly of RLC data units. The RLC layer formats and delivers traffic between the UE and the eNB. It provides three reliability modes for data transport, namely acknowledged mode (AM), unacknowledged mode (UM), or transparent mode (TM). The UM mode is suitable for delivery of time-sensitive real-time (RT) services. The AM mode is appropriate for non-RT (NRT) services such as file downloads. The TM mode is used when the length of the PDU is known *a priori*, such as in broadcasting system information. The RLC layer also provides in-sequence delivery of service data units (SDUs) to the upper layers and eliminates duplicate SDUs. It may also segment the SDUs depending on channel conditions.

The packet data convergence protocol (PDCP) provides robust header compression (RoHC) of user-plane IP packets and encryption, as well as radio resource control (RRC) functionality in the control plane (e.g., radio resource measurement, admission control, and scheduling). The RoHC is able to compress the headers of UDP (8 bytes) and RTP (12 bytes) to about 3–4 bytes. It is able to compress the 40-byte IPv4 and 60-byte IPv6 TCP headers to 8 bytes. This represents a 2.5% reduction in IPv4 TCP segment size (1500–1468 bytes). Due to the smaller payload, a more significant reduction can be achieved for VoIP packets (typical payload of 20 bytes) and TCP ACKs. The RRC layer in the eNB makes handover decisions based on adjacent cell measurements sent by the UE, pages for a UE, broadcasts system information, controls UE measurement reports (e.g., channel quality), and allocates a temporary cell identifier to an active UE.

4.2.9   Channel Functions and Mapping

The eNB provides two sets of reference signals for the UE to initiate cell search, acquire network timing, demodulate data signals coherently, and estimate channel quality. Each set of reference signals occupies a fixed pattern for each antenna in the LTE system. For example, the reference symbols in each set are separated by five data subcarriers and three data symbols, giving four reference symbols in a RB. Figures 4.9 to 4.11 show the location of the reference symbols in a RB for 1, 2, and 4 antennas. Unused symbols are required when two or more antennas are used in order to support concurrent transmissions from multiple antennas. In this case, when a reference symbol is transmitted from one antenna, the other antenna is idle. For example, four unused symbols are required in a two-antenna system, giving rise to an additional overhead of $4/84 \times 100\%$ or 4.8%. This overhead doubles to 9.6% for four antennas. The 4×2 configuration is primarily targeted for low Doppler spread deployments. Hence, there are fewer reference symbols for antennas 3 and 4.

The eNB transmits primary and secondary synchronization signals (twice every 10 ms on predefined slots for the UE to acquire timing information. These

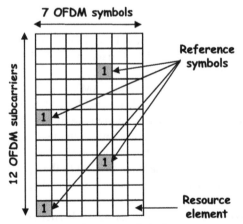

Figure 4.9   Reference symbols for a one-antenna LTE system.

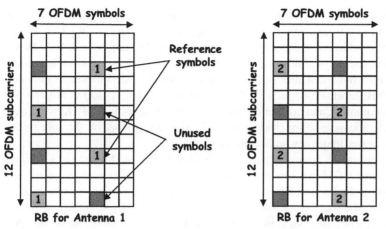

Figure 4.10   Reference symbols for a two-antenna LTE system.

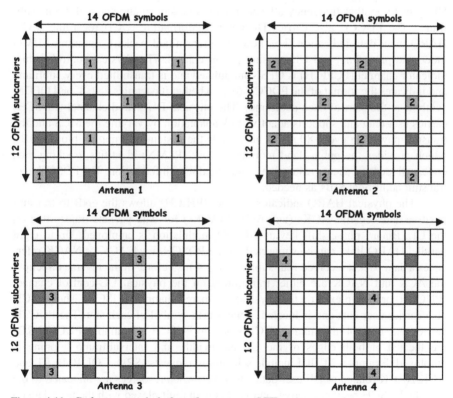

Figure 4.11   Reference symbols for a four-antenna LTE system.

signals provide a total of 504 physical layer cell identities via 168 secondary cell identity groups (N_s), each group containing three primary identities (N_p). Cell identities are computed as $3N_s + N_p$, where $N_s = 0, 1, 2, \ldots, 167$ and $N_p = 0, 1, 2$. As the name implies, the physical broadcast control channel (PBCH) on the DL is broadcast over the entire radio cell and carries the master information block that provides essential eNB information, such as the system bandwidth, HARQ configuration, system frame number, and number of active transmit antennas. It is QPSK modulated with cell-specific scrambling and is transmitted every 40 ms on 72 subcarriers around the carrier frequency within the first subframe.

The physical DL control channel (PDCCH) is QPSK modulated and is located in the first few OFDM symbols of a subframe and is primarily used to convey RB assignments on the UL and DL. A resource indication value (RIV) may be used to inform the UE the starting RB and the number of allocated RBs. The UE may hop from one frequency allocation in a slot to another within a subframe or between subframes. In intersubframe hopping, the frequency allocation may change from one subframe to another. For small RB allocations, intrasubframe hopping (also called inter-slot hopping) can be applied to achieve frequency diversity. In this case, the UE hops to another frequency allocation from one slot to another within one subframe. The PDCCH also carries DL control information (DCI) regarding MCS, HARQ, new data indication (i.e., new or retransmitted packet), optimum MIMO mode and precoding, and UL transmit power control. The physical control format indicator channel (PCFICH) is QPSK modulated and is transmitted every subframe and indicates the format of the PDCCH (e.g., whether it occupies 1, 2, 3, or 4 OFDM symbols at the start of each subframe). The physical DL shared channel (PDSCH) carries user data and supports adaptive link adaptation by varying the MCS and the transmit power. In every 80 ms, it also sends the system information block (SIB), which provides the public land mobile network (PLMN) identities, tracking area code, cell identity, access restrictions, and scheduling information. The PDSCH may transmit additional SIBs as needed.

The physical HARQ indicator channel (PHICH) allows the eNB to transmit positive and negative ACKs (i.e., ACKs/NACKs) for data packets transmitted on the UL. Binary phase-shift keying (BPSK) modulation is employed. Up to eight parallel HARQ links can be supported on the PDSCH. Each ACK or NACK references a data packet received four subframes or 4 ms before. In the reverse direction, the ACK and NACK are either transmitted on the physical UL control channel (PUCCH) or multiplexed on the physical UL shared channel (PUSCH). The PUSCH also carries user data and is located in the fourth symbol of each slot. The PUCCH carries user control information (UCI) when no PUSCH is available. PUCCH RBs are located at both edges of the UL bandwidth, and intersubframe hopping is employed. Examples of UCI include scheduling requests (SRs), CQI, PMI, RI, and ACK/NACK related to DL user data. However, when resources are allocated to the UE, the PUSCH becomes active, and UCI is then multiplexed with the user data on the PUSCH. Thus, the PUCCH is never transmitted simultaneously with PUSCH data. The PUCCH formats are shown in Table 4.4. Formats 1, 1a, and 1b carry three reference symbols per slot, whereas formats 2, 2a, and 2b carry two reference symbols per slot, each slot using a normal CP. The modulated data symbol is

TABLE 4.4 PUCCH Formats and Contents

PUCCH format	Modulation	Type	Number of bits per subframe
1	N/A	SR	N/A
1a	BPSK	ACK/NACK, SR	1
1b	QPSK	ACK/NACK, SR	2
2	QPSK	CQI, PMI, RI, ACK/NACK	20
2a	BPSK + QPSK	CQI, PMI, RI, ACK/NACK	21
2b	BPSK + QPSK	CQI, PMI, RI, ACK/NACK	22

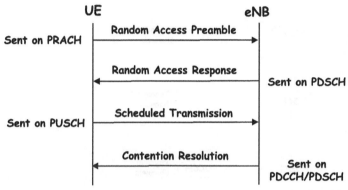

Figure 4.12   Random access procedure.

multiplied with a base cyclic sequence, which can be shifted to produce different orthogonal versions between symbols and slots to identify different UEs.

The physical random access channel (PRACH) on the UL carries a random access preamble to allow initial UE access, handover, or UL resynchronization (Figure 4.12). Two reference signals are transmitted on the UL, namely the demodulation reference signal (DRS) and the sounding reference signal (SRS). The DRS enables UL channel estimation and data demodulation (located in the fourth symbol in each slot with normal CP), whereas the SRS provides UL channel quality information. The SRS conveys the scheduling decisions and is normally transmitted in the first or last symbol of the subframe when there is no user data.

DL and UL physical channels are mapped into transport and logical channels by the MAC layer, as shown in Figures 4.13 and 4.14. Note that the number of transport channels is lower than the logical channels because the transport channels are shared. Some transport channels include the broadcast control channel (BCCH), the paging control channel (PCCH), the dedicated traffic channel (DTCH), the dedicated control channel (DCCH), and the common control channel (CCCH). The BCCH is a DL channel for broadcasting system control information. The PCCH is a DL channel that provides paging information and changes in system information.

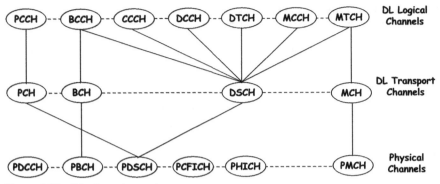

Figure 4.13   DL channel mapping.

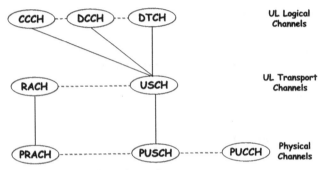

Figure 4.14   UL channel mapping.

This channel is used for paging when the network does not know the location cell of the UE. The DTCH is a point-to-point channel that is dedicated to a UE for the transfer of user information. It may exist in both UL and DL. The DCCH is a bidirectional point-to-point channel that allows control information to be exchanged between a UE and the network. It is used by a UE that has a RRC connection with the network. The CCCH is a channel for transmitting control information between UEs and the network. It is used by a UE that does not have a RRC connection.

The PUSCH power level (in dBm) is derived from Equation 4.1. There is a delay of four subframes between the power level adjustment command (issued by the eNB on the PDCCH) to the actual adjustment on the PUSCH. The measured power can also be used as a basis to initiate a handover (Figure 4.15).

$$P = \min\{P_{\max}, 10\log_{10} M + P_{\mathrm{o}} + xP_{\mathrm{loss}} + P_{\mathrm{a}} + T_{\mathrm{f}}\}, \tag{4.1}$$

where

$P_{\max}$ = Maximum allowed UE power

M = Number of PUSCH RBs

P_{o} = RRC configured cell-UE component

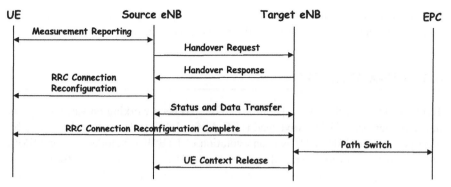

Figure 4.15   LTE handover.

x = RRC configured cell site parameter

P_{loss} = DL path loss estimate

P_a = Power adjustment via transmit power control command

T_f = PUSCH transport format.

4.2.10   Power Saving Modes

Like HSPA, LTE allows discontinuous reception (DRX) and discontinuous transmission (DTX), both reducing the UE transceiver duty cycle while in active operation. The off duration for DRX can be configured by the RRC via the PCCH. The discontinuous transmission and reception feature improves battery life by allowing the user device to turn off the transmitter when there is no data to send or receive. This benefits applications that send and receive bursty traffic (e.g., Web browsing and voice traffic).

4.2.11   Multimedia Broadcast Multicast Service

E-MBMS is a point-to-multipoint DL service of LTE. The multicast control channel (MCCH) carries E-MBMS control information to support DL transmission using one or more multicast traffic channels (MTCHs). In the single-cell mode, E-MBMS traffic is transported over the PDSCH. In the multicell mode, eNBs across different cells may be synchronized and configured as part of a Multicast/Broadcast Single Frequency Network (MB-SFN) so that identical transmissions can be broadcast to a group of users. This capability enables low-cost delivery of common content, such as popular multimedia content. The coordinated transmissions from the cells allow the UE to combine the RF signal energy from multiple transmitters, improving data rates and reducing power consumption via macrodiversity. In addition, a UE can roam between cells and maintain service continuity without the need for handovers. Transmissions via the MB-SFN employ a 7.5 KHz subcarrier spacing and a long CP in order to mitigate the signal delay from two or more adjacent eNBs. For example, a UE located close to an eNB may experience a longer delay between signals transmitted by adjacent eNBs than when the UE is located at the cell edge.

4.3   LTE-ADVANCED

To further improve the LTE capabilities, 3GPP has been working on various aspects in the framework of LTE-Advanced (LTE-A) for 3GPP Release 10, which may be deployed around 2015. LTE-A is an evolution of LTE that includes denser MIMO (e.g., up to eight layers or spatial streams for the DL, up to four layers for the UL), carrier aggregation (multi-channel operation using the same or different frequency bands) to support wider channel bandwidths (50 and 100 MHz), and new deployment strategies based on a heterogeneous network (HetNet) topology that blends macrocells with picocells, "metrocells," relays, and femtocells. Peak DL data rates are 100 Mbit/s (high mobility) and 1 Gbit/s (low mobility), whereas peak UL data rates are 50 Mbit/s (high mobility) and 500 Mbit/s (low mobility). There is potential for LTE-A to significantly improve the data rates through range expansion, ICIC, and user terminal interference cancellation. Like 802.16m, LTE-A exceeds the requirements of the International Telecommunication Union (ITU) 4G radio standard known as IMT-Advanced (IMT-A). ITU-R (ITU Radio Standardization Sector) has specified the minimum requirements and evaluation criteria for IMT-A in several technical areas as listed in Table 4.5.

4.3.1   Carrier Aggregation

Carrier aggregation (CA) creates wider bandwidth by using multiple aggregated carrier components (CCs). LTE-A enables multiple CCs to be aggregated to increase the bandwidth to up to 100 MHz. The aggregated CCs must be compatible with the channel bandwidths supported by LTE, as shown in Table 4.6. This allows seamless

TABLE 4.5   IMT-Advanced Requirements

Peak spectral efficiency (bit/s/Hz)	15 on DL (4×4), 6.75 on UL (2×4)
Cell spectral efficiency (bit/s/Hz/sector)	2.2 on DL (4×2), 1.4 on UL (2×4)
Cell edge user spectral efficiency	0.06 on DL (4×2), 0.03 on UL (2×4)
Bandwidth	Up to 40 MHz with carrier aggregation
Latency	100 ms for control plane (idle to active)
	10 ms for user plane
Link-level mobility efficiency (bit/s/Hz)	0.55 at 120 km/h, 0.25 at 350 km/h
Handover interruption time	Intrafrequency: 27.5 ms
	Interfrequency: 40 ms within a band, 60 ms between bands
VoIP capacity	40 (4 transmit antennas at eNB, 2 receive antennas at UE)

TABLE 4.6 UE Categories and Rates

UE category	Maximum DL rate	DL layers supported	Maximum UL rate
1	10.296 Mbit/s	1	5.160 Mbit/s
2	51.024 Mbit/s	2	25.456 Mbit/s
3	102.048 Mbit/s	2	51.024 Mbit/s
4	150.752 Mbit/s	2	51.024 Mbit/s
5	302.752 Mbit/s	4	75.376 Mbit/s[a]

[a]Requires 64-QAM to be supported on UL.

migration into LTE-A by reusing LTE resources. There are three types of CA. In intraband contiguous CA, a contiguous bandwidth wider than 20 MHz is used. In interband noncontiguous CA, multiple noncontiguous bands can be combined. For example, two frequency bands in the 2 GHz and 800 MHz bands can be used to increase the data throughput and improve transmission reliability using two spatial paths, each path operating on a different frequency band. In intraband noncontiguous CA, a noncontiguous band in the same band is used. This mode is useful for operators with fragmented spectrum in one band, as well as for operators that need to share the same cellular network.

4.3.2 HetNet Topology

Since improvements in spectral efficiency per link are approaching theoretical limits, the next step is to improve spectral efficiency per unit area in terms of bit/s/Hz/km^2. Relaying can further improve coverage and reduce deployment cost. Coordinated multipoint transmission and reception (CoMP) is an extension to ICIC and further improves cell-edge performance. By adopting a HetNet topology, LTE-A seeks to provide a uniform broadband experience, which is in contrast to a traditional cellular system, with eNBs employing similar power levels and antenna patterns. HetNets supports flexible and low-cost deployments with the use of different power eNBs to provide coverage and capacity, where it is needed the most. In a HetNet, the macro coverage area is overlaid with multiple smaller pico and femto cells, with eNBs that transmit at a substantially lower power than the macro eNB and may be deployed in an unplanned manner. These smaller cells bring the network closer to the users, thus ensuring better QoS for the users.

Small cells can be deployed to eliminate coverage holes in the macro network and improve capacity in hotspots, such as shopping malls, sports stadiums, and enterprises, effectively offloading traffic from the macro network. More importantly, these benefits may be realized without the expenses associated with macrocell installations. Allocating new spectrum or expanding macro networks may not be sufficient to keep up with traffic demand. HetNets are therefore becoming an increasingly attractive proposition. In contrast to 3G systems, where small cells made a relatively late entry, LTE specifications included small cells as an integral building block right from the beginning.

One of the key objectives of LTE is to increase the spectral efficiency and overall signal-to-interference-noise ratio (SINR) of the system. Two key elements in achieving the coverage and capacity objectives of HetNets are ICIC and range expansion. UL ICIC can be supported through high-interference and overload indicators. UL usage can be restricted at the cell edge for certain parts of the available bandwidth, for example, channels that are already under high usage. DL ICIC can be coordinated using the relative narrowband transmit power (RNTP) indicator. Again, the key idea is to reduce transmit power for certain parts of the bandwidth. Range expansion is a concept that allows more UEs to benefit directly from being served by low-power, small-cell eNBs, such as pico and femto eNBs in the presence of strong macrocell signals.

Consider the scenario where an LTE network is deployed using a combination of macro and small cells. Due to the large difference in transmit power, the DL coverage of a small cell is much smaller than that of a macrocell. If the UE selects the eNB to connect based on the DL received signal strength, many UEs will be attracted towards the macro eNB. As a result, the small cell may only be serving a few UEs. Furthermore, the macro eNB may run out of resources to serve all its UEs. This load imbalance may lead to the uneven distribution of data rates and adversely affect user experience. It is therefore highly desirable to balance the load between macro and small cells by expanding the coverage or range of the small cells. This requires mitigating the DL interference caused by macrocells to the UEs served by the small cells. To accomplish that, interference cancellation at the individual UEs can be employed. Alternatively, or additionally, resource partitioning or coordination may be performed among macro and small cells in the time, frequency, or spatial domains.

The resource partitioning may be fixed or adaptive. Time domain partitioning (e.g., slots or subframes) requires a synchronous network. This requires a network of macro- and small cells to be synchronized. Frequency domain partitioning may be more suited for asynchronous networks. Range expansion can be achieved by using the minimum path loss, as the eNB selection criterion, rather than the maximum received signal strength. As a result, more UEs will associate with the small cells, enabling a fairer distribution of wireless link resources to each UE.

Adaptive ICIC provides smart distributed resource allocation in the time, frequency, and spatial domains among interfering cells and improves intercell fairness in a heterogeneous network. A basic approach calls for the interfering macro eNB to give up use of some of its resources (e.g., RBs) in each radio frame to enable small cells to serve their users. The resource allocation may be based on each scheduling interval, the number of users served by the macro- and small cells, and their specific data rate requirements. More generally, transmit powers and spatial beamforming may be coordinated among interfering eNBs.

In another approach, resource allocation is negotiated over timescales much longer than the scheduling intervals. Such a slow-adapting scheme seeks to coordinate transmit powers, as well as the time, space, and frequency resources so that the total utility of the network is maximized in terms of user data rates, latency, or fairness. The hypothetical case is a central entity that has access to all network information before making these decisions. In practice, the decision making is distributed and affects only a subset of eNBs in the network, such as the macro and small cells contained within that macrocell's coverage area. The coordination is done by the macrocell via over-the-air messages or via the backhaul. For the latter, LTE-A specifies the X2 interface. Load information and resource partitioning requests, as well as responses, are sent via this low-latency (<200 ms) interface using standard messages. Having a standard X2 interface also helps enable intervendor interoperability in the RAN, which is not prevalent in legacy systems.

Interference cancellation at the UE is an important aspect of LTE-A. Some channels must be broadcast to enable legacy UEs to function. These channels are not subject to resource partitioning and must therefore be cancelled by the UEs to eliminate the interference they cause.

4.3.3 MIMO Modes

LTE supports up to four layers of DL multiplexing but the UE only supports a single layer on the UL. LTE-A provides SU-MIMO of up to eight layers for the DL and four layers for the UL. Theoretically, a peak spectral efficiency of 30 bit/s/Hz for the DL and 15 bit/s/Hz for the UL can be achieved. Thus, a 20 MHz channel can potentially achieve up to 600 Mbit/s on the DL. The MU-MIMO technique for both UL and DL can further improve the spectral efficiency by allowing up to four UEs share the same channelization codes using different midamble shifts.

4.3.4 Coordinated Multipoint Transmission/Reception

CoMP is a DL/UL technique to improve system capacity and cell edge user throughput. One approach is decentralized control based on an independent eNB architecture, and the other is centralized control based on a remote radio equipment (RRE) architecture. The RRE may be a femtocell eNB. In the independent eNB architecture, CoMP is performed by signaling between eNBs. This technique can utilize legacy cells, but the disadvantage is signaling delay and other overheads. In the RRE technique, the eNB can control all radio resources by transmitting baseband data directly between eNB and RREs on fiber connections. There is less signaling delay or other overheads using this technique. However, using fibers requires higher deployment cost. In addition, the centralized eNB must be able to accept a higher load.

4.4 FEMTOCELLS

Femtocells are simplified cellular eNBs typically deployed in homes, enterprises, or hotspots to extend cellular coverage in indoor environments. The Home Node B (HNB) is used in femtocells. The HNB is equivalent to the femto access point (FAP). Femtocells operate on licensed spectrum and transmit at low power (100 mW or less). They provide voice and Internet service to cellphone users by connecting them to the Internet and operator's core network using a broadband connection (e.g., DSL, cable) as a backhaul. They also provide capacity gains through spectrum reuse. Thus, femtocell users may experience better voice call quality and data rates due to improved wireless coverage. Cellular operators benefit from reduced infrastructure deployment costs that are otherwise needed for network evolution, including capacity upgrades and coverage improvements. In addition to coverage extension, femtocells may offload traffic from the macrocell to a wired broadband network, allowing macro users to achieve higher throughputs.

4.4.1   Deployment

To minimize deployment costs, femtocells are typically deployed with minimal or no RF planning. This requires femtocells to be able to coexist with macrocells. Thus, HNBs must be self-configuring and able to choose the appropriate frequency or time slot of operation and RF output power such that they minimize interference to nonfemto users (i.e., macro users and users belonging to other femtocells) [4]. In addition, a handset must be intelligent enough to connect to the HNB in the presence of a macro signal. New handsets can have smart features to distinguish between femto and macro signals, own femto versus office femto, and accordingly prioritize their search. They can use additional information, such as geographical location, to search for the desired femto.

Enterprise femtocells may differ significantly from residential femtocells in terms of coverage, interference management, self-configuration, and mobility management due to the use of multiple femtocells. The coverage provided by a femtocell in an enterprise is typically in the range of 5000–7000 sq ft, and contiguous coverage involving multiple femtocells may be needed. Macro signal strength in an enterprise can vary by 20–30 dB.

4.4.2   Interference Management

Femto users may experience DL interference from femtocells when femtocells and macrocells share the same RF frequency channel. On the UL, femto users at the edge of femtocell coverage transmit at a higher power and can create interference on the macro UL. Thus, an unusual situation arises in femtocells in that interference can occur when operating on licensed frequency bands. Both DL and UL interference issues can be mitigated by limiting the coverage of each femtocell and increasing the number of femtocells. Interference management also requires a level of coordination to restrict the transmit power imbalance between neighboring femtocells.

4.4.3   Traffic Offload Using Femto HNBs

The Selected IP Traffic Offload (SIPTO) for the HNB provides seamless access for a UE connected via a HNB to a wired IP network (e.g., the Internet). When the UE is under wireless LAN coverage, the operator can offload some traffic to the wireless LAN. At the same time, it may be beneficial to keep some traffic (e.g., VoIP traffic) for cellular access. With SIPTO, UE bandwidth is maximized at the lowest possible cost and without any service disruption or interruption. Additionally, it may be possible to provide a limited nonseamless wireless LAN offload via a transient IP connection to a wireless LAN (referred as Direct IP Access). This implies that the UE uses the wireless LAN IP address, and no IP address preservation is provided between the wireless LAN and the LTE network.

4.5   ANTENNA DESIGN CHALLENGES FOR 4G SMARTPHONES

Contributed by Frank M. Caimi, PhD, SkyCross, Inc.

The proliferation of mobile wireless devices with the promise of high data rate connectivity afforded by 4G continues to provide exciting economic opportunities, as well as new paradigms for just about every facet of life. Streaming video, location-based services, access to massive databases, and so on, are now expected and commonplace functions available to anyone using a smartphone or tablet device. The trend in mobile wireless devices has been to provide faster access, improved processors, more memory, brighter and higher resolution screens, additional connectivity with Wi-Fi, GPS, 3G, and 4G world access—all with longer battery life in thinner and sleeker packages. Compound this with the desire of mobile operators to expand their available band allocations, and what results is a difficult industrial design arena, where suppliers are vying for physical space within the confines of a smartphone or similar device to accommodate necessary components. One such component is the antenna—essentially a transducer that converts time-varying electrical current to radiated energy, and often considered a last-minute addition to the physical structure. Now with the introduction of 4G and 1×2 or 2×2 MIMO communications protocols, not one but two antennas are needed, each with specific requirements typical of any transducer—bandwidth, efficiency, cross-coupling, size, and cost. With newer 3GPP definitions for carrier aggregation, simultaneous band coverage will also be necessary. So what are the limitations and challenges associated with antenna design for 4G?

4.5.1   Physical Considerations

A number of factors affect MIMO antenna performance in a handheld mobile communication device. While these factors are related, they generally fall into one of three categories: antenna size, mutual coupling between multiple antennas, and device usage models.

4.5.1.1   Antenna Size   The size of an antenna is dependent on three criteria: bandwidth of operation, frequency of operation, and required radiation efficiency. Bandwidth requirements have obviously increased, as they are driven by FCC frequency allocations in the United States and carrier roaming agreements around the world. Different regions use different frequency bands, now with over 40 E-UTRA band designations. Many of these bands are overlapping but require world-capable wireless devices to typically cover a frequency range from 698 to 2690 MHz.

A simple relationship exists between the bandwidth, size, and radiation efficiency for the fundamental or lowest frequency resonance of a physically small antenna [5–9].

$$\frac{\Delta f}{f} \propto \left(\frac{a}{\lambda}\right)^3 \eta^{-1}. \tag{4.2}$$

Here a is the radius of a sphere containing the antenna and its associated current distribution. Since a is normalized to the operating wavelength λ, the formula may be interpreted as "fractional bandwidth is proportional to the wavelength normalized modal volume." The radiation efficiency η is included as a factor on the right side of Equation 4.2, indicating that greater bandwidth is achievable by reducing the efficiency. Note that currents exist not only on the antenna element, but also on the attached conductive structure or "counterpoise." For instance, mobile phone antennas in the 698–960 MHz bands use the entire printed circuit board (PCB) as a radiating structure so that the physical size of the antenna according to Equation 4.2 is actually much larger than what appears to be the "antenna." The "antenna" may be considered a resonator that is electromagnetically coupled to the PCB so that it excites currents over the entire conductive structure or chassis. Most smartphones exhibit conductive chassis dimensions of approximately 60×110 mm, which from an electromagnetic modal analysis predicts a fundamental mode somewhat over 1 GHz. This suggests that performance bandwidth degrades progressively at lower excitation frequencies. The efficiency-bandwidth trade-off is complex, requiring electromagnetic simulation tools for accurate prediction [10, 11]. Results indicate that covering the frequency range of 698–960 MHz (E-UTRA bands 5, 8, 12, 13, 17, 18, 19, 20) with a completely passive antenna of desirable antenna size and geometry becomes difficult without making sacrifices in radiation efficiency.

Factors determining the achievable radiation efficiency are not entirely obvious, as the coupling coefficient between the "antenna" and the chassis, radiative coupling to lossy components on the PCB; dielectric absorption in plastic housing, coupling to coexisting antennas, as well as losses from finite resistance within the "antenna" resonator structure, all play a part. In most cases, the requirements imposed by operators suggest minimum radiation efficiencies of 40–50%. Thus, meeting a minimum total radiated power (TRP) requirement (a measure for the spherical radiated power of the antenna) essentially requires tradeoffs between the power amplifier (PA) output and the achievable antenna efficiency. In turn, poor efficiency at the antenna translates to less battery life, as the PA must compensate for the loss.

4.5.1.2 *Mutual Coupling between Multiple Antennas*   With the advent of 4G standards, such as WiMAX and LTE, more than one antenna is required to operate over a single frequency range. Currently, the requirement to support the LTE offerings of AT&T and Verizon in North America require operation in the 700 MHz frequency range per the recent FCC UHF TV band reallocation. These new frequency bands (bands 13 and 17) have been the nemesis of antenna design engineers. Since the longitudinal dimension of the chassis is a primary contributor to the fundamental chassis resonance mode, an antenna located at either end on top of the phone will excite nearly the same current modal distribution, and therefore produce virtually the same radiation pattern. Since both antennas are coupled to the same mode, they experience mutual coupling so that the power introduced into one

antenna is partially coupled to the opposite antenna's source resistance, and is subsequently lost as heat. Figures 4.16 and 4.17 illustrate the similarity of the radiated far-field patterns for antennas at opposite ends of the phone chassis and the resulting effects on mutual coupling.

One solution is to relocate one antenna to excite a different radiation mode on the chassis. This can result in smaller bandwidth and efficiency depending on the chassis width-to-length ratio. For single-band 1×2 MIMO configurations rolling out now, this approach can meet the -3 or -6 dB relative efficiency requirement for the diversity antenna relative to the main antenna as imposed by operators. However, it is not suitable for 2×2 MIMO in current smartphone designs for bands 13 and 17 due to the reduced efficiency/bandwidth tradeoff, and subsequent power loss. There are, however, alternative approaches for reducing mutual coupling effects through the use of various chassis modifications combined with antenna placement, but these can be more difficult for industrial designers to accommodate. Several recently developed antenna design methods reduce mutual coupling through the use of a single structure antenna that optimally excites different radiation modes [13].

The method reduces mutual coupling between the antenna's two feedpoints, resulting in a minimum antenna pattern envelope correlation coefficient (APECC), as supported by a detailed electromagnetic analysis. Since only one antenna structure is needed, industrial design constraints and concerns over meeting SAR are minimized. A comparison of Figures 4.16 and 4.17 shows a more favorable result for radiation efficiency and correlation coefficient. These techniques have been used in WiMAX 2×2 applications with great success and will likely benefit future smartphone designs.

4.5.1.3 Correlation Coefficient MIMO communications systems require channel independence to achieve additional information capacity. Typically, a signal correlation metric, the complex envelope correlation coefficient, is specified according to:

$$\rho_e = \frac{E\left[\left(r_1 - \overline{r_1}\right)\left(r_2 - \overline{r_2}\right)\right]}{\sqrt{E\left[\left(r_1 - \overline{r_1}\right)^2\right] E\left[\left(r_2 - \overline{r_2}\right)^2\right]}}, \tag{4.3}$$

where the signals r_1 and r_2 are complex valued envelope signals for the two received signals r_1 and r_2 shown in Figure 4.18. The operator E represents the expected value.

In order for the communications system to fully exploit the independence of signal channels, it is commonly believed that the transmitter and receiver must utilize multiple transducers (antennas) having certain characteristics. Since the signals may arrive from different directions and with different phase, the antennas should exhibit characteristics that allow for highest independence of the received signals. One such characteristic is antenna pattern independence—measured by the antenna pattern correlation coefficient. The pattern correlation coefficient ρ_p can be calculated from full sphere antenna pattern measurements and is given in terms of the electrical field components by [14]:

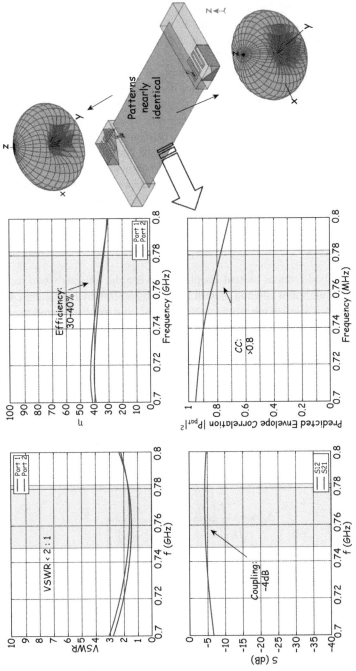

Figure 4.16 Coupling and radiation patterns resulting for antennas at opposite ends of the phone chassis.

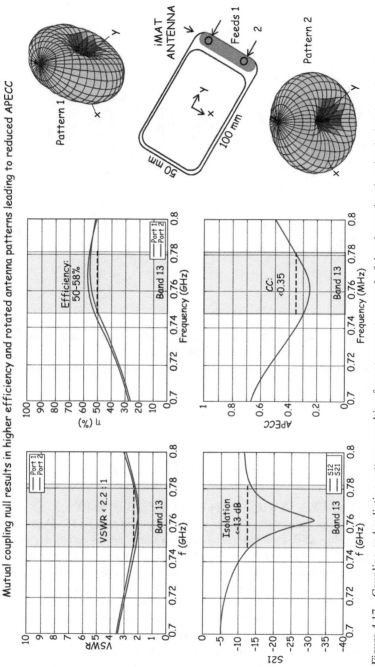

Figure 4.17 Coupling and radiation patterns resulting for antennas at one end of the phone chassis using isolated mode antenna (iMAT) [12].

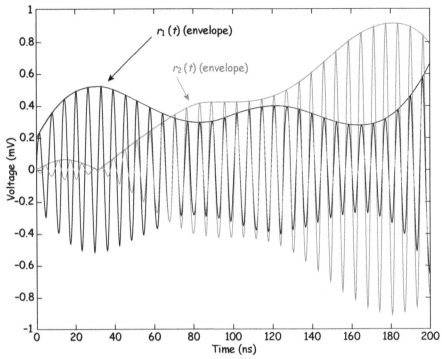

Figure 4.18   Received signals from two antennas and corresponding envelope waveforms $r_1(t)$ and $r_2(t)$.

$$\rho_p = \frac{\int_0^{2\pi} \int_0^{\pi} A_{12} d\Omega}{\sqrt{\int_0^{2\pi} \int_0^{\pi} A_{11} d\Omega} \sqrt{\int_0^{2\pi} \int_0^{\pi} A_{22} d\Omega}},\qquad(4.4)$$

where $A_{mn}(\theta, \varphi) = XE_{mv}(\theta, \varphi)E_{nv}^*(\theta, \varphi) + E_{mh}(\theta, \varphi)E_{nh}^*(\theta, \varphi)$ and $X = S_v/S_h$.

The subscript indices (m, n) refer to either antenna port 1 or 2. Depending on their integer value, $\Omega(\theta, \phi)$ is the spatial angle in steradians as commonly depicted, and $E_{1,v}$, $E_{1,h}$, $E_{2,v}$, and $E_{2,h}$ are the complex envelopes of either the vertical or horizontal field components resulting from excitation of either antenna port 1 or 2, respectively. The cross polarization power ratio X is defined as the ratio of the mean received power in the vertical polarization (S_v) to the mean received power in the horizontal polarization (S_h). The formula as shown represents a uniformly weighted measure of correlation, but the signal power can also take into account the characteristics of the propagation environment by weighing the field components by P_θ and P_ϕ (the probability distributions of the power incident on the antenna in the vertical and horizontal polarizations), yielding a revised form of Equation (4.5) where

$$A_{mn}(\theta, \varphi) = XE_{mv}(\theta, \varphi)E_{nv}^*(\theta, \varphi)P_v(\theta, \varphi) + E_{mh}(\theta, \varphi)E_{nh}^*(\theta, \varphi)P_h(\theta, \varphi).\qquad(4.5)$$

Perhaps most interesting is that the correlation coefficient depends primarily on the phase difference between the two antenna patterns—indicating that good correlation performance requires careful consideration of the antenna placement and design in the case where multiple single antennas are used in close proximity. The APECC, that is, the correlation coefficient between the signals envelope in Rayleigh distributed multipath channel is approximately given by the square of the complex correlation [15]. Thus,

$$\text{APECC} = \rho_p^2. \tag{4.6}$$

It is not necessary to have complete signal decorrelation, that is, APECC = 0. Rather, a limit is imposed, with APECC below 0.5 the most desirable for LTE systems proposed currently.

Operator requirements for APECC can be difficult to achieve or to measure in real devices. Some operators are firm in requiring measured APECC below 0.5, while others can accept values below 0.7. The APECC shown in Figure 4.16 is greater than 0.8 and would be unacceptable in either case. These are free space test results, and can vary significantly depending on how the testing is done. Furthermore, these effects change with the phone usage model due to perturbations of the antenna near field radiation and proximity detuning. In summary, it should be noted that:

- The receive envelope correlation is a function of both the environment (or channel model), and the antenna parameters.
- If the antenna patterns are known, then the envelope correlation can be predicted by choosing a statistical channel model.

Test methods for determining the overall MIMO performance due to the combined effects of the antenna pattern, APECC, and channel model are current topics in 3GPP, since MIMO system performance depends on both the antenna configuration and the channel model for the specific usage environment. Test methods are now being developed that allow antenna patterns to be measured in an anechoic test chamber as is the current practice for SISO systems, and these are used independently to determine the system performance from an appropriate statistical channel model. Other test methods use a reverberation chamber specially configured to match the environment from which direct measurement of the signal correlation can be made.

4.5.1.4 *Device Usage Models*

Modern smartphones and tablets have a wide variation in usage models compared with conventional handsets. The device must work as a phone when held to the head, but also needs to work for a wide variety of handheld and body worn positions. For gaming applications, it may be necessary to grip the ends of the device with both hands instead of holding it by the middle with one hand, yet other applications do not require the device to be held at all. Increased display area and variations in grip are making it increasingly more difficult to find a good location for the antenna where it will not be influenced by the display or by proximity to the body. Even if a place can be found to locate the antenna within a given industrial design, the various usage models create additional challenges for obtaining adequate antenna performance.

4.5.2 Current Handset Antenna Configurations and Challenges

Prior to concerns over carrier aggregation, wireless devices operated on one band at a time with the need to change when roaming. Consequently, the required instantaneous bandwidth would be considerably less than that required to address worldwide compatibility. Take a 3G example for instance, where operation in band 5 (824–894 MHz) compared to operation in bands 5 plus 8 (824–960 MHz). Then, add the requirements for band 13 and band 17, and the comparison becomes more dramatic: 824–960 versus 704–960 MHz. This becomes problematic, as legacy phone antennas support pentaband operation, but only bands 5 and 8. Given Equation 4.2, several choices exist. The most obvious would be to increase the antenna system size, (i.e., the antenna and phone chassis footprint) and/or to reduce the radiation efficiency. Since 4G smartphones require two antennas, neither approach is necessarily desirable from an industrial design standpoint, although it is possible to cover the 704–2170 MHz bands with a completely passive antenna in a space allocation of $6.5 \times 10 \times 60$ mm [16]. Some solutions also involve splitting the system architecture into low-band and high-band sections and using separate antennas for each section, thereby further doubling the number of antennas. Various alternative antenna configurations motivated by industry are the following:

- Limit the antenna(s) instantaneous bandwidth within the current antenna space allocations to allow use of two antennas without compromising industrial design (antenna supplier motivation).
- Make the antenna(s) smaller to achieve a compact and sleek device with greater functionality by limiting the instantaneous bandwidth with the same or improved antenna efficiency (OEM motivation).
- Improve the antenna efficiency, and therefore the network performance by controlling the antenna instantaneous frequency/tuning (operator motivation).
- Make the antenna agile to adapt to different usage models (OEM/user/operator motivation).
- Combinations of the above.

Perhaps the simplest approach would be to limit the instantaneous operation to a single band to satisfy the protocol requirements for a single region. To satisfy the roaming requirements, the antenna could be made frequency agile on a band-by-band basis. This approach represents the most basic type of "state-tuned" antenna.

An illustration of the performance gain achievable with state tuning was conducted for a 3G mobile phone equipped with a legacy multiband antenna covering low frequency bands 5 and 9 (824–960 MHz), as well as high frequency bands 1 and 2 (1850–2170 MHz). As shown in Figure 4.19, substandard radiation efficiencies of 15–23% were measured across the low band frequency range. A retrofit of

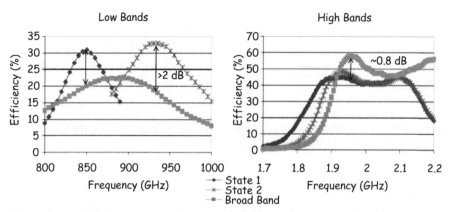

Figure 4.19   Radiation efficiency improvement using state-tuned antenna in comparison to legacy broadband antenna.

the phone antenna using a two-state tuning method to separately cover bands 5 and 9 achieved measured performance gains of up to 2 dB without changing the antenna size or location.

Since typical antennas used in mobile devices are not just single mode resonators, they can support simultaneous high-band operation, allowing for carrier (band) aggregation depending on the aggregation plan and the specifics of the antenna design. However, not all combinations for simultaneous band usage are possible without compromises in antenna performance. The example shown in Figure 4.19 is illustrative of single low-band and multiple high-band aggregation compatibilities. The high band radiation efficiency in this case remains essentially the same independent of the low band tuning state.

Not surprisingly, the industry is considering a variety of possibilities to provide "adaptive" or "smart" antennas. Adding requirements for adaptability for different usage models or to accommodate different industrial designs or manufacturing tolerances, would suggest using more tuning states, or reconfiguring the antenna structure in response to sensors already available in the wireless device. The industry is also looking to further reduce the antenna instantaneous bandwidth for subband operation through the use of closed-loop feedback. In this way, the antenna can be optimized for the frequency or frequencies actively in use. These approaches involve the use of active components, which introduce challenges of nonlinearity, response time, stability, power handling capability, and test methodology.

4.5.3   Antenna Implementation

A variety of methods exist for "tuning" an antenna, and can be implemented using various components depending on the number of tuning states required, the specific antenna design, the frequency range to be covered, harmonic generation requirements, and control interface capability. The example of Figure 4.19 uses a direct tuning method. This is accomplished by changing the structure of the antenna— so-called "aperture tuning," where the antenna fundamental resonance frequency is

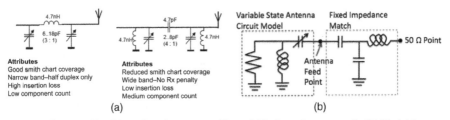

Figure 4.20 (a) Fixed broadband antenna with variable impedance match. (b) Variable state antenna with fixed feed point matching.

changed in response to a control signal. In contrast, "feedpoint matching" relies on dynamically tuning the antenna matching network to transform the antenna terminal impedance to the usual 50-Ω feedpoint. Both methods have their advantages—the former being a highly efficient method in concept, and the latter being easier to implement as a circuit add-on. The disadvantages of each method are more difficult to generalize. Both schemes are illustrated in Figure 4.20.

Aperture tuning may use a simple switch to select different load elements on the antenna structure to vary the antenna's electrical length, thus causing a resulting in a change in resonant frequency. Tuning of the matching network is typically implemented using variable capacitance elements or variable inductance using switches in a pi or tee configuration. Conventional components such as varicap diodes, barium strontium titanate (BST) devices, PIN diodes, and CMOS or gallium arsenide switches allow some flexibility for a limited number of states. Digitally addressable chip-level products that allow control of terminal capacitance C_t are in various stages of development. These integrated circuits are being designed for support of standard and proprietary control interfaces, such as I2C, SPI, and MIPI-RFFE (http://www.mipi.org/DigRF_RFFE) and generally employ a minimum of 5-bit resolution over the range from C_{tmax} to C_{tmin}. The capacitance range can be configured from roughly 3:1 for some devices to more than 10:1 for others. Fine-scale resolution is desirable to allow compensation for head and hand proximal detuning of the antenna impedance or frequency, and can also be used for fine-scale tuning of the antenna or matching network during the manufacturing process. Ultimately, the tuning will be accomplished under closed-loop control based on feedback from sensors onboard the user platform, which account for the user's usage model and information obtained from onboard processors. Power handling, ability to hot switch with applied RF power, harmonic generation, and electrostatic discharge (ESD) compliance are also necessary design concerns in any implementation.

To mitigate the effects of various usage models, it may be necessary to combine the benefits of state tuning and mode isolation. Modal isolation reduces mutual coupling and correlation coefficient, while state tuning allows the operation of the antenna over the necessary range of frequencies for a given space allocation and modal volume according to Equation 4.2. Figure 4.21 shows the measured average efficiency for a single-structure, variable state isolated mode antenna (iMAT) covering multiple frequency bands using six tuning states. The antenna uses two

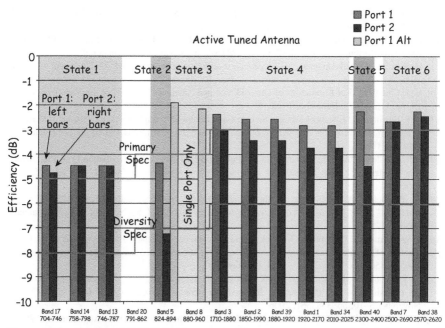

Figure 4.21   State tuned iMAT single antenna structure covering all 4G and 3G applications with two MIMO antenna ports.

feedpoints, which can support either balanced or imbalanced gain/efficiency with an APECC less than 0.5 for typical smartphone geometric form factors.

4.5.4   Conclusion

Significant challenges exist for antenna designers to meet the needs of modern wireless devices that will rely on the implementation of MIMO protocols over an increasing spectral range. Requirements to support lower frequencies associated with newer band allocations in the United States, as well as the need to add additional antennas, place significant design constraints on existing smartphone industrial designs. Usage models will likely drive more advanced designs that reconfigure to meet specific operational modes. Advanced design techniques, like state tuning and modal isolation, may answer the challenge allowing device designers the flexibility to implement sleek, feature-rich devices without sacrificing network-level performance.

REFERENCES

[1] LTE Deployments, http://ltemaps.org.

[2] 3GPP Releases, http://www.3gpp.org/releases.

[3] B. M. Popovic, "Generalized Chirp-like Polyphase Sequences with Optimal Correlation Properties," *IEEE Transactions on Information Theory*, Vol. 38, pp. 1406–1409, July 1992.

[4] Femto Forum, Interference Management in UMTS Femtocells, December 2008, http://www.femtoforum.org/femto/Files/File/FF_UMTS-Interference_ Management.pdf.

[5] H. Wheeler, "Small Antennas," *IEEE Antennas and Propagation*, Vol. AP-23, No. 4, pp. 462–469, 1975.

[6] L. J. Chu, "Physical Limitations of Omni-Directional Antennas," *Journal of Applied Physics*, Vol. 19, pp. 1163–1175, December 1948.

[7] R. F. Harrington, "Effects of Antenna Size on Gain, Bandwidth, and Efficiency," *Journal of Research of the National Bureau of Standards—D, Radio Propagation*, Vol. 64D, No. 1, pp. 1–12, January–February 1960.

[8] J. S. McClean, "A Re-Examination of the Fundamental Limits on the Radiation Q of Electrically Small Antennas," *IEEE Transactions on Antennas and Propagation*, Vol. 44, pp. 672–675, May 1996.

[9] S. R. Best, "A Discussion of the Quality Factor of Electrically Matched Small Wire Antennas," *IEEE Transactions on Antennas and Propagation*, Vol. 53, No. 1, pp. 502–508, January 2005.

[10] P. Vainikainen, J. Ollikainen, O. Kivekäs, and I. Kelander, "Resonator-Based Analysis of the Combination of Mobile Handset Antenna and Chassis," *IEEE Transactions on Antennas and Propagation*, Vol. 50, No. 10, pp. 1433–1444, October 2002.

[11] C. T. Famdie, W. L. Schroeder, and K. Solbach, "Numerical Analysis of Characteristic Modes on the Chassis of Mobile Phones," *1st European Conference Antennas Propagation*, Nice, France, November 6–10, pp. 358–361, 2006.

[12] F. M. Caimi, "Antenna Solutions for MIMO in LTE Handsets (with Other Applications)," *IWPC MIMO Summit Conference Proceedings*, San Diego, California, November 11–12, 2008.

[13] F. M. Caimi and M. Montgomery, "Dual Feed, Single Element Antenna for WiMAX/MIMO Application," *International Journal of Antennas and Propagation*, Article ID 219838, 2008.

[14] S. D. Parson, *The Mobile Radio Propagation Channel*, 2nd edition, John Wiley, 2000.

[15] R. G. Vaughan and J. B. Andersen, "Antenna Diversity in Mobile Communications," *IEEE Transactions on Vehicular Technology*, Vol. 36, No. 4, pp. 149–172, November 1987.

[16] C. W. Chiu, C. H. Chang, and Y. J. Chi, "A Meandered Loop Antenna for LTE/WWAN Operations in a Smartphone," *Progress in Electromagnetics Research C*, Vol. 16, pp. 147–160 , 2010.

[17] A. Larmo, et al., "The LTE Link-Layer Design," *IEEE Communications Magazine*, pp. 52–59, April 2009.

HOMEWORK PROBLEMS

4.1. Wireless technologies can be deployed in challenging situations. Describe how you would design a broadband wireless access network to provide connectivity within (a) a large aquarium with millions of gallons of water and serving thousands of visitors per day; (b) a busy airport concourse of over 200,000 sq ft with moving walkways, trams, and visitors; (c) a federal reserve bank building with marble floors and walls throughout; (d) a slowly revolving restaurant on top of America's tallest hotel of 72 floors; (e) an island with a 2-mi span—half the island is occupied by residential homes, hotels, and high-rise condominiums, while the remaining space houses a naval military base; (f) a city surrounded by scenic mountainous parks—one park is located 2000 ft above lakes and rivers, whereas a large concentration of natural stone arches can be found in another park; and (g) a large state covering one-fifth the size of continental United States with many natural disasters, such as wildfires, volcanoes, and floods, and these emergencies may require interoperable digital two-way wireless communications using the Project 25 standard. In (a), (b), (d), (e), and (f), include a wireless capability in the network to allow visitors to instantly upload high-quality photos or videos captured on their smartphones to family or friends.

4.2. The subcarrier spacing in the LTE standard is 15 KHz, giving an FFT period of 66.7 μs. The CP for this FFT period is 4.6875 μs. For 802.11, the CP is 0.8 μs. Explain why a shorter CP is used for 802.11. Note that since the number of samples taken within the FFT period is 128, the sampling interval is 66.7 μs/128 or 0.52 μs, giving a sampling frequency of 1.92 Msamples/s.

4.3. Why is the OFDM symbol duration normally fixed in many wireless standards (e.g., 4 μs for 802.11, 66.7 μs for LTE)? Justify the choice of a 15 KHz OFDM subcarrier spacing in LTE versus a 312.5 KHz OFDM subcarrier spacing in 802.11n.

4.4. Unlicensed mobile access (UMA) (http://www.umatechnology.org/overview) provides a set of open specifications that aim to make 3GPP cellular services available over unlicensed networks (e.g., Wi-Fi and Bluetooth). It was initiated by Nokia in September 2004 and defines transparent roaming and handoffs for connections to be routed through cellular network even though they may travel over an 802.11 network (using VoIP). This reduces the cost and dependence on landline office phones, and extends the cellular coverage to indoor public networks (e.g., café, airplane, and airport), as well private networks (e.g., home, office, and campus). In the United States, T-mobile offers residential and business UMA service. Explain why dual-mode UMA handsets may deplete batteries faster than cellular handsets. What is the advantage of UMA compared with the femtocell approach?

4.5. One of the ways to bring cellular coverage inside a building is to employ distributed antennas. A single femto AP can be deployed in the office premise, and that AP is connected to a wired hub, which connects to a number of antennas distributed among several locations in the building (see figure below). For example, an entire floor can be covered with one dominant femto source (a femtocell). This is in contrast to the costly deployment of multiple femto APs to expand service coverage, which may require complex handoffs between different coverage areas. Coverage areas of 10,000 sq ft or more can be achieved with the distributed antenna solution. The antennas communicate with the AP using RF signals over cable via the hub. The AP typically operates at very low power (1–10 mW range) to prevent interference with the regular cellular network. Since all antennas operate on the same frequency, if one of the antennas encounters excessive interference, will this impact the other antennas? Will link adaptation affect users located outside of the femtocell? How can MIMO antennas be included? Note that in an in-building deployment, the need to support mobility is less critical since most communications are conducted when the user is stationary, hence minor dead spots may be tolerated (e.g., moving a few feet for better signal reception is generally acceptable compared with moving to the nearest window). This applies even to voice connections since the person will normally complete the call before moving elsewhere. However, coverage area overlaps may be inevitable, and they lead to low SNR because the handset detects multiple sources. What is the advantage of backhauling the cellular traffic via the femto AP to a T1 line provided by the carrier? A similar concept was adopted by some 802.11 switches to extend wireless LAN signal coverage. In this case, the switch connects a number of "dumb" antennas at the front end, and wireless access is controlled by 802.11 radios located at the switch. This reduces the cost for deploying multiple APs at the front end, which also take up more space. However, the amount of wiring is not reduced. Unlike the femto approach, each antenna connected to the switch may operate on a different frequency channel. Explain whether the DCF (CSMA/CA) used by 802.11 will perform better than the PCF (polling access) in this situation.

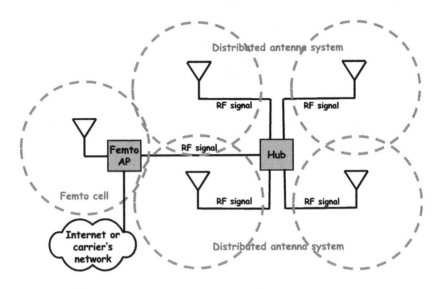

4.6. Consider an LTE femto AP that maintains simultaneous connections to the handset and LTE eNB using the same channel. Because the femto AP is not constrained by battery power and space, it can employ additional transmit power and/or antenna diversity to ensure good signal connectivity and data rates with the LTE eNB. This removes the need for a wired backhaul broadband connection, facilitating femto deployment significantly. The femto AP essentially acts as a repeater between the macro eNB and the handset. Although traffic cannot be offloaded from the eNB, the higher bandwidth available with LTE offsets this limitation. Describe how you would resolve self-interference issues when deploying such a network. Suppose 802.11 is used for the local connection between the femto AP and the handset. In this case, dual-mode handsets are employed since the LTE and 802.11 radios are built into the femto AP. Describe the technical issues in enabling a seamless handoff from a LTE connection to 802.11 for better indoor coverage.

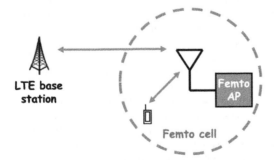

4.7. In a typical femtocell deployment, a wireless handover needs to be negotiated between the eNB and the femto AP. Equally important, especially for VoIP calls, is a wireline handover from the wired connection connecting the eNB to a new wired connection connecting the femto AP. Explain how such a wireline handover can be accomplished seamlessly and in tandem with the wireless handover. How does the complexity of such a handover compare with the case when the femto AP is completely wireless? Note that the femto AP may not be capable of performing a regular handover similar to an eNB.

4.8. In addition to widely varying interfering power levels, hidden collisions can also arise in a wireless network. A user device can be hidden from the transmitter if it is situated behind a large obstacle (e.g., a building) that absorbs most of the RF energy. Explain whether hidden collisions can occur in an LTE network.

4.9. Explain how the MB-SFN concept in LTE can be extended to 802.11. Note that 802.16m provides a similar feature as the MB-SFN system under the E-MBS framework. Describe the advantages and drawbacks of the MB-SFN system compared with a mobile broadcast digital TV system, such as the ATSC-M/H system.

4.10. Explain how interference resulting from coexisting macro and femtocells can be effectively mitigated using MU-MIMO. Is precoding useful even when one spatial stream (i.e., SISO) is used? ■

CHAPTER 5

ATSC DIGITAL TV AND IEEE 802.22 STANDARDS

Consumer desire for media content anywhere and anytime continues to tax cellular networks' ability to deliver video via their one-to-one architecture. As a result, cellular operators are imposing data limits on services that have raised concerns among content providers and consumers. On the other hand, broadcasters' one-to-many architecture has an edge in terms of delivering video to mass audiences with few constraints, an advantage that could serve the industry well in terms of meeting consumer expectations. The migration of analog to digital TV promises free yet high-quality broadcast TV transmission. In addition, mobile digital TV is gaining momentum with the ratification of the ATSC M/H standard in September 2009. Access to spectrum is critical to development of a wireless broadband platform. The ability for a carrier to share TV band spectrum with other providers makes the most sense in rural areas where underutilization is more often the case than in highly populated metropolitan areas. Small service providers would be able to lease spectrum from larger carriers. This chapter covers the ATSC digital TV standard and the emerging two-way 802.22 standard that utilizes the TV band white space for unlicensed operation.

5.1 DIGITAL TV FREQUENCY CHANNELS

The number of digital TV (DTV) channels in the United States is increasing steadily since the demise of the analog channels in June 2009. There were originally 68 digital channels, where each channel can transmit more than one signal. Some of the 6 MHz channel frequencies in the United States are shown in Table 5.1. With the migration to DTV, the antenna must be able to receive both very high frequency (VHF) channels (channels 2–13) and ultra-high frequency (UHF) channels (channels 14–69). The 700 MHz band covers the frequency range of 698–806 MHz (channels 52–69), formerly occupied by analog TV channels. In the United States, some of the channels are currently assigned to DTV channels (e.g., channels 53, 54, 55, 59, 68, and 69), whereas others are reserved for public safety (e.g., channels 63, 64). In Atlanta, for example, the WUPA TV station broadcasts over channel 69. The rest of the bandwidth (62 MHz) is available for auction, and this is subdivided into 5 blocks:

Broadband Wireless Multimedia Networks, First Edition. Benny Bing.
© 2013 John Wiley & Sons, Inc. Published 2013 by John Wiley & Sons, Inc.

TABLE 5.1   Some DTV channel frequencies in the United States

TV channel number	Frequency band (MHz)	TV channel number	Frequency band (MHz)	TV channel number	Frequency band (MHz)	TV channel number	Frequency band (MHz)
2	54–60	19	500–506	36	602–608	53	704–710
3	60–66	20	506–512	37	608–614	54	710–716
4	66–72	21	512–518	38	614–620	55	716–722
5	76–82	22	518–524	39	620–626	56	722–728
6	82–88	23	524–530	40	626–632	57	728–734
7	174–180	24	530–536	41	632–638	58	734–740
8	180–186	25	536–542	42	638–644	59	740–746
9	186–192	26	542–548	43	644–650	60	746–752
10	192–198	27	548–554	44	650–656	61	752–758
11	198–204	28	554–560	45	656–662	62	758–764
12	204–210	29	560–566	46	662–668	63	764–770
13	210–216	30	566–572	47	668–674	64	770–776
14	470–476	31	572–578	48	674–680	65	776–782
15	476–482	32	578–584	49	680–686	66	782–788
16	482–488	33	584–590	50	686–692	67	788–794
17	488–494	34	590–596	51	692–698	68	794–800
18	494–500	35	596–602	52	698–704	69	800–806

- Block A: 12 MHz (698–704/728–734 MHz)
- Block B: 12 MHz (704–710/734–740 MHz)
- Block C: 22 MHz (746–757/776–787 MHz)
- Block D: 10 MHz (758–763/788–793 MHz)
- Block E: 6 MHz (722–728 MHz).

As can be seen, with the exception of Block E, the other blocks are split into two subblocks with the same bandwidth in each subblock, and this can facilitate simultaneous uplink (UL) and downlink (DL) communications (i.e., two-way full-duplex transmission). The 700 MHz auction netted a record of over $19 billion, with AT&T ($7 billion) and Verizon ($9 billion) taking up the C-block at a total cost of $16 billion [1]. The D-block netted $472 million from Qualcomm. AT&T subsequently purchased Qualcomm's lower D and E block spectrum licenses for $1.93 billion, which was approved by the FCC in December 2011.

5.2   DIGITAL TV STANDARDS

The FCC's landmark Notice of Proposed Rule Making issued in May 2004 [2] opens up a significant portion of TV spectrum for unlicensed use. This was motivated by the transition from analog to digital TV in the United States. By U.S. law, since June 12, 2009, terrestrial TV transmission is based on Advanced Television Standards Committee (ATSC) DTV standard. A significant amount of TV white space (i.e., available or unoccupied TV channels that have been allocated to primary users such as broadcasting stations) can be exploited by secondary or cognitive devices. FCC voted to open up the white spaces from 512 to 698 MHz on November 2008 [3]. Dell plans to integrate white space radios into future products [4]. Devices operating in unused TV channels are defined by FCC as TV band devices. Currently, FCC has decided to allow only fixed TV band devices.

5.2.1 Overview of Advanced Television Systems Committee

The ATSC (http://www.atsc.org) is an international, nonprofit organization developing voluntary standards for DTV. Over 195 ATSC members represent the broadcast, broadcast equipment, motion picture, consumer electronics, computer, cable, satellite, and semiconductor industries [5]. The ATSC was formed in 1982 with the task of examining and documenting new technologies for terrestrial TV broadcasting. Initial efforts focused on analog technologies quickly led to digital architectures, culminating in the publication of the ATSC DTV standard in 1996. From that time to the present, an enormous international effort has achieved a complex transition from analog to digital broadcasting in the United States and elsewhere.

5.2.2 ATSC DTV Standard

The ATSC DTV standard allows one-way, over-the-air broadcast of standard-definition (SD) and 720p/1080i high-definition (HD) channels to fixed, portable, and mobile devices (e.g., TV, PC, laptop, and smartphone), without the need for a proprietary customer premise equipment (CPE), such as a set-top box (STB). It replaces the NTSC analog standard. ATSC DTV channels are currently broadcast in over 200 U.S. cities by over 1500 TV stations and can be accessed free of charge by any user with an antenna and device with a DTV tuner. ATSC uses 188-byte MPEG transport stream (TS) packets to carry data, similar to cable networks. Since July 2008, it has adopted the H.264/MPEG-4 advanced video codec (AVC) and is defined as two parts: A/72 part 1 (Video System Characteristics) and A/72 part 2 (Video Transport Subsystem Characteristics). Lightweight and portable USB ATSC DTV tuners (and recorders) are also available for laptops.

5.2.3 Digital Video Broadcast-Terrestrial 2

Digital Video Broadcast-Terrestrial (DVB-T) is the European counterpart of ATSC. Digital Video Broadcast-Terrestrial 2 (DVB-T2) is the latest platform for high-definition TV (HDTV) transmission in Europe. The FEC coding cascades Bose Chaudhuri Hocquenghem (BCH) cyclic error-correcting codes with LDPC to avoid undetected errors at low BER while maintaining a high code rate (i.e., 3/5, 2/3, and 3/4). Data services are separated into logical channels called physical layer pipes (PLPs), where FEC and interleaving are applied separately to each data stream. The

PLPs occupy bandwidth slices in a time-frequency domain (OFDM symbols and subcarriers). The standard supports larger-size FFT OFDM (1K, 2K, 4K, 8K, 16K, and 32K FFT sizes) and QPSK, 16-, 64-, and 256-QAM modulations. Rotated constellation is an optional feature to introduce additional diversity. There are two separate mechanisms for reducing the peak-to-average power ratio (PAPR). The active constellation extension increases the amplitude of some outer constellation points, thereby improving the operating range with no throughput loss. The reserved-carrier PAPR reduction reserves some carriers to synthesize a peak-canceling waveform (at the expense of a small reduction in throughput). Multiple antenna capabilities are available. Optional transmit diversity is based on the Alamouti scheme. Since this is the only scheme that allows independent decoding, backward compatibility to single antenna systems is possible.

5.3   MOBILE TV

Many wireless subscribers now watch TV and video on a handheld device. This is driven by the proliferation of video-capable smartphones and tablets, and online TV/video websites (e.g., Amazon, Hulu, Netflix, and YouTube). However, mobile TV is not a substitute for traditional TV. The most obvious difference is in the usage patterns and the length of the viewing sessions. In mobile TV, audio quality and audio synchronization with the video are particularly important since this helps viewers to follow the plot in situations when image quality is degraded. Mobile TV network deployment is largely based on overlay solutions. For example, a 700 MHz TV band network can be overlaid over a cellular network. To reduce the need for new infrastructures, existing broadcast standards are preferred to support mobile TV. Terrestrial DTV standards, such as ATSC and DVB-T, can be modified to support mobile TV broadcast.

5.3.1   Mobile ATSC Standard

ATSC is partnering the Open Mobile Video Coalition (OMVC) to launch mobile DTV, which will allow viewers to watch TV programs on portable and mobile devices. OMVC (http://www.openmobilevideo.com) is an alliance of U.S. commercial and public broadcasters formed to accelerate the development and rollout of mobile DTV products and services. The partnership between ATSC and OMVC is supported by more than 800 TV stations. Consumer devices, transmission technologies, and data services include smartphones, accessory receivers for Wi-Fi phones and laptops, in-vehicle devices, DVD players, and USB receivers for laptops that are all equipped with mobile DTV reception capability. Electronic service guide (ESG) suppliers can also take advantage of mobile DTV's two-way capability, which also allows upload of personalized information, content consumption, user behavior data, and emergency alert system (EAS) service. More importantly, DTV broadcasters can offer user interactivity and more choices to viewers.

The original ATSC DTV standard was designed for a fixed reception (using highly directional fixed antennas) and employs 8-vestigial sideband or 8-VSB modulation. It is not robust enough against Doppler shift and multipath RF interference caused by mobile environments. To address these issues, additional channel coding mechanisms are introduced in ATSC-Mobile/Handheld (ATSC-M/H) devices, which is a U.S. standard for mobile DTV. The standard is an extension of the terrestrial ATSC DTV standard, and is therefore compatible to the terrestrial standard but does not cause interference in legacy reception. This is achieved by flexibly dividing the bandwidth into a variable-size mobile DTV part and a legacy DTV part. ATSC-M/H employs three layers namely presentation layer (audio/video coding and close captioning), management layer (transport, streaming, and non–real-time file transfer, ESG), and physical layer (RF transmission and FEC). To date, 45 broadcast TV stations in the United States have tested mobile DTV signals using the ATSC-M/H standard, allowing mobile DTV service to feature programs similar to fixed DTV. The A/153, ratified in October 2009, standardizes the characteristics of the emitted M/H signal and the functionality within the signal. Receiver manufacturers have developed dozens of devices, ranging from video-capable smartphone receivers to laptop computer receivers to in-vehicle receivers. The ATSC broadcast system with the main and mobile services is shown in Figure 5.1. The mobile service employs IP transport and uses a portion of the total available bandwidth (roughly 19.39 Mbit/s).

ATSC has developed a standard to enhance and extend the mobile service [5]. The Scalable Full Channel Mobile Mode (SFCMM) system is intended for specialized applications, where maximum use of the available digital bandwidth is required. A separate effort within ATSC has focused on the delivery of non–real-time (NRT) services over the legacy DTV system and the mobile DTV system. Many TV programs do not need to be delivered in real time. They can be downloaded overnight, or at some other time, and presented when the viewer wants to see them. NRT is especially attractive for mobile services, since most mobile viewing is done on an on-demand basis. For example, the consumer may want to watch something while waiting at the doctor's office. Because of the generally unpredictable nature of mobile viewing, the concept of "appointment viewing" may not always be practical

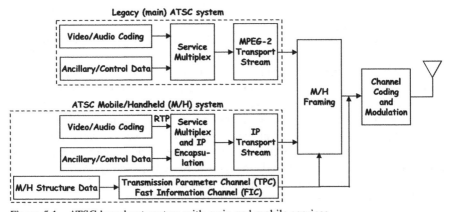

Figure 5.1 ATSC broadcast system with main and mobile services.

for mobile TV. NRT solves this problem by allowing the consumer to select what they want to see from a menu, with the program or service preloaded on their mobile device. NRT is one element of the emerging ATSC 2.0, which targets to offer new services and features for the fixed viewing environment. Other possible features of ATSC 2.0 include new advanced coding technologies, Internet connectivity, enhanced service guides, audience measurement, and conditional access.

5.3.2   Digital Video Broadcast-Handheld

DVB-Handheld (DVB-H) is one of the earliest mobile TV and IP datacast standards (http://www.dvb-h.org). It was adopted by the European Telecommunications Standards Institute (ETSI) in November 2004 for broadcasting TV transmissions to handsets. It has the ability to receive up to 15 Mbit/s in an 8 MHz channel in the 700 MHz band and can be tailored to work with a 5 MHz bandwidth in L-band (1.670–1.675 GHz). The handset transmit power is limited to less than 100 mW. This allows the receiver to power off in inactive periods, leading to a significant reduction of battery power consumption. DVB-H employs OFDM and OFDMA. It transforms digital TV into IP data packets, which are transmitted in short 100 ms time slots. DVB-H is backward compatible with DVB-T. It is essentially DVB-T with additional error coding and interleaving. It employs the same modulation methods as DVB-T, although DVB-H may use a larger constellation (e.g., 64-QAM) to service bigger payloads, as shown in Table 5.2. Like ATSC-M/H, DVB-H must deal with higher Doppler spread and the limited antenna size and power consumption.

5.3.3   Digital Multimedia Broadcasting

Digital multimedia broadcasting (DMB) is a digital broadcast standard for sending data, radio and TV to mobile devices, including cellphones. It can operate via satellite (S-DMB) or terrestrial (T-DMB) transmission. DMB is based on the digital audio broadcast (DAB) standard. It has some similarities with DVB-H. T-DMB is an ETSI standard.

TABLE 5.2   DVB-H
Modulations and Code Rates

Modulation	Code rate
QPSK	1/2
QPSK	2/3
16-QAM	1/2
16-QAM	2/3
64-QAM	1/2
64-QAM	2/3

TABLE 5.3   Terrestrial Analog TV Broadcast Standards

Standard	Modulation	Frequency bands (MHz)	Channel bandwidth (MHz)
NTSC	Analog (AM)	54–88, 174–216, 470–806	6
PAL	Analog (AM)	54–88, 174–216, 470–806	6,7,8
SECAM	Analog (FM)	54–88, 174–216, 470–806	6,7,8

TABLE 5.4   Terrestrial DTV Broadcast Standards

Standard	Modulation	Frequency bands (MHz)	Channel bandwidth (MHz)
ATSC	8-VSB	54–88, 174–216, 470–806	6
DVB-T	64-QAM^a	174–216, 470–806	6,7,8

^{a}Most common field deployments.

TABLE 5.5   Mobile TV standards

Standard	Modulation	Frequency bands (MHz)	Channel bandwidth (MHz)
ATSC-M/H	8-VSB	54–88, 174–216, 470–806	6
DVB-H	QPSK^a	470–806	6,7,8

^{a}Most common field deployments.

5.3.4   Comparison of TV Standards

Tables 5.3–5.5 provide a comparison of the analog TV, DTV, and mobile TV standards. OFDM is by far the most popular PHY transmission method because it is able to deal with delay spreads as high as 30 µs.

5.4   THE IEEE 802.22 STANDARD

IEEE 802.22 is the newest working group (WG) [6, 7] of the IEEE 802 committee that was formed in October 2004. The formation of the WG is motivated by FCC's proposed rule making in 2004. The WG focuses on point-to-multipoint wireless regional area networks (WRANs) that takes advantage of the favorable transmission characteristics of the VHF and UHF TV bands to provide two-way broadband wireless access over a large area. The WRAN comprises a fixed base station (BS) with fixed and portable user terminals (CPEs) operating in the VHF/UHF TV broadcast bands between 54 MHz to 862 MHz. The IEEE 802.22-2011 standard was published on July 2011 [8]. This is the first IEEE 802 standard that operates in TV white spaces. In general, urban areas have much less white space than low-population rural areas. As shown in Figure 5.2, the standard provides much-needed rural broadband

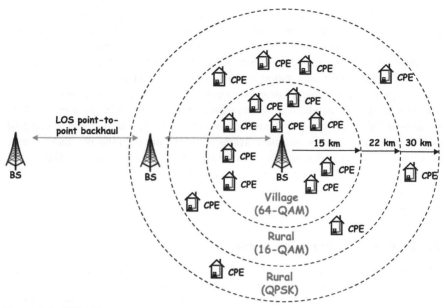

Figure 5.2   WRAN coverage.

connectivity, targeting unserved, underserved, and hard-to-reach areas, thus helping to bridge the digital divide. An important cost advantage of a WRAN is that network backhaul can be completely wireless using the minimum number of BS sites. Another advantage of a WRAN is that it can support mobility of up to 114 km/h with no handoff.

The typical operating radius is 30–40 km with a maximum range of up to 100 km. Long round-trip propagation delays of up to 37 ms are compensated by the OFDM PHY, and additional delays are absorbed by the MAC layer. The standard is able to support up to 255 CPEs with outdoor antennas located 10 m above ground level. The BS may employ sectorized or omnidirectional antennas for data transmission, whereas the CPE employs directional antennas. A different set of antennas is used for sensing and geolocation. Although the sensing antenna is omnidirectional in coverage, it requires horizontal and vertical polarization sensitivities to sense TV and microphone signals, respectively. 802.22 supports 6, 7, and 8 MHz TV channels with 5.6241, 6.5625, and 7.4944 MHz signal bandwidths. Sampling frequency, carrier spacing, symbol duration, signal bandwidth, and data rates will be scaled by the channel bandwidth for worldwide operations. A total bandwidth of 282 MHz is available in the United States with 47 TV channels, ranging from 54–60, 76–88, 174–216, 470–608, and 614–698 MHz. A minimum peak throughput is achievable at the cell edge: 1.5 Mbit/s for the DL and 384 Kbit/s for the UL. Depending upon the country of deployment, each WRAN can deliver 22–29 Mbit/s without interfering with reception of existing TV broadcast stations. This is because 802.22 incorporates advanced cognitive radio capabilities, including dynamic spectrum access, incumbent database access, accurate geolocation techniques, spectrum sensing, regulatory domain dependent policies, spectrum etiquette, spectrum management, and coexistence for optimal use of the available spectrum. Besides employing sensing and database access for incumbent protection, a specially designed beacon was also developed for this purpose. GPS-based geolocation is mandatory, but terrestrial geolocation based on triangulation methods is also supported.

Cognitive radio allows low-power license-exempt 802.22 CPEs to share radio spectrum in the TV bands. It minimizes interference to the primary users, such as incumbent TV broadcasting systems and low-power wireless microphones. The goal is to dynamically identify and use portions of spectrum not in use by other systems in a given location and time. The maximum output transmit power in the United States is 4 W EIRP for fixed CPEs and BSs (higher power BSs are allowed in other countries). The transmit power for the CPE is the highest among all IEEE 802 wireless standards.

5.4.1 Physical Layer Overview

The 802.22 PHY is optimized for long channel response times and highly frequency selective fading channels. Three primary functions include data transfer, spectrum sensing, and geolocation-enabled cognitive sensing. The 802.22 PHY employs OFDM on the DL and OFDMA on the UL. The air interface comprises 2048 subcarriers spread across the channel to increase frequency diversity. There are 368 guard, 1440 data, and 240 pilot subcarriers. The guard interval/symbol interval ratio is the same as 802.16-2004 (i.e., 1/4, 1/8, 1/16, 1/32). 802.22 employs outdoor directional antennas for communication with the BS. Multiple antennas are not supported due to the physical size of antennas and omnidirectional antennas are used for sensing. DL bandwidth is allocated progressively across subchannels in the frequency domain to simplify the DL bandwidth allocation, and reduce overhead and decoding latency. The fixed access operation implies channels change slowly. Thus, minimal time diversity gain is achievable with burst allocation in the time domain. The UL bandwidth is allocated progressively across OFDMA symbols in the time domain to reduce the number of subchannels used by the CPE. This reduces the transmit power, and, hence, the interference to incumbent systems.

5.4.2 Adaptive Modulation and Coding

Adaptive modulation and coding schemes (MCSs) are used by the BS to minimize the CPE transmit power. As shown in Table 5.6, 16 transmission modes are supported. Modes 1–4 are used for control signaling, whereas the remaining modes are used for data transmission.

- Mode 1 is used for CDMA opportunistic ranging bursts.
- Mode 2 is used for superframe control header (SCH) transmission.
- Mode 3 is used for coexistence beacon protocol (CBP) transmission.
- Mode 4 is used for frame control header (FCH) transmission.
- Modes 5–16 are used for data transmission.

TABLE 5.6   MCSs and PHY Rates in 802.22

PHY mode	Modulation	Code rate	Peak data rate for 6 MHz[a] bandwidth (Mbit/s)	Spectral efficiency (bit/s/Hz)[b]
1	BPSK	Uncoded	NA	NA
2	QPSK	1/2 and repeat: 4	NA	NA
3	QPSK	1/2 and repeat: 3	NA	NA
4	QPSK	1/2 and repeat: 2	NA	NA
5	QPSK	1/2	4.54	0.76
6	QPSK	2/3	6.05	1.01
7	QPSK	3/4	6.81	1.13
8	QPSK	5/6	7.56	1.26
9	16-QAM	1/2	9.08	1.51
10	16-QAM	2/3	12.10	2.02
11	16-QAM	3/4	13.61	2.27
12	16-QAM	5/6	15.13	2.52
13	64-QAM	1/2	13.61	2.27
14	64-QAM	2/3	18.15	3.03
15	64-QAM	3/4	20.42	3.40
16	64-QAM	5/6	22.69	3.78

[a]For 7 and 8 MHz channels, figures will scale proportionally.

[b]Net spectral efficiency ranges from 0.624 to 3.12 bit/s/Hz.

The spectral efficiencies are comparable with 802.16-2004 and are computed based on a continuous stream of 1440 data subcarriers for a given MCS (i.e., assuming no time gaps and superframe/frame headers). Unlike 802.16-2004, however, 802.22 supports a code rate of 5/6 for higher efficiency. Convolutional coding is mandatory. In this case, data are encoded with 1/2 rate and a constraint length of 7. The different code rates are obtained by puncturing the output of the convolutional coder. Optional FEC modes include block interleaving, which employs Turbo-based iterative interleaving. There are two variants of Turbo codes, namely duobinary convolutional Turbo code (CTC) and shortened block Turbo code (SBTC). Optional low-density parity-check (LDPC) codes are also available.

5.4.3   Preambles

There is a different preamble for the superframe, frame, and CBP. Each preamble comprises one symbol with a 1/4 guard/symbol ratio and enables burst detection, synchronization, and channel estimation. All CPEs are synchronized to the BS using the superframe preamble that comprises four repeated short training sequences (STSs). The frame preamble comprises two repeated long training sequences (LTSs). The CBP preamble uses the same structure as the superframe preamble but with a different STS to ensure low cross-correlation between the CBP and superframe STSs.

5.4.4   Bandwidth Resource Allocation

The basic tile comprises one subcarrier by seven symbols for the UL and DL. In 802.16e, a tile comprises four contiguous subcarriers, giving 420 tiles for 2048-FFT. The smallest unit of resource allocation is the subchannel, which comprises 28 subcarriers with 24 data and 4 pilot subcarriers. There are 60 subchannels in each symbol. In the DL, all data subcarriers in the 60 subchannels are interleaved with a block size of 24 × 60 (or 1440) before transmission to improve frequency diversity. On the UL, two subchannels are reserved for ranging, bandwidth allocation message (MAP) transmission, or urgent coexistence situation (UCS) notification. The remaining subchannels are interleaved with a block size of 28 × 58, which includes both pilot and data. The same frequency interleaving method is applied to the UL and DL. Each symbol is divided into subchannels of 28 subcarriers (24 data and 4 pilot subcarriers). Pilot symbols are inserted in the UL to ensure the BS receives a pilot symbol for every seven symbols (or every seven subcarriers) for robust channel estimation and frequency tracking. The pilot's position is rotated over a period of seven symbols and seven subcarriers. Pilots are omitted on the DL to allow faster channel estimation at the CPE using less than 7 symbols. This is useful for delay-sensitive applications. No frequency domain interpolation for channel estimation is required.

5.4.5   Spectral Awareness

The spectrum manager provides spectrum sensing and geolocation functions, which allow the BS to control channel usage and the CPE transmit power. This prevents harmful interference to licensed incumbents (i.e., analog TV in some countries, DTV, and licensed devices, such as wireless microphones) and allows coexistence with 802.22 systems. The CPEs may be forced to stop transmission or change to a different channel to prevent interference. Geolocation/database and spectrum sensing are two mandatory methods employed. The BSs and CPEs are required to use satellite-based geolocation technology to locate cognitive radio devices and a database of licensed transmitters. Neighboring networks can be synchronized to a global timing source. All 802.22 CPEs must be installed in fixed location so that the BS may know their locations. The CPE must first determine its location (by locking to the satellites) before associating with the BS. The BS location is accurate within a 15 m radius whereas the CPE location is accurate within a 100 m radius. The BS must have access to an accurate/up-to-date incumbent database service. Additional 802.22 operations data and backup channels in the area of interest may also be included in the database. Backup channels allow current operating channels to be rapidly vacated to protect the incumbents.

5.4.6   Spectrum Sensing Function

The BS and CPEs implement the spectrum sensing function (SSF), which allow them to observe and identify the TV spectrum for a set of signal types used by licensed devices and report the results of this observation. There are MAC

TABLE 5.7   Sensing Thresholds for Key Signal Types

Signal type	Signal power (dBm)
WRAN	−93
ATSC	−114
NTSC	−114
Wireless microphone	−114
802.22.1 Sync burst	−116

management frames that allow the BS to control the operation of the SSF within each of the CPEs. Spectrum sensing comprises per-channel sensing, quiet periods, standardized reporting, and implementation independence. Multiple channels can also be sensed at the same time. The sensing thresholds for the key signal types, including incumbent signals, are shown in Table 5.7. In-band and out-of-band spectrum measurements are performed by the CPE. In-band measurement consists of sensing the actual channel that is being used by the BS and the CPE. Out-of-band measurement consists of sensing the rest of the channels. In addition to these sensing measurements, fast and fine sensing are also allowed. Fast sensing allows sensing at speeds below 1 ms per channel and can be performed by the CPE or the BS. Fine sensing incurs more time (25 ms per channel or more), and it is used based on the outcome of the previous fast sensing mechanism. To perform reliable sensing, quiet times are allocated and no data transmission is permitted during these intervals. Such periodic interruption of data transmission may impair the performance of cognitive radio systems. This can be addressed by a proposed operation mode called dynamic frequency hopping (DFH), where data is transmitted in parallel with spectrum sensing without any interruption.

5.4.7   Medium Access Control Overview

Just like 802.16, all 802.22 MAC services are connection oriented. These connections can be created dynamically and are mapped to a 12-bit connection ID (CID). A frame comprises an integral number of fixed size OFDM slots. Each slot consists of one symbol and one subchannel. The UL subframe may comprise contention intervals scheduled for:

- CPE association (initial ranging between BS and CPE)
- CPE link synchronization, power control, and geolocation (periodic ranging)
- Bandwidth request
- UCS notification
- Quiet period resource adjustment.

The UL and DL transmission scheduling is controlled by the BS. This means that subscribers in the same wireless cell do not hear each other's transmission. The

802.22 MAC supports time division duplexing (TDD), which removes the potential problems associated with frequency division duplexing (FDD). For example, with TDD, sensing/reciprocity is possible when a system operates using one channel, thereby increasing the probability of spectrum availability. However, FDD may be supported in future. The DL subframe comprises a single PHY PDU (PPDU). The UL subframe may have a number of PPDUs from different CPEs. The time buffer of one symbol is included before and after the self-coexistence window. This is used for absorbing the propagation delay and initial ranging. The 802.22 MAC employs a superframe structure with 375 superframes in 60 seconds. This facilitates inter-BS synchronization (e.g., with GPS).

The various types of QoS services supported by 802.16 are also supported in 802.22. These include UGS, NRTPS, RTPS, and best effort service. Bandwidth requests are submitted by the CPEs on a contention basis. Although ARQ is supported, HARQ is not supported. Unicast, multicast, and broadcast services are supported. The cognitive functions are achieved via the following:

- Dynamic and adaptive scheduling of quiet periods to balance the QoS requirements of users and the need for spectrum sensing. Quiet periods range from 1 symbol (about 0.33 ms) to one superframe (160 ms).

- CPEs may alert the BS the presence of incumbents via dedicated UCS messages or low priority MAC messages.

- BS can request one or more CPEs to move to another channel using the FCH or dedicated MAC messages.

5.4.8 MAC Frame Format

The MAC superframe structure in the self-coexistence mode is shown in Figure 5.3. A long superframe of 160 ms is required to absorb a potentially long signal propagation delay associated with TV band transmission. The TDD superframe is made up of multiple frames and contains a SCH and a preamble. The frames are sent by the BS in channels that will not cause interference. The durations of the DL and UL subframes can be adjusted. When a CPE is turned on, it will sense the RF spectrum to locate available channels and will receive all necessary information to connect to the BS. 802.22 allows linear bandwidth allocation where subchannels are progressively used. The minimum UL burst size on a given subchannel is seven symbols to account for the longer signal propagation delay. The self-coexistence window (SCW) has a duration of five symbols, which includes a three-symbol CBP interval, if scheduled. A comparison of the key features of 802.22, 802.16m, and LTE is shown in Table 5.8.

5.4.9 Coexistence Beacon Protocol

The CBP mitigates mutual interference among colocated 802.22 WRAN systems and reduces cochannel operation, which may degrade performance significantly. The CBP employs beacon transmissions in a SCW that are scheduled at the end of some frames. The SCW comprises a preamble, an SCH, and a CBP MAC protocol data

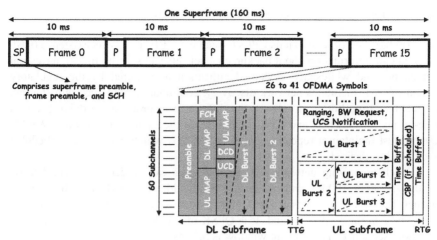

Figure 5.3   802.22 MAC frame format. CBP, coexistence beacon protocol; DCD, downlink channel descriptor; FCH, frame control header; MAP, bandwidth allocation message; RTG, receive–transmit transition gap; SCH, superframe control header; SCW, self-coexistence window; TTG, transmit–receive transition gap; UCD, uplink channel descriptor; UCS, urgent coexistence situation.

TABLE 5.8   802.22, 802.16m, and LTE Comparison

	802.22	802.16m	LTE
Air interface	OFDM for DL, OFDMA for UL	OFDMA for UL/DL	OFDMA on DL, SC-FDMA on UL
FFT Size	2048 subcarriers for all bandwidths	2048 subcarriers for 20 MHz bandwidth	2048 subcarriers for 20 MHz bandwidth
Duplexing	TDD	TDD/FDD	TDD/FDD
Channel bandwidth	6, 7, 8 MHz	5, 10, 20 MHz	1.4, 3, 5, 10, 15, 20 MHz
Resource allocation	Linear	Two dimensional	Two dimensional
Subcarrier permutation	Distributed with enhanced interleaver	Distributed (permutated) or grouped (optional)	Grouped (localized) with channel dependent scheduling
Multiple antenna techniques	None	Single or multiuser MIMO and beamforming	Single or multiuser MIMO and beamforming
Superframe size	160 ms	20 ms	None
Frame size	10 ms	5 ms	10 ms
Subframe size[a]	None	0.617 ms	1 ms
Coexistence with incumbents	Supported	None	None
Self-coexistence	Dynamic spectrum sharing	Master frame assignment	Master frame assignment
Internetworking	IP	IP	IP

[a]Refers to segments of a frame, not to be confused with the UL/DL TDD partitions.

243

unit (MPDU). The BS commands CPEs to send out CBPs for self-coexistence (CBP bursts contain information about backup channel sets and sensing times), terrestrial geolocation, and whitespace device identification as required by the regulatory domain rules. It facilitates network discovery, coordination, and spectrum sharing, and includes time buffers to absorb the difference in propagation delay between close-by and distant BSs and CPEs operating on the same channel. Each WRAN system must maintain a repeating pattern of SCWs, which can also be scheduled on-demand as desired.

A self-organizing network of beacon devices can be deployed. The beacon duration is about 100 ms and occupies a bandwidth of 77 KHz in a 6 MHz TV channel. The beacon operates at a maximum power of 250 mW in the UHF TV band and 50 mW in the VHF TV band. It is QPSK-modulated with an eight-chip pseudorandom sequence. The sensing window is 5.1 ms.

5.4.10 Security

802.22 data confidentiality, privacy, and authenticity (integrity) is based on the 128-bit Advanced Encryption Standard in Galois Counter Mode (AES-GCM), which is used for encryption and authentication. GCM is an authenticated encryption algorithm for symmetric key 128-bit block ciphers that has been widely adopted because of its efficiency and performance. In 802.22, GCM is used to compute the integrity check vector (ICV), and pseudorandom sequence numbers are appended to each packet. Network authorization is based on the Rivest, Shamir, and Adleman (RSA) and elliptic curve cryptographic (ECC) X.509 certificates, which are used for mutual authentication and network entry authorization. Signals, such as the wireless microphone beacon and CBP, are authenticated using ECC digital signatures. No encryption is provided for these packets. The secure control and management protocol is used for key management. All management messages (except for broadcast, initial ranging, and basic CID) are protected. For device security, the Trusted Computing Group's Trusted Platform Module (TPM) specifications are recommended to enable tamper-proof hardware and software. In terms of spectrum access authorization, the BS may deauthorize a CPE at any time using the sensing and incumbent database services. In addition, the spectrum manager may prohibit a CPE from registering if it does not have adequate sensing capabilities.

5.4.11 IEEE 802.22.1

Although some channels are not used for TV broadcasts, low-power, licensed devices, such as wireless microphones operated by broadcasters, may use these channels, and are entitled to protection by regulation to avoid disrupting incumbent services. 802.22.1 is an air interface of devices that broadcast warning beacons to protect the operation of low-power licensed devices operating in the TV bands from harmful interference. The beacons transmit identifiable synchronization bursts to signal the presence of the wireless microphones. In addition, optional information about locations and operational parameters of the protected devices may be included. WRAN devices must be able to sense the 802.22.1 beacon and decode the information. 802.22.1 employs a superframe structure that repeats without interruption on a given TV channel occupied by an incumbent licensed wireless microphone or other licensed low-power devices. A large portion of the superframe structure is divided over two logical channels, which are transmitted in parallel. The synchronization logical channel consists of a succession of synchronization bursts, whereas the beacon logical channel consists of the PPDU, which mostly encompasses the MAC beacon frame. The MAC beacon frame contains information, such as the physical location of the beaconing device and the estimated duration of TV channel occupancy. The superframe consists of a succession of 31 slots. Each slot is comprised of 32 differential QPSK symbols, with a symbol duration of 1/9609.1 seconds. Each slot contains one synchronization burst, as well as a fixed number of PPDU bits. The protecting device (PD) is a beaconing device that is protecting a low-power, licensed device. Each PD monitors its TV channel for a random number of superframes to determine the presence or absence of a primary PD (PPD). If no PPD is heard, the PD will act as a PPD and initiate its own beacon. However, if one or more PPDs are detected, the PD may either act as a PPD (and initiate beaconing) or as a secondary PD (SPD) and attempt to contact a PPD.

5.5   WHITESPACE ALLIANCE

The Whitespace Alliance or WSA (http://www.whitespacealliance.org) promotes the development, deployment, and use of standards-based products and services using the TV bands. The WSA will also help put in place interoperability, conformance, and compliance testing.

REFERENCES

[1] FCC 700 MHz Auction, http://wireless.fcc.gov/auctions/default.htm?job=auction_factsheet&id=73.
[2] FCC Notice of Proposed Rule Making, May 25, 2004, http://hraunfoss.fcc.gov/edocs_public/attachmatch/FCC-04-113A1.pdf.
[3] FCC Second Report and Order and Memorandum Opinion and Order, November 14, 2008, http://hraunfoss.fcc.gov/edocs_public/attachmatch/FCC-08-260A1.pdf.
[4] N. Srivastava and S. Hanson, Expanding Wireless Communications with White Spaces, http://www.dell.com/downloads/global/vectors/white_spaces.pdf.
[5] J. Whitaker, "Future Directions for Digital Television," *IEEE Wireless Communications Industry Perspectives*, June 2011.
[6] C. Stevenson, et al., "IEEE 802.22: The First Cognitive Radio Wireless Regional Area Network Standard," *IEEE Communications Magazine*, pp. 130–138, January 2009.
[7] IEEE 802.22 Working Group, http://www.ieee802.org/22.
[8] IEEE 802.22 Standard Download, http://standards.ieee.org/about/get/802/802.22.html.

HOMEWORK PROBLEMS

5.1. Due to the longer operating wavelength of 802.22, the antenna size can be large compared with the 2.4 and 5 GHz frequencies. Based on this reasoning, why are small antennas used by AM/FM radios that operating at even lower frequencies (and hence longer wavelengths)?

5.2. Compare the guard interval/symbol interval ratio for 802.11, 802.16m, and 802.22. What can you conclude about the overheads due to the guard interval? How would you modify 802.16 and 802.11 to operate in the TV bands? Since LTE does not support superframes, is this a disadvantage? Why is OFDMA not needed for the DL operation in 802.22? Explain how you can ensure the same operating range for sensing and reception in 802.22. What are the consequences of not having equal sensing and reception range?

5.3. If an 802.22 WRAN has been operating for some time, how can the BS or CPE sense the presence of a primary user that has become active suddenly? Will this imply that some form of interference is inevitable? Can signal-specific and feature-based sensing algorithms help to alleviate the problem? Comment on the reliability of long-range sensing and the impact on the response time.

5.4. Beacons may further generate undesirable interference for 802.22 WRANs. How can the use of beacons be minimized or avoided?

5.5. Some 802.11 receiver sensitivities are listed in the following tables. Explain why the sensitivities increase (corresponding to the ability to detect a lower signal power) as the data rate decreases and as broader channels (e.g., 40 MHz) are used. The cognitive radio mechanism in 802.22 requires a high detection sensitivity of -114 dBm (or -19 dB SNR) due to the large coverage area in WRANs. This corresponds to a power of $10^{-11.4}$ mW. Justify the need for such a high sensitivity in wide area networks covering tens of miles. Will this sensitivity virtually eliminate the presence of any white space in the 700 MHz band?

802.11b/g

Data rate (Mbit/s)	Minimum sensitivity (dBm)
1	−93
2	−91
5.5	−88
6	−86
9	−85
11	−85
12	−84
18	−83
24	−79
36	−77
48	−72
54	−70

802.11n

Modulation	Code rate	Minimum sensitivity for 20 MHz channel (dBm)	Minimum sensitivity for 40 MHz channel (dBm)
BPSK	1/2	−82	−79
QPSK	1/2	−79	−76
QPSK	3/4	−77	−74
16-QAM	1/2	−74	−71
16-QAM	3/4	−70	−67
64-QAM	2/3	−66	−63
64-QAM	3/4	−65	−62
64-QAM	5/6	−64	−61

5.6. An extension to cognitive radio is the use of wideband compressive sensing techniques to allow the capture of accurate RF information when sweeping through sparsely populated bands. Explain whether such bands exist in practice when receivers are equipped with a minimum detection sensitivity of −114 dBm.

5.7. For licensed TV band devices (i.e., the primary users), is spectrum sensing needed when channel access can be arbitrated via a BS? If incumbents who do not use the licensed bands consider leasing or auctioning these bands, will this remove the need for spectrum sensing?

5.8. The PHY performance of 802.22 WRANs is shown in the table below. Justify the use of 64-QAM with a 1/2 code rate.

Modulation	Code rate	Net data rate	Net efficiency	SNR
		Cyclic prefix = 1/8		
QPSK	1/2	3.74	0.624	4.3
	2/3	4.99	0.832	6.1
	3/4	5.62	0.936	7.1
	5/6	6.24	1.04	8.1
16-QAM	1/2	7.49	1.248	10.2
	2/3	9.98	1.664	12.4
	3/4	11.23	1.872	13.5
	5/6	12.48	2.08	14.8
64-QAM	1/2	11.23	1.872	15.6
	2/3	14.98	2.496	18.3
	3/4	16.85	2.808	19.7
	5/6	18.72	3.12	20.9

5.9. The 700 MHz UHF band is a prime RF band reserved for first broadband wireless application—TV broadcasting. Signal propagation in this band enjoys certain advantages when compared with higher frequencies, enabling broadband connectivity even in rural areas. Since signal attenuation is lower than microwave frequencies, deeper wall penetration in buildings and houses is possible, and signal impairments due to rain, snow, and multipath are less significant. This also leads to wider coverage, longer range, and fewer handoffs—an omnidirectional coverage of at least 20 km from a well-sited BS is typical. For example, a trial broadband network in Washington, DC at 700 MHz covers the entire metro area with 10 deployment sites, compared with 400 sites for 4.9 GHz. Thus, the band can support many interesting deployments, including high-bandwidth and high-speed vehicular applications (see N. Itoh and K. Tsuchida, "HDTV Mobile Reception in Automobiles," *Proceedings of the IEEE*, January 2006, pp. 274–280). NASA's new onboard Electra UHF relay transceiver provides faster data rates required for all future orbiters, landers, and rovers. A 700 MHz 802.11 wireless LAN prototype has been reported. Describe the engineering challenges in these 3 deployments.

5.10. Software defined radio (SDR) is a key technology to reusing RF hardware, thereby reducing wireless deployment costs. Software-defined cellular base stations have been deployed to provide flexibility, allowing easy and less costly upgrades. Explain whether SDR is necessary for the deployment of cognitive radio technology in 802.22.

5.11. Verizon plans to use all the 700 MHz spectrum it won in the FCC auction for deploying LTE 4G services. More than 40 cities have been granted or seeking waivers from the FCC to build broadband networks in the 700 MHz band under the National Broadband Plan. In addition, the same frequency band has been designated for building a nation-wide public-safety broadband network that will eventually carry voice, video, and data for first responders. Since both utilities and public safety agencies are typically the first on the scene of an emergency of disaster, sharing spectrum and infrastructure in this band is critical. LTE is the platform of choice because it offers interoperability between the infrastructures and can accelerate response times, improve situational awareness, and increase the safety of all personnel. Clearly, deployment, usage, and business models will determine the success of these new rollouts. Explain whether long-range wireless operation using the 700 MHz band is more suited for broadcast operations than two-way communications. Describe the advantages and disadvantages of using LTE instead of 802.22 when operating at 700 MHz frequencies. ∎

CHAPTER 6

MESH, RELAY, AND INTERWORKING NETWORKS

Despite advances in wireless technologies that allow improvement in data rates and transmission reliability, it is challenging for cellular base stations (BSs) to provide adequate coverage in cluttered urban areas where signal obstructions caused by tall structures are common. Mesh and relay networks serve to bridge that gap in coverage with wireless routers deployed at street level. These mesh routers allow end-user devices to access the network by forwarding data traffic wirelessly. Unlike BSs, the mesh or relay routers can be deployed in strategic locations (e.g., in places experiencing high traffic volume) to minimize power consumption. This in turn benefits battery-powered user handsets and enables energy-efficient network deployment. This chapter focuses on the 802.11s, 802.16j, and 802.11u mesh, relay, and interworking amendments.

6.1 INTRODUCTION

Simple methods of extending wireless coverage include the use of a signal repeater or a single-hop packet relay. Femtocells employ such relays that forward incoming traffic from the BS to the handset, whereas 802.11 range extenders employ signal repeaters. Unlike these methods, a mesh network comprises a concatenation of wireless hotspots that may require packet routing. Each hotspot constitutes a hop, and may serve a number of stations within a home, a classroom, an office, a street café or even a restaurant. These hops form a wireless backhaul network, which ultimately connects to the Internet service provider. Hence, a fixed network infrastructure is still required to provide service to mesh networks. A wireless mesh network is typically organized as a flat network involving only the PHY and MAC layers. A hierarchical structure relies on higher networking layers and applies to the interworking portion of the backhaul, which can be connected to other types of wireless access networks, such as 802.16 and LTE. In public mesh networks, static wireless routers are typically mounted on streetlights or telephone poles. In community (grassroots) mesh networks, these mesh routers are directly mounted on the rooftop of a house. The mesh router may forward the packets across several hops via a predetermined path or route. This requires consideration on the tradeoff between network overheads,

Broadband Wireless Multimedia Networks, First Edition. Benny Bing.
© 2013 John Wiley & Sons, Inc. Published 2013 by John Wiley & Sons, Inc.

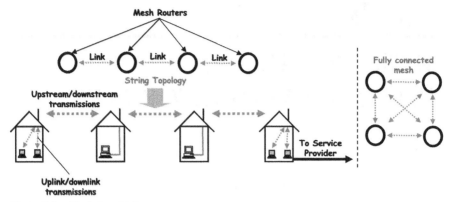

Figure 6.1   Mesh topologies.

maintenance cost, density of network design, and path redundancy. For example, packet forwarding and routing takes up additional bandwidth resources compared with a direct transmission between the source and destination stations. Although each additional hop adds latency, this can be compensated by the higher router bandwidth due to the shorter range and spatial frequency reuse. However, a bottleneck may occur on the slowest or most congested link in the path.

Network topologies may range from a string to a fully connected mesh, as shown in Figure 6.1. The string topology generates the least interference and does not require any routing overheads. This is because there is only one path (i.e., no path redundancy), and a mesh router simply filters the MAC address of the incoming packet and forwards it to its neighbor(s) if a station with a matching MAC address is not found within its subnet. The fully connected mesh network is costly to deploy, but it is the most reliable since each router is connected to every other router and path redundancy is maximized. A distinction is made between uplink/downlink (UL/DL) and upstream/downstream (US/DS) transmissions. An UL transmission is relevant within a single hop and refers to the transmission from the station to the router, whereas an US transmission refers to router-to-router transmission such that a packet is forwarded to another router located closer to the destination station. In a mesh network, an UL transmission can be directed US or DS depending on the location of the destination station.

6.1.1   Mesh Radio Transceivers and Channels

Mesh technology evolved from one radio transceiver per router to multiradio routers that provide high network throughput, high scalability, and multimedia support (Table 6.1). Each radio transceiver employs a single frequency channel, which implies that a radio can only operate in the half-duplex mode (i.e., transmit or receive but not at the same time). Single-radio systems are limited in throughput due to the half-duplex nature of operation and the presence of hidden stations. For example, the throughput of a three-hop topology is about one-fourth of a single hop since the intermediate mesh router can either transmit or receive at any time (Figures 6.2 and

TABLE 6.1 **Evolution of Mesh Networks**

	1st generation	2nd generation	3rd generation
Density	1 radio	2 radios	≥3 radios
Scalability	Very limited	Limited	High
Latency over hops	High	Medium	Low
Throughput over hops	Very low	Low	High
Real-time application support	Limited	Limited	High

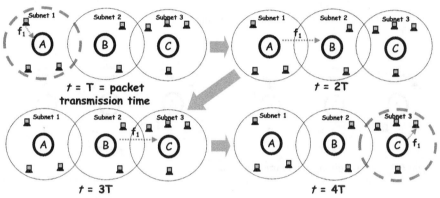

Figure 6.2   Single-radio upstream transmission.

6.3). Thus, four packet transmission periods are needed to relay one packet from router A to router C. This can be considered the worst-case throughput since a station may transmit a second packet to router A at time $t = 3T$, if it is located at a sufficient distance from router B, and hence, does not cause interference. Alternatively, it can employ directional beamforming antennas instead of broadcast antennas. On the other hand, if each data packet requires an acknowledgment (ACK) from the destination station, the useful network throughput will be degraded further. In addition, many protocols broadcast control packets to all routers (a process known as flooding) in order to discover the appropriate path to send a packet. Thus, the overheads will increase if the paths connecting the routers need to be updated frequently. Note that a single-radio router need not be restricted to using the same frequency channel all the time. With channel switching, cochannel interference (CCI) can be minimized because a router further US or DS can change to a different frequency channel. For example, if router B uses channel 2 instead of channel 1 at time $t = 3T$, it will allow any station served by router A to transmit a new packet without causing interference to router B. However, since channel switching involves coordination between two or more routers, this may incur additional overheads.

A two-radio system allows the mesh router to employ different frequencies for simultaneous transmissions with one neighboring router, as well as stations

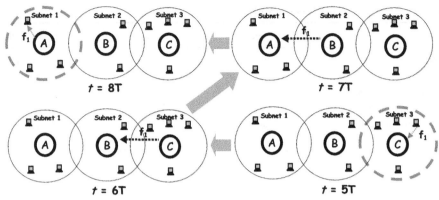

Figure 6.3   Single-radio downstream transmission.

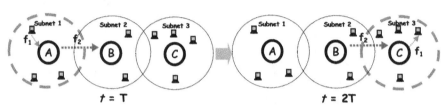

Figure 6.4   Dual-radio upstream transmission.

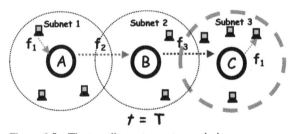

Figure 6.5   Three-radio upstream transmission.

within the router's subnet. In this case, a router may concurrently receive and transmit information on two different frequency bands, but one band is reserved for communications with stations in the subnet served by the router. However, the link connecting two adjacent routers is limited to half-duplex operation because US and DS communications cannot be conducted at the same time. As shown in Figure 6.4, when router B is receiving a packet from router A, router B cannot transmit to router C on the US or to router A on the DS.

A three-radio system, as shown in Figure 6.5, allows concurrent US transmissions. If the router is not an endpoint, two radios are used for US transmission while the third radio is used for communication with stations in its subnet. However, concurrent US and DS transmissions cannot be achieved. Note that a two-radio

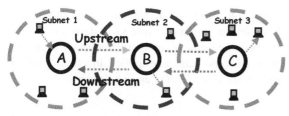

Figure 6.6   A multiradio mesh network.

system can also be configured in a similar manner but relay routers are employed instead. For instance, mesh router B may use only two radios but strictly performs forwarding functions and cannot serve any station in Subnet 2 using a third frequency band.

A multiradio system involving more than three radios per mesh router can support full-duplex transmissions with adjacent routers (Figure 6.6), as well as with stations. It is the most scalable way to deploy a mesh network since the constraints associated with the network topology are removed, allowing the routing overhead and path redundancy to be balanced efficiently.

6.1.2   Advantages of Mesh Networks

Mesh networks provide greater convenience in network access. For example, home wireless networks can be extended to include connections from virtually any open space (e.g., swimming pool, and backyard). These networks also allow efficient traffic distribution, which may include peer-to-peer traffic. When path redundancy is available, local interference, denial of service (DoS) attacks, and traffic bottlenecks can be avoided. For the operators, network maintenance becomes easier since the residential network can be accessed away from the customer's premise. In addition, scalable network deployment is possible. An operator can start with the minimum number of mesh routers, and as the number of users and traffic load increase, more routers can be installed. This is in contrast to cellular networks, where operators need to justify the cost for building a BS by evaluating the number of potential subscribers it can serve. Large-scale indoor or outdoor mesh networks can be created quickly in a plug-and-play fashion since no network cable is required. The shorter hops of these networks reduce the effects of the Fresnel zone that impact cellular deployments. Thus, low-lying outdoor areas do not pose problems in network deployment.

6.1.3   Packet Routing

Packet routing is required only when redundant paths to the destination are available. Because the ability to move packets along multiple paths improve network reliability, good quality of service (QoS) can be achieved in a mesh network even when operating on unlicensed frequency bands. Mesh routing protocols should be designed with some unique capabilities in mind. For example, routing protocols designed for an ad-hoc wireless network with no topology assurance may not be suitable for mesh networks with structured topologies. Similarly, protocols designed for the Internet where instances of network congestion occur more frequently than link outages (due to the global span of the network) cannot be applied directly to wireless mesh networks that usually provide network access for up to a few miles.

6.1.4 Public Mesh Networks

The success of public mesh networks may depend on the ownership of the network rather than on technology issues. For example, municipal 802.11 mesh networks have enjoyed only limited success due to the difficulty in obtaining right of way in the public space even though these networks can save substantial operational costs for the government agencies and benefit residents, businesses, and tourists. Successful municipal 802.11 deployments in the United States include Portland (http://www.wifipdx.com), Tempe (http://www.tempe.gov/wifi), and Minneapolis (http://www.ci.minneapolis.mn.us/wirelessminneapolis).

6.2 802.11 MESH NETWORKS

A mesh capability in 802.11 may transform both private and public 802.11 networks by removing the wiring needed to connect access points (APs). In addition, the same PHY and MAC layers can be employed throughout the span of the network. If these networks are used to deliver broadband access services to residences and businesses, this may see the distinction between access and local area networks blurring for the first time in the history of computer networking. Transmit power is not an issue with static mesh APs since they are individually powered. These APs become layer 2 mesh routers that provide a structured network infrastructure and serve as "checkpoints" in frame forwarding, thereby ensuring some measure of reliability in data transfer. Mesh stations involving user devices and handsets may also provide routing capability, further improving the robustness of the network against router failures and offering the ability to deploy completely wireless ad hoc networks without any mesh APs. The One Laptop per Child [1] project employs this mode of operation. However, this operational mode should be evaluated carefully due to the battery power limitations and security issues associated with the mesh station. Unlike a mesh AP, which is normally placed in an elevated location, a user device acting as a mesh station may not be aware of all other stations within the wireless subnet. This may potentially lead to a higher occurrence of hidden collisions. Thus, the deployment of such a network is probably justified for a small number of stations.

6.2.1 802.11s Amendment

The 802.11s Task Group (TG) was formed in July 2004 and defines interoperable mesh networking protocols for the Extended Service Set (ESS). The 802.11s amendment was ratified in September 2011 after 12 draft revisions. An open source implementation of the ratified amendment is available in Reference [2]. 802.11s improves the reach of 802.11 networks by allowing mesh stations to configure as a wireless mesh network. The 802.11s mesh station encompasses both user devices and APs, and acts like a mesh router. A mesh gate connects these stations, and hence the wireless mesh network, to the wired distribution system (DS). Thus, the mesh gate comprises an 802.11s radio interface, as well as a wired network interface. The interconnected mesh stations will route data frames over the best available path. Path selection is based on MAC-address layer 2 routing, which can be different from layer 3 IP routing. 802.11s allows vendors to employ interoperable and proprietary path selection protocols. It uses the existing 802.11 address format to define a mesh network with autoconfiguration capabilities. There are new end-to-end multihop security mechanisms. Mutual authentication among mesh stations creates secure associations. Each mesh station may act as a supplicant or authenticator for neighboring mesh stations. Reauthentication can occur rapidly for roaming stations to preserve the network session. Other capabilities include topology discovery, frame forwarding, congestion control, channel allocation, and traffic and network management.

6.2.2   Mesh Discovery

A basic service set (BSS) that forms a self-contained network of mesh stations is called a mesh basic service set (MBSS). An MBSS may contain zero or more mesh gates. A mesh station performs either active scanning or passive scanning to discover an active MBSS and its mesh profile. The mesh profile consists of the mesh ID (i.e., the identity of an MBSS), active path selection protocol, active path selection metric, congestion control mode, synchronization method, authentication protocol, and emergency service support. Based on the result of the scan, the mesh station may establish a new MBSS or become a member of the existing MBSS.

Since an MBSS has multihop capability, it appears that all mesh stations are directly connected at the MAC layer even if the stations are not within range of each other. This is in contrast to an independent BSS, where stations cannot communicate if they are not within range of each other. Unlike the independent BSS, an MBSS may have access to the DS by connecting to one or more mesh gates. Thus, the MBSS can be used as a distribution system medium (DSM). APs, portals, and mesh gates may employ the MBSS as a DSM to provide the distribution system service (DSS). This allows different infrastructure BSSs to be integrated with the MBSS to form an ESS. An AP may identify the infrastructure BSS that it forms. However, in an MBSS, no such central entity exists. While an infrastructure BSS requires the ESS and thus the DS, the MBSS network may not require access to a DS. However, if an MBSS has one or more mesh gates providing access to the DS, the MBSS may exist in disjointed areas and yet form a single network.

Each mesh station periodically transmits a beacon frame and responds with a Probe Response when a Probe Request is received from a neighboring mesh station that is performing mesh discovery. The beacon and Probe Response carries the mesh ID. Thus, the mesh station's beacon helps other mesh stations detect and join the network. In an MBSS, multiple mesh stations transmit beacon frames periodically, and these stations may be located out of range of each other but employ the same frequency channel. This suggests that mesh beacon frames may suffer from hidden collisions, and because these frames are transmitted in the broadcast mode, no return

ACKs are possible, and hence, repeated hidden collisions may occur. A mesh beacon collision avoidance (MBCA) protocol is defined to mitigate hidden collisions of beacon frames. MBCA is composed of beacon timing advertisements, target beacon transmission time (TBTT) selection, and TBTT adjustment. The mesh station uses the beacon timing information of its neighbor for its TBTT selection and TBTT adjustment. For example, when a beacon frame is received from one of its neighbors with which the mesh station maintains synchronization, the mesh station will calculate the TBTT of the received beacon frame as follows:

$$T_{\text{TBTT}} = T_r - (T_t \text{ modulo } (T_{\text{BeaconInterval}} \times 1024)), \tag{6.1}$$

where

T_{TBTT} is the calculated TBTT

T_r is the beacon frame reception time

T_t is the value of the timestamp field in the received beacon frame

$T_{\text{BeaconInterval}}$ is the value of the beacon interval field in the received beacon frame.

The mesh station will calculate the time difference between the TBTT of the received beacon frame and the time predicted from the past TBTT as follows:

$$T_{\text{Delta}} = \left| T_{\text{TBTT,c}} - (T_{\text{TBTT,p}} + (T_{\text{BeaconInterval}} \times N_{\text{Count}})) \right|, \tag{6.2}$$

where

T_{Delta} is the absolute time difference

$T_{\text{TBTT,c}}$ is the TBTT calculated from the received beacon frame

$T_{\text{TBTT,p}}$ is the TBTT calculated for the first time after the latest status number update

$T_{\text{BeaconInterval}}$ is the value of the beacon interval field in the received beacon frame

N_{Count} is the number of TBTTs since $T_{\text{TBTT,p}}$ has been calculated.

The status number is set to a value from a modulo-16 counter, starting at 0 and incrementing by 1 for each transmission of a beacon frame after the mesh station encountered any of the following events:

- It starts or stops maintaining synchronization with a neighbor station.
- It receives a beacon frame from a neighbor station with which it maintains synchronization and the calculated T_{Delta} is greater than 255 µs.
- It completes the TBTT adjustment procedure, which requires the mesh station to check that it does not transmit beacon frames concurrently with beacon transmissions from other stations within a two-hop range.

Within an MBSS, direct communication between neighboring mesh stations is allowed only when they become peer mesh stations. After mesh discovery, two neighboring mesh stations may establish mesh peering. A mesh station can establish mesh peering with multiple neighbors. Mesh peering management (MPM) facilitates mesh peering establishment and deactivation.

6.3   HYBRID WIRELESS MESH PROTOCOL

The Hybrid Wireless Mesh Protocol (HWMP) is a mandatory layer 2 path selection protocol in 802.11s. Unlike layer 3 routing protocols, such as those developed for ad hoc wireless networks and the Internet, a layer 2 routing protocol is more responsive to changes in the physical link conditions. HWMP employs a combination of the proactive tree-based distance vector protocol to support static mesh deployments and elements of the reactive ad hoc on-demand distance vector (AODV) protocol [3] to support mobile mesh deployments. Thus, HWMP enables efficient path selection in a wide variety of mesh networks (with or without access to a network infrastructure). Both modes of operation employ a single set of primitives, which allows the operator to apply the most efficient routing protocol for a given deployment or application.

A mesh gate can announce its presence in the MBSS by sending a gate announcement (GANN) frame. When a mesh gate has access to IEEE 802 stations that lie outside the MBSS, it can act as a proxy for the IEEE 802 station. Such a mesh gate is called a proxy mesh gate. When a root mesh station exists, which may include a root mesh gate, the mesh network can employ distance vector routing to establish paths to the root beforehand. In this case, traffic to and from the root follows a tree structure as shown in Figure 6.7. Because the shortest path is chosen from the transmitting mesh station to the root, and all connections terminate at the root, spanning tree routing is the most efficient and guarantees loop-free connectivity. Clearly, the root mesh station will require more bandwidth resources since more traffic is directed to it. Providing fairness by eliminating spatial bias ensures that mesh stations closer to the root mesh station do not achieve a higher throughput than mesh stations requiring a higher number of hops.

The on-demand mode is always available, independent of whether a root mesh station is present or not. It allows mesh stations to communicate using peer-to-peer paths. The on-demand mode is enhanced using the proactive tree building mode. This can be performed by configuring a mesh station as a root mesh station using either the proactive Path (Set-up) Request (PREQ) or root announcement (RANN) mechanism. The PREQ mechanism creates paths from the mesh stations to the root, using only group-addressed communication. The RANN mechanism creates paths between the root, and each mesh station using acknowledged communication. The RANN is transmitted in an HWMP Mesh Path Selection frame. The on-demand and proactive modes may be used concurrently because the proactive mode is an extension of the on-demand mode.

Unlike the native AODV layer 3 protocol, there are modified rules for processing layer 2 AODV primitives and parameters in 802.11s. Examples include Sequence Numbers and Time-to-Live (TTL). The freshness of the path is provided to prevent forwarding loops and duplications. As shown in Figure 6.7, to establish an on-demand path, mesh station 4 will broadcast PREQ for Address B. Besides

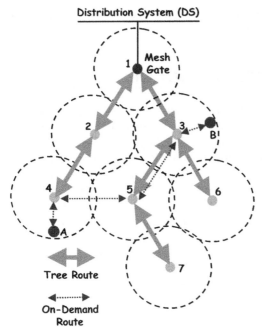

Figure 6.7   On-demand path establishment.

discovering a path to one or more targets, the PREQ can also be used to build a proactive (reverse) path selection tree to the root mesh gate. When the PREQ arrives at mesh station 3 (proxy of B), the station will send a unicast Path (Set-up) Reply (PREP) back to mesh station 4 to complete the "direct" path. The PREP is used to establish a forward path to a target and to confirm that a target is reachable. The PREP includes the hop count information, and this return message passes through the intermediate mesh stations; therefore, these stations know how many hops are needed to reach the destination. The reverse routing path from B to A is established after mesh station 3 receives the PREQ. The desired routing path from A to B is created after A receives the PREP. A path error (PERR) will be issued if a broken link for the next hop of an active path is detected or when a frame destined for a mesh station for which it has no (active) path is received.

6.3.1   Frame Forwarding Function

Once a path connecting the pair of source and destination mesh stations has been identified using HWMP, mesh stations along the path can forward the data frames, or more specifically, the MAC service data units (MSDUs). Thus, the MSDUs are forwarded by all intermediate mesh stations in the path, even if they are not neighbors. There are two frame forwarding methods in 802.11s. Intramesh forwarding is concerned with MSDUs originating and directed within the mesh network. Just like a regular 802.11 wireless LAN, this usually requires four addresses: receiver,

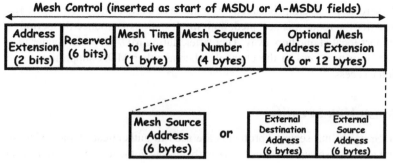

Figure 6.8   802.11s mesh control.

transmitter, destination, and source. Proxy forwarding is concerned with MSDUs originating from or destined to stations located outside the mesh network. This requires six addresses: receiver, transmitter, destination, source, external destination, and external source. The external addresses are carried in a Mesh Address extension, which contains two additional parameters: TTL and Sequence Number, as shown in Figure 6.8. 802.11s extends data and management frames by adding a mesh control field. When the mesh stations communicate over a single hop, their frames do not carry the mesh control field. The 2-bit address extension field indicates the presence of up to three additional MAC addresses in the mesh control field. Standard management frames have three addresses only. The mesh destination address is included as Address 3 in the standard 802.11 frame header, whereas Address 4 is included in the mesh control field. Once a candidate peer has been identified, a mesh station establishes a peer link with another mesh station. Even when the physical link breaks, mesh stations may keep the peer link status to allow faster reconnection.

The discovery of the proxy mesh station is achieved by requesting a path to the proxied entity and receiving a reply by the corresponding proxy mesh station. For unicast frames without a valid on-demand path, such frames may be sent to the root mesh station first, but this is not mandatory. The Mesh Sequence Number is the primary mechanism that detects duplicate reception in order to avoid duplicate retransmission. TTL is backup mechanism for avoiding infinite transmission. Thus, the TTL and sequence number fields combine to prevent frames from looping forever. The main causes for unreliable transmissions include missing ACKs (which result in undetected collisions) and mesh stations that the sender does not know about. Because mesh stations that have switched to a new channel cannot detect the status of the network allocation vector (NAV) on the new channel, collisions are likely to occur. Thus, channel switching is not included.

6.3.2   Mesh Deterministic Access

Mesh deterministic access (MDA) is an optional access method that minimizes contention. The MDA-supporting mesh station requests a lower-contention period called MDA opportunity (MDAOP). The MDA-supporting neighbor accepts the request if there is no overlap between the requested period and other registered MDAOPs. After the request is accepted, both mesh stations announce the period to other neighbors located one hop away. Thereafter, during MDAOPs, the sender may have higher priority access to the channel, whereas other MDA-supporting mesh stations defer access.

6.3.3 Mesh Link Security

In an MBSS, mesh link security protocols are used to authenticate a pair of mesh stations and to establish session keys between them. The authenticated mesh peering exchange (AMPE) protocol relies on a shared pairwise master key (PMK) between two mesh stations to establish authenticated peering and derive session keys. To ensure secure mesh association between two neighboring mesh stations, the mesh key distributor (MKD) and mesh authenticator (MA) are required. The MKD is responsible for authentication of a booting mesh station. The authentication result is reusable and cached on the MKD. Authentication is a two-step process: the MA authenticates the mesh station, and keys to encrypt the traffic are derived. The EAP exchange provides the shared secret key PMK, which is valid for the entire session. Once authentication is successful, the booting mesh station performs a four-way handshake with neighboring MAs to exchange security keys (Figure 6.9). This mesh station may become an MA to serve other new mesh stations. The four-way handshake is establishes another key called the pairwise transient key (PTK). The PTK is generated by concatenating the PMK, MA nonce, mesh station nonce, MA MAC address, and mesh station MAC address. The nonce values are arbitrary, but can only be used once. The product is then encrypted using a cryptographic hash function. The handshake also yields the group transient key (GTK) that is used to decrypt multicast and broadcast traffic. Because the IEEE 802.1X authentication process can be time-consuming, the mesh station can invoke preauthentication independently of association with the MA. However, authentication is required before an association can be established.

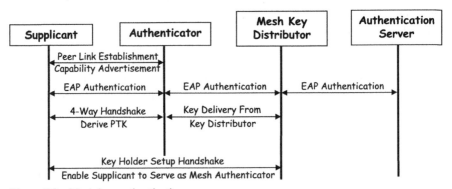

Figure 6.9   Mesh key authentication.

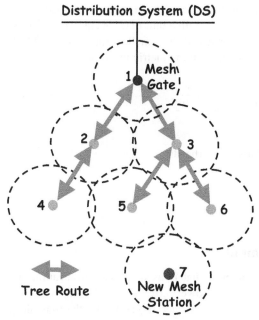

Figure 6.10   Mesh authentication.

6.3.4   Secure Peer Link Establishment

Secure peer link establishment is a symmetrical process between peers that is performed before routing. It improves the security of routing protocols by employing bilateral open and confirm primitives, which is different from conventional 802.11 networks (Figure 6.10). Secure peer link establishment is completed in two exchanges if all necessary security parameters (e.g., PMKs) are available to the two mesh stations. For example, if mesh station 6 is not able to authenticate mesh station 7, mesh station 6 reverts to 802.1X supplicant role, with mesh station 7 acting as an authenticator. Although mesh station 7 is not a member of the mesh network, it sees mesh stations 4 and 6. The link budget between mesh station 6 and mesh station 7 is better than between mesh station 4 and mesh station 7. The number of hops to reach the mesh gate is the same. Mesh station 7 is able to authenticate candidate mesh station 6, and hence, mesh station 7 establishes a peer link with mesh station 6.

6.3.5   Airtime Metric

HWMP uses the airtime metric to assess a path that connects the destination. The metric reflects the amount of bandwidth resources consumed by transmitting the frame over a particular link. As shown in Equation 6.1, this radio-aware metric considers the link quality (e.g., frame error probability, data rate) and link usage. Frame losses due to exceeding the mesh TTL are not included in the F_{loss} estimate as they are not correlated with link performance. The metric is encoded as an unsigned integer in units of 0.01 TU, where the time unit (TU) is equal to 1024 μs. The destination mesh station selects the path with the smallest radio metric. Mesh stations continuously monitor their links to keep the metric up-to-date.

$$A = \frac{1}{1 - F_{\text{loss}}} \left[O + \frac{F}{R} \right], \tag{6.3}$$

where

O = Network access and protocol overheads

F = Frame length (bits)

R = Transmission rate (Mbit/s)

F_{loss} = Frame loss ratio.

6.3.6 Mesh Power Management

All beacon frames provide a time reference that is used for synchronization and power saving. A mesh station can manage the power activity level of its links. Power-saving mesh stations are either in light- or deep-sleep mode. Under the light sleep mode, the mesh station must switch to full power to listen to beacons from peer mesh stations. In the deep sleep mode, the mesh station only wakes up to transmit its own delivery traffic indication map (DTIM) beacons. The power-saving mesh station can be informed of buffered traffic during the awake period that follows the beacon. To do this, it relies on an active mesh station to perform mesh power mode tracking for each of its neighboring peer mesh stations, and deliver buffered traffic.

6.3.7 Layer 2 Congestion Control

Unlike APs attached to high-speed wired networks, the overall performance of a wireless mesh network may be dictated by the speed of the slowest link in the path. 802.11s mesh stations normally select the appropriate modulation and coding scheme (MCS) before the link is established. Clearly, the available buffer space in the mesh station should cater for the highest order MCS (corresponding to the maximum wireless rate) for a specified time period. 802.11s employs intramesh congestion control to provide layer 2 flow control and minimize unnecessary wireless bandwidth usage (e.g., packet retransmissions) caused by buffer overflow at the mesh stations. Intramesh congestion control consists of three main mechanisms: local congestion monitoring and congestion detection, congestion control signaling, and local rate control. A mesh station may send a congestion control request (CCR) message to inform its neighbors to select a transmission rate that is lower than the one specified in the CCR message. Consider the mesh topology in Figure 6.6. Suppose mesh station A sends a CCR message to mesh station B to lower the transmission rate. If mesh station B lowers its transmission rate to mesh station A, but maintains a higher input rate from mesh station C, mesh station B may potentially experience a buffer overflow because it cannot output relay packets at a rate that is fast enough. Thus, the lowest rate CCR should be sent to all neighboring mesh stations in the path.

6.3.8   Mesh Coordination Function

A mesh station may employ the optional mesh coordination function (MCF) for channel access. MCF consists of contention-based channel access (based on EDCA) and MCF controlled channel access (MCCA). Because MCCA is reservation based, it allows mesh stations to access the wireless medium at selected times with less contention than EDCA. There are two types of transmit opportunities (TXOPs) in MCF: EDCA TXOPs and MCCA TXOPs. The EDCA TXOP is obtained by a mesh station succeeding in EDCA contention. The MCCA TXOP is obtained by a mesh station gaining control of the wireless medium during an MCCA opportunity (MCCAOP), which is an interval of time reserved for data frame transmissions, by exchanging MCCA frames.

Mesh stations may employ MCCA management frames to make reservations for transmissions. The mesh station transmits an MCCA Setup Request frame to initiate a reservation and becomes the MCCAOP owner. Receivers of the MCCA Setup Request frame are the MCCAOP responders. The MCCAOP owner and the MCCAOP responders alert their neighbors of this impending MCCAOP reservation via an MCCAOP advertisement. During its MCCAOP, the MCCAOP owner obtains a TXOP by successfully contending for the wireless medium using EDCA. Because of its reservation, the MCCAOP owner experiences no interference from other neighboring mesh stations, which refrain from transmission during the reserved time periods. In order to use MCCA, a mesh station maintains synchronization with its neighboring mesh stations. Mesh stations that use MCCA employ a DTIM interval with a duration of $2^n \times 100$ TUs, where n is a non-negative integer less than or equal to 17. In addition, a mesh station will need to track the reservations of its neighboring mesh stations.

6.3.9   Mesh Channel Switching

The mesh channel switch may be triggered by the need to avoid interference to a detected radar signal or to reassign mesh station channels to maintain MBSS connectivity. When a mesh station switches the operating channel, it uses a vendor-specific channel switch protocol prior to channel switch execution. A mesh station may make use of the information in Supported Channel and Supported Regulatory Classes elements, as well as measurement results undertaken by the mesh stations to assist the selection of the new channel. When an MBSS switches from 20 to 40 MHz or from 40 to 20 MHz, a mesh station may need to perform path maintenance to find an optimized path. A mesh station that receives a channel switch announcement may choose not to perform the switch but move to a different MBSS. When a mesh station accepts a channel switch, it will adopt information received in the Channel Switch Announcement and Mesh Channel Switch Parameters element.

6.4   802.16 RELAY NETWORKS

Centralized wireless mesh networks are also known as wireless relay networks. The 802.16j task group was formed on March 30, 2006 to define a multihop relay (MR) network specification for operating frequencies below 11 GHz. Only the time division duplex (TDD) mode is supported. The specification was ratified on May 2009 after nine draft revisions. The 802.16j amendment is an optional deployment that allows relay stations (RSs) to extend the reach and coverage or enhance the performance of an 802.16 wireless access network via multihop layer-2 frame forwarding. It simplifies the complexity of the BS design considerably, thereby reducing cost and enabling rapid 802.16 network deployment. Because traffic is aggregated at the RS before being forwarded, signaling overheads are reduced, and bandwidth efficiency is improved.

In MR networks, the BS may be replaced by a multihop relay BS (MR-BS) and one or more RSs. The superordinate station can be either an MR-BS or RS, whereas a subordinate station can be either an RS or subscriber station (SS). The 802.16j specification also supports relay mobility and employs multiple antennas to enhance the spectral efficiency of the relay link. Each RS is under the supervision of an MR-BS. Traffic and signaling between the SS and MR-BS are relayed by one or more RSs. The SS may also communicate directly with the MR-BS. The specification supports fixed and mobile relays. The RS can support QoS, hybrid ARQ (HARQ), and mobile station (MS) network entry/handover. Virtual relay is achieved via a multicast RS group, which is controlled by the superordinate station. This employs multiple antennas to enhance the spectral efficiency of the relay link. A backward compatible frame structure supports both relay frames and legacy frames. The MS can benefit from such relay operation without awareness of the presence of the RS.

6.4.1   PHY and MAC Layer Extensions

The PHY includes extensions to OFDMA for transmission of PHY protocol data units (PPDUs) across the relay link between the MR-BS and the RS. The MAC layer includes signaling extensions to support relay functions, such as network entry (of an RS, and of an SS through an RS), bandwidth request, forwarding of PDUs, connection management, and handover. Two different security modes are defined. The centralized security mode is based on key management between an MR-BS and an SS. The distributed security mode incorporates authentication and key management between an MR-BS and a nontransparent RS and between an RS and an SS. An RS may be configured to operate either in the normal connection ID (CID) allocation mode, where primary management, secondary, and basic CIDs are allocated by the MR-BS, or in the local CID allocation mode, where the primary management and basic CID are allocated by the RS. The network management of RS employs a secondary management connection.

6.4.2   Scheduling Modes

Two different scheduling modes (centralized and distributed) are specified for controlling the allocation of bandwidth for a station or RS. In the centralized scheduling mode, the bandwidth allocation for an RS's subordinate stations is determined by the MR-BS. Conversely, in the distributed scheduling mode, the bandwidth allocation of an RS's subordinate stations is determined by the RS, in cooperation with the MR-BS. The MAC layer includes signaling extensions to support functions such as network entry of an RS, and of a station through an RS, bandwidth request, forwarding of PDUs, connection management, and handover.

6.4.3   Relay Modes

The standard defines the transparent and nontransparent relay modes with different transmit zones for the RS and BS. In general, these transmit zones require time gaps to partition the UL and DL subframes into orthogonal sets (e.g., BS to RS, RS to MS). A nontransparent RS can operate in both centralized and distributed scheduling mode, whereas a transparent RS can only operate in centralized scheduling mode. Nontransparent RSs transmit the DL frame-start preamble, FCH, MAP, or channel descriptor (DCD/UCD) messages. To do this, these relays employ a common transmit zone. The transparent mode is used to increase capacity. This is because all stations in the network are dependent on the BS for framing and synchronization information. In this case, a frame received by the RS from the BS is forwarded by the RS to the station in the next frame interval. Thus, this mode supports only two hops. On the other hand, the nontransparent mode can support two or more hops, and hence, can be used to improve the wireless coverage area. However, it is more complex since the RS is allowed to transmit synchronization information in the frame headers, which may interfere with BS synchronization. Clearly, a transparent RS may communicate with superordinate and subordinate stations using a single carrier frequency, whereas a nontransparent RS may employ one or more carrier frequencies. A dual-radio nontransparent RS can reduce the complexity of a single-radio nontransparent RS by allowing the RS to communicate with the BS using one radio and other stations using another radio. At the same, it can also increase the capacity of the system due to the additional bandwidth from a second radio.

6.4.4   Cooperative Relays

Cooperative relaying is supported but optional. It is possible to achieve diversity by sending appropriately coded signals across different MR-BS and RS transmit antennas during the transmission of a data burst to subordinate stations. The three modes of operation for cooperative relaying are cooperative source diversity, cooperative transmit diversity, and cooperative hybrid diversity. Cooperative source diversity employs multiple sources with the same signals and coordinated by an MR-BS, as shown in Figure 6.11. In this case, the transmitting antennas simultaneously transmit the same signal using the same time–frequency resource. Cooperative transmit diversity relies on multiple sources, each with a different space–time coded (STC) signal. The STC-encoded signals are transmitted across the transmitting antennas using the same time–frequency resource (Figure 6.12). Cooperative hybrid diversity is identical to cooperative transmit diversity except that at least one value for the antenna assignment is assigned to multiple physical antennas. In all cases, the timing

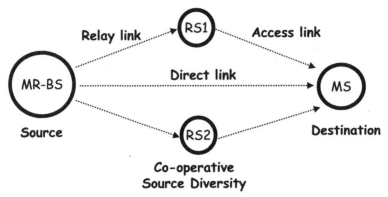

Figure 6.11   Cooperative source diversity.

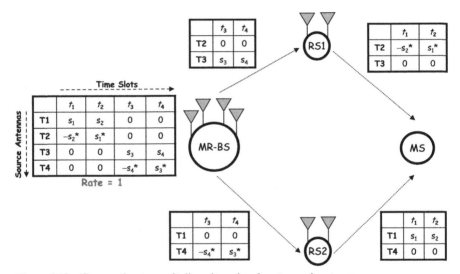

Figure 6.12   Cooperative transmit diversity using four transmit antennas.

difference between sources must be less than the duration of the OFDM cyclic prefix (CP). In addition, frames sent by the MR-BS and RS at a given frame time should arrive at the MS within the CP interval.

6.5 802.11 INTERWORKING WITH EXTERNAL NETWORKS

Like femtocells, 802.11u interworking is simpler than mesh or relay networks since it typically employs a single wireless hop. More importantly, 802.11u allows 802.11

wireless networks to be integrated with wireline networks in a coherent manner. In 802.11u, an AP can interact with external networks using a Subscription Service Provider Network (SSPN) interface for the purpose of authenticating users and provisioning services. The exchange of authentication and provisioning information between the SSPN and the AP passes transparently through the portal. This information is stored in the AP management information base (MIB), but the protocol to exchange this information is not defined. The SSPN interface provides the means for an AP to authenticate and authorize a non-AP station (e.g., a wired Ethernet switch or an IP router) from an external network. The SSPN also allows the non-AP station to access services in destination networks (DNs) other than the SSPN (e.g., Internet access via an IEEE 802 LAN).

The 802.11u network discovery and selection functions are:

- Discovery of suitable networks through the advertisement of access network type, roaming consortium, and venue information
- Selection of a suitable 802.11 infrastructure using advertisement services in the BSS or in an external network reachable via the BSS
- Selection of an SSPN or external network with its corresponding 802.11 infrastructure.

Other 802.11u interworking functions include:

- Emergency services
- Emergency call and network alert support at the link level
- QoS map distribution
- SSPN interface service between the AP and the SSPN.

The generic advertisement service (GAS) can be used by a station to support the network selection process. It also allows a non-AP station to communicate with other information resources in a network before joining the wireless LAN. The GAS enables stations to discover the availability of information related to desired network services. This may include information provided in local access services, available SSPNs, and other external networks. GAS uses a generic container to advertise network service information over an 802.11 network. Public Action frames are used to transport this information. 802.11u also supports emergency services. It is able to identify a traffic stream used for emergency services. The interworking service provides QoS mapping for SSPNs and other external networks. Since each SSPN or other external network may have its own end-to-end layer 3 packet marking practice, the QoS map service remaps the layer 3 service to a common over-the-air service. The SSPN interface supports service provisioning and transfers user permissions from the SSPN to the AP.

REFERENCES

[1] One Laptop per Child, http://one.laptop.org.
[2] Open 802.11s, http://open80211s.org.

[3] C. Perkins, E. Belding-Royer, and S. Das, "Ad-Hoc On-Demand Distance Vector (AODV) Routing," IETF RFC 3561, July 2003.

HOMEWORK PROBLEMS

6.1. Spatial frequency reuse in a wireless mesh network operating on unlicensed frequency bands is challenging because the links may have to be assigned dynamically depending on current interference conditions. Assuming a linear topology with 10 single-radio mesh routers, design a channel assignment strategy for all links that connect the routers. The assignment should minimize CCI between the links. The routers are connected sequentially in time. In other words, the first router connects to the second router before the third router is powered and connects to second router. Suppose one of the mesh routers encounters congestion and requires more bandwidth (e.g., 40 MHz instead of 20 MHz). How would you modify your channel assignment algorithm to accommodate this? Repeat for the case when a mesh router detects excessive interference or radar transmission and needs to switch to a different channel. If a single-radio mesh router switches to a different channel, are all other routers in the network required to switch channels too? Suppose all mesh routers are powered simultaneously and connectivity is required. How would you redesign your channel assignment strategy? How will your strategy scale to hundreds or thousands of routers?

6.2. The star-mesh topology possesses in-built path reliability. As shown below, the path redundancy is effective, but not excessive as in a fully connected mesh network. However, the center mesh router in the star mesh topology requires more channels and possibly more radios. For N mesh routers, compute the number of bidirectional links (served by two radios in each router) for the fully connected mesh and the star mesh topologies. Explain whether packet routing protocols are needed for these two topologies. How can packets be broadcast efficiently to all other mesh routers using a single radio?

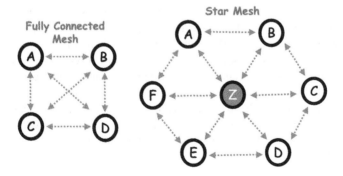

6.3. For the linear mesh topology, derive the minimum number of radios for each mesh router to enable full duplex US and DS transmissions in the mesh backhaul as well as concurrent communications with user devices in each wireless subnet. Will this number apply to 2-hop linear mesh topologies? CCI should be minimized. How can a mesh router decide whether to forward a packet to the US or DS mesh router? Repeat for the fully connected mesh and the star mesh topologies.

6.4. Figure 6.6 assumes LOS point-to-point US and DS transmissions between routers using directional beamforming antennas. Modify the figure to show omnidirectional point-to-multipoint wireless coverage in the US and DS transmissions and deduce the number of radios needed to provide full-duplex US/DS transmissions. When mesh router B transmits to user devices within its subnet (communications within a subnet is usually done by broadcasting), is it necessary to use a power level that is high enough to include subnets serviced by routers A and C? If an unstructured mesh network is composed of mesh stations involving user devices and no backhaul mesh routers, how many radios per station are required to enable full duplex communications while minimizing hidden collisions?

6.5. Consider a linear mesh topology with three mesh routers. List the pros and cons of establishing such a network as opposed to setting up two routers with a rate fall back to achieve the same range as the linear mesh network. Note that fewer user devices are typically served by each router if more routers are used. How effective are the two networks in defending against man-in-the-middle attacks?

6.6. In Figure 6.4, how will the network throughput be affected if each router in the two-radio mesh network needs to receive an entire packet correctly before forwarding to the neighboring router? What can you conclude about the usefulness of dual-radio systems compared with single-radio systems?

6.7. The figure below shows the benefit of packet fragmentation and pipelining in achieving concurrency on transmission over multiple packet routers.

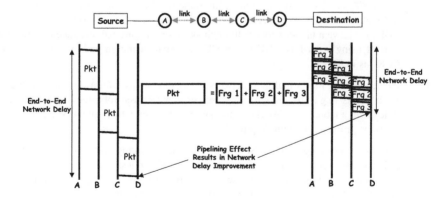

Taking into account the additional fragment headers and assuming no ACKs or other network overheads, verify the end-to-end delay in the figure below for varying number of routers and fragments. What can you conclude about the optimum number of fragments for a fixed number of routers?

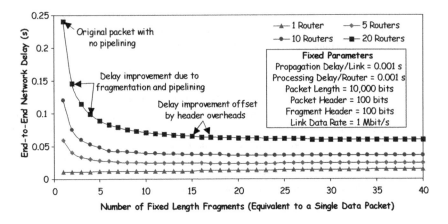

Suppose there is only one hop between the source and destination (i.e., no packet forwarding is needed). Compute the end-to-end delay for this case. Would you expect this delay to be always lower than the fragmented case? From this delay, compute the delay overhead in the packet forwarding process (for the two additional hops). Show that packet forwarding takes up significant bandwidth resources compared with the single hop case, even though packet routing overheads have been omitted in this computation.

6.8. Consider the use of the user datagram protocol (UDP) and the transmission control protocol (TCP) in a mesh network. TCP requires ACKs to be transmitted in the reverse direction to data packets, but these transport layer ACKs are different from those sent by the 802.11 protocol at the MAC layer. On the other hand, UDP does not require ACKs to be sent by the receiver, but corrupted UDP packets may still be recovered by the 802.11 MAC layer. Assuming the same number of data packets is transmitted and that there is no access contention at the MAC layer, would you expect TCP or UDP to achieve a higher network throughput? Using Figure 6.6, evaluate the pros and cons of using a per-link ACK (as in 802.11) versus using an end-to-end (multilink) ACK (as in TCP).

6.9. Would you say that a multihop network of mesh stations with a diversity of 802.11 amendments (e.g., 802.11a/g/n) possesses greater intrinsic robustness against a DoS attack than a mono (e.g., 802.11g only) system? Explain why rogue APs may not pose a serious problem in a multihop mesh network.

6.10. The open shortest path first (OSPF) is a link-state IP routing protocol (defined in RFC 2328) that was employed in earlier deployments of wireless mesh networks. The protocol is classified as link state because every router knows the topology of the entire network and needs to maintain a topology database (each router using itself as the root) to keep it up-to-date. Clearly, OSPF works for a structured mesh network where user devices are not involved in packet routing. OSPF is based on the Dijkstra routing algorithm. Consider the following mesh network with symmetrical US and DS links. The cost for each link is indicated. The routing table after three iterations is shown below, where *C* indicates the cost for the path. The goal is to locate the paths with the lowest cost. At the start of each iteration, the router that provides a new least cost path in the previous iteration is added to the list *L*.

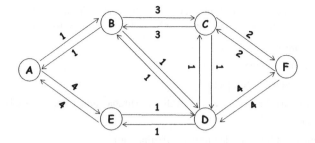

Iter.	L	P_{AB}	C_{AB}	P_{AC}	C_{AC}	P_{AD}	C_{AD}	P_{AE}	C_{AE}	P_{AF}	C_{AF}
0	A	A,B	1	–	∞	–	∞	A,E	4	–	∞
1	A,B	A,B	1	A,B,C	4	A,B,D	2	A,E	4	–	∞
2	A,B,D	A,B	1	A,B,D,C	3	A,B,D	2	A,B,D,E	3	A,B,D,F	6
3	A,B,D,C	A,B	1	A,B,D,C	3	A,B,D	2	A,B,D,E	3	A,B,D,C,F	5

The shortest paths from router A to all other routers are shown below. Derive the minimum number of radios in each router for full-duplex communications, assuming that links with a cost of two or more do not cause interference between subnets. Although the airtime metric can be used to compute the cost for each link in the mesh network, evaluate the pros and cons of using the physical distance to determine the cost instead.

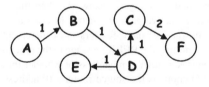

6.11. The AODV packet routing algorithm is a routing protocol designed for mobile ad hoc networks (RFC 3561). The key characteristics of AODV include:

- Finding routes only as needed
- Using monotonically increasing sequence numbers to track path freshness and prevent loop prevention
- Only keeping track of the next hop for a path instead of the entire path
- Discarding paths when they expire (which reduces the overhead for path maintenance)
- Implementing only a single path when multiple paths between source and destination are available.

AODV uses periodic HELLO messages to track neighbor routers and initiates the path discovery process when the source router sends a Route Request (RREQ) message. A Route Reply (RREP) is received from the destination router to confirm the path. For each valid path maintained by a router as a routing table entry, the router also maintains a list of precursors that may be forwarding packets on this path. These precursors are immediate neighbors of the router. The routing table fields used by AODV are:

- Destination IP Address
- Destination Sequence Number
- Valid Destination Sequence Number flag
- Other state and routing flags (e.g., valid, invalid, repairable, and being repaired)
- Network Interface
- Hop Count (number of hops needed to reach destination)
- Next Hop
- List of Precursors
- Lifetime (expiration or deletion time of the path).

The next hop in the path entry is assigned to be the router from which the RREP is received, which is indicated by the source IP address field in the IP header. When a router wishes to send a packet to a destination router, it checks its routing table to determine if it has a current path to the destination. If there is a path, it forwards the packet to the next hop router. If not, it initiates a path discovery process. Thus, the AODV protocol is only used when two endpoints do not have a valid active path to each other. AODV assumes all routers in the network are trusted.

Forwarding routers increase the sequence number for a given destination when forwarding RREP messages. Routers originating RREQ messages must increment their own sequence number before transmitting the RREQ. Destination routers increment their sequence numbers when the sequence number in the RREQ is equal to their stored number.

A Route Error (RERR) message is initiated to notify the network when the router detects a broken link for the next hop of an active path or when it receives a packet destined for a router for which it has no (active) path. Routers must increment the destination sequence numbers before transmitting the RERR message to the precursors. Routers must also mark affected path entries as invalid regardless of whether they are transmitting and/or receiving. The RREQ contains the source router's IP address and current sequence number, as well as the destination IP address and the destination sequence number. It also contains the broadcast ID number that is incremented each time a source router sends an RREQ. The broadcast ID and source IP address form a unique identifier for the RREQ. Broadcasting of control packets is done via flooding. The sequence numbers help to avoid the possibility of forwarding the same packet more than once.

Apply the AODV routing protocol to the fully connected mesh and star mesh networks, assuming that router A is the source and router D is the destination. Explain whether collisions are possible if the RREQ message is replicated as it travels along different paths, and these messages arrive at a common destination router at about the same time.

6.12. Explain whether there is a need to distinguish between relay packets (forwarded by one mesh router to another) and packets that are received by the router from user devices. Since a mesh station's total input rate is the sum of the relay and packet rates, should intra-mesh congestion control be applied to all packets in addition to relay packets? Explain whether the CCR message should be sent to all mesh stations in the path, and not just neighboring stations.

6.13. Evaluate the use of beamforming antenna arrays to reduce the number of radios and the interference levels in wireless mesh networks. Assuming a two-radio mesh network as depicted in Figure 6.4, will carrier sensing improve the performance of the beamforming antennas?

6.14. Peer-to-peer (p2p) communications is a scalable way of delivering broadcast content. It decreases the load on congested servers, minimizes distribution cost, and reduces downloading time for users. There are essentially two kinds of p2p network nodes. Peers are computers that download and upload data. Trackers are servers assisting peers to find other peers. Peers typically interact with the tracker using a simple protocol that resides on top of HTTP. BitTorrent (BT) is a leading open source p2p distribution protocol for p2p file sharing. To initiate a BT session, a Web server, a centralized tracker, and a seed node are activated. A .torrent file is generated at the server. A new peer joins the session by downloading the .torrent file and then contacting the tracker for a list of other participating peers. The peer then selects a set of neighboring peers from the list of peers provided by the tracker and then establishes connections with them to exchange file pieces. BT uses file fragmentation and pipelining to keep the HTTP connections at full capacity. Explain how p2p technology can be used to alleviate network congestion in a wireless mesh network.

6.15. Consider a ring of mesh APs that are interconnected via directional antenna beams in a wireless mesh network. Each AP serves a number of stations in a wireless subnet. Frame transmission is unidirectional, as shown in the figure below. As such, a routing protocol is not required, and in addition, only one US or DS channel is needed, thereby reducing overheads. A token is circulated around the ring. An AP that wishes to transmit replaces the token with a packet. The packet is repeated by all other APs in a round-robin fashion until it reaches the AP that is a recipient of the packet. This AP replaces the packet with an ACK. Again, the ACK is repeated by all other APs until it reaches the transmitter, which replaces the ACK with the token. Compare the advantages and disadvantages of such a network compared with using 802.11s to interconnect the APs. Explain whether the packet needs to include the addresses of the mesh APs. Suggest methods to optimize network performance by minimizing links and maximizing buffers.

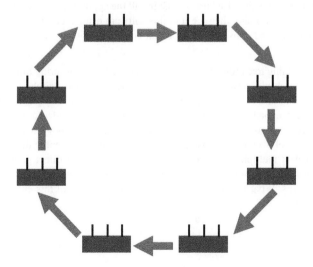

6.16. The 802.11 switch acts as a central wireless LAN controller for all connected APs. It validates all association requests, controls each AP's power and channel setting, and automatically detects any AP failures (and instructs nearby APs to compensate for the loss by adjusting transmit power and channel settings). New APs can be automatically discovered by the switch, which uploads the appropriate power and channel settings. Unauthorized (rogue) APs connected to the switch can be detected and proactively shut down. Encryption and authentication procedures are offloaded from the APs to the switch. The switch can perform 802.1X authentication, access control, policy configuration/management, personal firewall creation for users to control Web access, and can be connected to an aggregation switch, such as a Gigabit Ethernet switch. Transparent reauthentication can be achieved quickly, and policies can be enforced even as users move among different APs. Access policies can be applied against a variety of criteria, such as specific user name, a group of users, location, or time of day. With these capabilities, network administrators can provide guest users with only Internet access (via http, https, or IPSec), whereas employees have access to a wider range of TCP ports and services. Note than a wireless LAN switch is different from a wired switch in that it does not provide management and control functions on a port-by-port basis and does not provide dedicated bandwidth to the end user. Some 802.11 switches allow the station to associate with a single MAC address located at the switch, as opposed to multiple MAC addresses in multiple APs. Since the MAC address never changes, the architecture makes roaming easy because reassociation is unnecessary, and existing connections can be maintained when the user moves from one AP to the next. Rogue APs cannot connect to the network because it does not have the same MAC address as the switch. Explain whether mesh APs should include such centralized management capabilities. If implemented, which mesh network topology would be most suitable for performing switching functions? The front-end AP that is connected to the switch can be classified as lightweight since heavy processing is offloaded to the switch. Explain whether these lightweight APs are cheaper than regular APs.

6.17. Consider a mobile IP method (based on RFC 3344) to enable mobility when integrating wireless mesh networks with LTE or WiMAX networks. The protocol allows the mobile station to change its location without the need to restart applications or terminate and reestablish an ongoing connection. End users must install the appropriate software onto the data services they will use and then log into this software using their chosen passwords. Evaluate the pros and cons of this approach compared with layer 2 mobility management.

6.18. The IEEE 802.1D spanning tree protocol provides redundancy while ensuring loop-free connectivity among multiple network switches. Applying 802.1D to wireless mesh networks may result in bandwidth efficiencies resulting from unnecessary packet forwarding since some direct paths may be blocked by the root switch in order to preserve the unique paths (see left figure below). This may result in higher latency because the packet may need to travel over a longer path. A possible solution is to employ the IEEE 802.1S protocol to elect multiple root switches. In this case, alternate direct paths are available (see right figure below). Describe the pros and cons of adopting 802.1S in wireless mesh networks. What network topology will you choose? Will this eliminate the need for AODV?

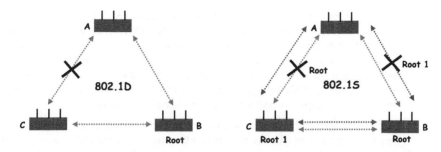

6.19. IP multicasting relies on a group management protocol (to establish and maintain multicast groups) and multicast routing protocols to route packets efficiently. Multicast routing protocols construct distribution trees in the network to forward multicast traffic. The paths defined by the distribution trees are based on source or shared trees. With a source tree, the root is located at the source, and branches form a spanning tree in the network. Source trees guarantee minimal latency but require more network resources since a separate branch is needed to reach every member of the multicast group. Shared trees use a single common root placed at some chosen point in the network. Multicast sources send their traffic to the root, which forwards traffic down the shared tree to all group members. Although shared trees make more efficient use of resources, optimal paths may not be used, which increases latency. How can the best of source and shared tree routing be adapted to densely and sparsely populated wireless mesh networks? Suppose a router receives multicast traffic that exceeds a predefined latency threshold. What can be done to improve the QoS in this situation? The optimized link state routing (OLSR) protocol (IETF RFC 3621) is a proactive routing protocol that uses the link state as a driving factor and includes a multicast capability: a subset of routers called multipoint relays provides anchors for neighbor routers. Link state information distribution can remain local, and multicasting is supported "naturally." The implementation of OLSR requires many different control messages, although it is more efficient than its flooding based protocols. Evaluate the pros and cons of employing a multipoint relay to distribute multicast traffic in a wireless mesh network.

6.20. Explain whether the path selection decision in a mesh network should be standardized or left to vendor implementation.

6.21. A user handset, such as a smartphone, is normally able to detect the signal from a BS and then perform necessary ranging operations so that the signal from the handset can reach the BS. Explain how outdoor 802.11 mesh networks may implement this feature, which is not available in the native standard.

6.22. Under what circumstances will an 802.11s mesh network require only three addresses in the data frame?

6.23. The IEEE 802.11 standard defines four authentication methods: open system authentication, shared key authentication, fast BSS transition (FT) authentication, and simultaneous authentication of equals (SAE). Open system authentication admits any station to the DS. Shared key authentication relies on WEP to demonstrate knowledge of a WEP encryption key. FT authentication relies on keys derived during the initial mobility domain association to authenticate the stations. SAE authentication uses finite field cryptography to prove the knowledge of a shared password. IEEE 802.11 authentication based on the robust security network association (RSNA) framework allows definition of new authentication methods. Thus, RSNA may support SAE authentication. RSNA also supports authentication based on preshared keys (PSKs) after open system authentication. RSNA does not support shared key authentication. SAE authentication can be used in an MBSS. SAE authentication is performed prior to association, allowing a mesh station to be authenticated by several APs simultaneously while still being associated to another AP. RSNA security can be established after association using the resulting shared key. Explain whether SAE can replace FT authentication.

6.24. Evaluate the pros and cons of using a layer 2 airtime metric instead of a layer 3 metric based on hop count in wireless mesh network routing. Since the native layer-3 AODV protocol uses a hop count metric, will its performance be degraded when the airtime metric is used instead? ■

CHAPTER 7

WIRELESS VIDEO STREAMING

Video traffic constitutes a significant percentage of mobile traffic, more than twice the volume of data and voice traffic. On-the-go video entertainment is rapidly becoming "cloud-based," allowing users to access on-demand video services at their convenience: anywhere, anytime, and on any device. Such services not only require higher capacities at the server and mobile backhaul, but also reliable network connectivity. Online TV (e.g., ESPN3) makes truly national or global events possible, reaching millions of consumers via handheld devices. For example, the 2012 Super Bowl attracted over 2.1 million unique viewers when the game was streamed online in the United States for the first time. All 302 events of the 2012 summer Olympics were also streamed live in the United States. The emergence of over-the-top service providers (e.g., Netflix, Hulu, and Amazon) offers more video choices to the consumer by providing replacement or supplementary TV services. Currently, Netflix has over 26 million subscribers, while Hulu handles over 30 million online users (over 2 million are paid subscribers) and over 1 billion video streams per month. Content distribution platforms to store, transcode, and deliver petabytes of video on commodity hardware are readily available. However, the popularity of mobile online entertainment creates many challenges for wireless service providers. In particular, 4G cellular networks will need to work in tandem with Wi-Fi hotspots in order to overcome the bandwidth crunch associated with video transmission. This chapter focuses on the techniques to efficiently transport high-quality videos across bandwidth-constrained wireless networks.

7.1 HIGH-DEFINITION AND 3D VIDEOS

YouTube has the largest library of both user-generated and premium videos. In May 2011, the number of views on YouTube hit 3 billion per day [1]. The first video posted on YouTube was a 19-second clip called *Me at the Zoo*. Today, more than 48 hours of video are uploaded every minute. These short-duration videos are perfect for small-screen smartphones. In addition to smartphones, many new tablets are able to capture 720p (1280 × 720 pixels) and 1080p (1920 × 1080 pixels) high-definition (HD) videos. For example, the new iPad tablet is able to display images at a resolution of 2048 × 1536 pixels, which is roughly 3.1 million pixels and over a million pixels more than a 1080p video. These tablets are like mini-TVs and are seeing much

Broadband Wireless Multimedia Networks, First Edition. Benny Bing.
© 2013 John Wiley & Sons, Inc. Published 2013 by John Wiley & Sons, Inc.

longer viewing times than smartphones. The use of autostereoscopic 3D displays is on the rise. The HTC Evo 3D smartphone now comes with an autostereoscopic 3D touchscreen [2]. The phone's 3D camcorder captures 720p videos (in addition to 2D 1080p videos), which are stored in full-resolution temporal format for easy preview and playback. These 3D videos can be uploaded and played back on YouTube. Many of these phones come with 802.11 and HDMI connectivity, allowing the HD videos to be streamed to a big screen TV or computer. Game consoles, such as Nintendo 3DS, now support autostereoscopic 3D displays [3]. In addition to entertainment and live captured 3D videos, Google launched anaglyph 3D videos in Google Street View, allowing users to see the streets in 3D.

7.2 VIDEO COMPRESSION

Video content is mostly compressed using lossy algorithms and then decompressed for viewing on a display. Sometimes, the decompression is performed for HD interfaces (e.g., HDMI). Lossy video compression results in higher efficiency (i.e., smaller compressed file size) compared with lossless compression (e.g., methods designed for data compression). The loss is essentially caused by a quantization process, which is determined by a quantization parameter (QP) in the video codec. A lower QP value reduces the loss, which results in a corresponding improvement in video quality (Figure 7.1). The amount of loss that can be tolerated depends on the resolution of the video and the loss perception threshold of the human visual system, which can vary from subject to subject.

If the original video (captured from the source) is compressed and then decompressed, the size of the decompressed video is the same as the original video. This is because the decompressed video is represented as 3 planes (i.e., Y, U, V) of color pixels (usually using 256 levels or 8 bits) for viewing, which is the same format as the original source video. However, the quality of the two videos may differ, since the decompressed video has undergone a lossy compression process. Many video capture devices, including professional video production equipment, will compress the source video to some degree. Thus, loss in video quality may be inevitable, although this loss is usually minimized by the capture device. Transport networks

(a) (b)

Figure 7.1   A lower QP leads to higher video quality. (a) QP = 20. (b) QP = 40.

may increase the loss further with more aggressive compression (e.g., using a higher QP value) in order to conserve bandwidth. On the other hand, HD interfaces, such as HDMI, send videos to the display in the decompressed (uncompressed) format.

It is important to distinguish between high-quality (HQ) and HD videos. An HD video may appear poorer than a HQ standard definition (SD) video if the quantization is very coarse. A HD video may be more difficult to compress than a SD video because more pixels need to be processed, even if the HD video is quantized with poor quality. Thus, many research articles have primarily focused on lower resolution videos because it takes a shorter time to generate the results. However, these results may not apply directly to HD videos. Since many touchscreens on smartphones and tablets are able to display HQ and HD videos, it is important to understand the effects of compression on these videos.

7.2.1 MPEG Standard

The ISO/IEC MPEG family of video coding standards are all based on the same general architecture, namely, motion-compensated temporal coding coupled with block-based discrete cosine transform (DCT) spatial coding. They employ hierarchical video coding to support interoperability between different services and to allow receivers to operate with different capabilities (e.g., devices with different display resolutions). The MPEG standard also defines requirements that are relevant to network transport. For example, random access of frames by the user at the receiver is provided with a maximum access time of 500 ms. This sets a limit on the round-trip network propagation delay.

MPEG exploits the spatial and temporal redundancy inherent in video sequences. The temporal sequence of MPEG frames consists of three types, namely intra-coded (I) frames, predicted (P) frames, and bidirectionally interpolated (B) frames. These three frame types aim to strike a balance between random access of frames and compression efficiency. I-frames provide reference points for random access but suffer from reduced compression efficiency when compared with the P- and B-frames. P-frames are coded using motion-compensation prediction techniques with reference to either I- or other P-frames. Because B-frames are interpolated from past and future reference frames, they provide the best compression efficiency. The interleaving of these frames in a video sequence is application dependent. For example, for video conferencing applications, fewer I-frames are used since there is little motion in the video. A sequence such as IBBPBBPBBI may generate a larger number of successive packets containing each I-frame, whereas the B-frames require the least number of packets for network transport. Motion compensation requires the frames to be divided into macroblocks (MBs), which is a group of adjacent pixels, usually 16 × 16 pixels. Motion prediction is performed based on motion translation (not motion rotation). Motion vectors (MVs) associated with each MB are represented using differential coding with respect to adjacent blocks. The bit requirements for representing the MVs are further reduced using variable-length coding. The MVs are computed using a MB matching algorithm. The motion prediction error (resulting in a MB mismatch) is coded using DCT and variable length codes. MPEG reduces spatial redundancy using smaller blocks of 8 × 8 pixels. DCT is used for block-by-block compression. Each block of image pixels is converted to the frequency domain using DCT, producing 64 transform coefficients. High frequency components, such as those produced by differential coding of P- and B-frames, are quantized coarsely. I-frames contain information in many frequencies and therefore require fine-grained quantization.

7.2.2 H.264/MPEG-4 AVC Standard

H.264 or MPEG-4 AVC is currently the most popular video coding standard, especially for online video streaming. It succeeds the equally popular MPEG-2 standard. H.264 was developed by the Joint Video Team (JVT), a partnership between the ITU-T (SG16/Q6) Video Coding Experts Group (VCEG), and the ISO/IEC (JTC 1/ SC 29/WG 11) Moving Picture Experts Group (MPEG). The standard is also known as MPEG-4 Part 10 Advanced Video Coding (AVC). Motivated by the success of H.264/AVC, MPEG and VCEG are working together again to develop a new H.265 or High Efficiency Video Coding (HEVC) standard. HEVC provides a significantly higher coding performance than H.264/AVC, especially for HD videos. It includes technologies to enable parallel encoding/decoding and to simplify implementation. Given its large and hierarchical block structure and significantly increased number of coding parameters, HEVC presents many new challenges, including coding optimization, mode decision, rate control, hardware design, and error concealment. The typical video coding efficiencies of the various standards are depicted in Figure 7.2.

7.2.3 Constant Bit Rate and Variable Bit Rate Videos

Compressed video can be further categorized under constant bit rate (CBR) and variable bit rate (VBR). In general, CBR encoding may result in variable video quality due to rate caps imposed by the encoder's output. These rate caps are usually constrained by the network requirements (e.g., a compressed CBR video stream can

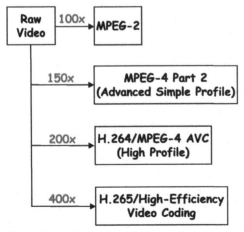

Figure 7.2 Efficiencies of video coding standards.

be capped to about 1.5 Mbit/s to fit into a T1 line rate of 1.536 Mbit/s). On the other hand, VBR encoding leads to constant video quality (due to a fixed QP value), but the bit rate of a single stream can be highly variable and bursty due to scene changes and motion that require more bits to be generated per frame than "talking heads" with little motion. Multiplexing of several VBR streams can help smooth the overall bit rate variation of the aggregated output.

7.3 VIDEO STREAMING INTERFACES AND STANDARDS

The 802.11aa amendment deals with multimedia streaming over 802.11 networks [4]. With the high data rates offered by 802.11n, there is a strong inclination to extend the convenience of wireless connectivity to multimedia streaming applications. In order for 802.11 to become a viable alternative to wired connections, several challenges need to be overcome in order to meet user expectations. Multimedia applications, in particular, video applications, require the network to satisfy two key requirements:

- Timely delivery of payload, which is usually large in size.
- Sufficient medium time for transmission and/or retransmission of the payload to cope with the inherent unreliability of the medium.

Multimedia streaming scenarios vary widely depending on the content (e.g., audio, video, voice, and data), the capabilities of the source and sink devices, and the location of these devices. In some cases, the content gets delivered to multiple sinks (video to TV display and audio to one or more speakers, some wired and others wireless). As a result, multimedia streaming is challenging and is much more complex than data and voice applications. The performance of a multimedia application depends on many subsystems end-to-end, and the 802.11 link is just one of them. Achieving the highest throughput may not always translate into the best multimedia performance. If a suitable data rate has been determined, minimizing packet loss and error rates and hence, the delay, may assume greater importance. 802.11aa focuses on the PHY and MAC layers to facilitate streaming applications. Some of the key features are described as follows.

7.3.1 Robust Multicast

802.11 multicast is not reliable since there is no acknowledgment (ACK) mechanism to ascertain that all members of the multicast group successfully received the transmission. 802.11aa specifies three multicast mechanisms: directed multicast, unsolicited retries, and extended block ACKs.

7.3.2 Prioritization

When there are multiple multimedia streams in a BSS, a mechanism is needed to prioritize between them. Within a compressed video stream, not all information carry the same importance. For instance, an I-frame is more important than a B-frame because the loss of an I-frame can have a higher impact on the quality of the rendered video. A mechanism is needed to tag packets within a stream as well as between streams to identify their relative importance. This priority information needs to be preserved end-to-end.

7.3.3   Overlapping BSS Management

802.11 provides two medium access mechanisms: enhanced distributed channel access (EDCA) and hybrid coordination function controlled channel access (HCCA). Devices contend for medium access with EDCA and are allocated contention-free medium access opportunities with HCCA. EDCA admission control is an extension of EDCA where 802.11 stations advertise their load and if enough resources are available, get corresponding medium time allocated by the AP. Both EDCA admission control and HCCA allocations are local to a BSS and is not shared with neighboring BSSs. As a result, resource allocations within a BSS get invalidated when one or more neighboring BSS operates in the same channel. Mechanisms to advertise BSS load and share the available medium time across BSSs are required.

7.3.4   Interworking with 802.1AVB

Multimedia streaming involves streaming content over a variety of wired and wireless networks. The IEEE 802.1 Audio Video Bridging (AVB) Task Group (http://www.ieee802.org/1/pages/802.1ba.html) produces a set of protocols (802.1Qat, 802.1Qav, 802.1AS, and 802.1BA) designed for end-to-end multimedia streaming. For example, 802.1BA is a suite of standards that enable audio and video streams over Ethernet LANs. Only 2 ms of buffering (maximum delay over seven hops) is required to support the most time-sensitive traffic, and network congestion will not cause dropping of stream packets. 802.1Qat deals with reservation and maintenance of QoS for a multimedia flow [5]. 802.1Qat deals with up to seven hops between the source and the sink, with up to two of the hops being 802.11. 802.1Qat employs a Stream Reservation Protocol (SRP) to reserve end-to-end resources. It enhances Ethernet by implementing admission control. SRP defines the concept of layer 2 traffic stream descriptors and a signaling mechanism for end-to-end management of the stream resources so as to guarantee quality of service (QoS). Unlike the resource reservation protocol (RSVP), SRP works for multicast streams. Some 802.11 mechanisms need to be extended to meet the needs of 802.1Qat SRP. Once the reservation is setup, periodic QoS maintenance reports are monitored to ensure that the required QoS is maintained.

7.3.5   Higher Layer Factors

A lack of sufficient buffering in layer 3 may cause packets to be dropped at the receiver despite the fact that they were delivered correctly by the lower layers. While the buffer sizes chosen for layer 3 may be appropriate for data/voice applications, the same setting may be suboptimal for multimedia applications. A mechanism is needed to determine and set this parameter to best fit multimedia applications. Additional buffering at the application layer can be used especially in cases where the streamed content is not real time. With such buffering, excessive delay and lost packets can be mitigated. Mechanisms exist to determine this parameter when operating over a wired connection. These mechanisms need to be adapted for operation over 802.11 links.

Video decoders can conceal many packet errors, including lost packets. This is because there is redundancy in the compressed video frames. If the payload size is set to 1500 bytes, seven 188-byte multimedia transport stream (TS) packets can be packed into a single 802.11 packet. With 802.11n, more TS packets can be aggregated. Aggregation in 802.11 is a key feature to reduce channel access overheads and improve channel utilization. Error concealment methods need to be revisited in the context of aggregation, and optimal aggregation for multimedia applications should be determined. In addition, enabling robust coding schemes, such as low-density parity-check (LDPC) coding, can allow longer packets to be sent, thereby improving multimedia performance.

Timely feedback from the decoder to throttle the encoder improves the quality of the rendered content, especially for real-time streams. The encoder may be generating a certain output stream based on some assumptions about the conditions at the decoder. Feedback on changes will help the encoder adapt to generate the best output given the conditions in the near-past.

Application layer tagging may be needed. Some multimedia applications do not take advantage of the MAC layer prioritization. Unless the 802.11 MAC performs some deep packet inspection to determine packets from a multimedia stream, the appropriate priority is not assigned to the packet. As a result, multimedia packets compete with time-insensitive application packets, which give rise to poor multimedia performance.

7.3.6 Digital Living Network Alliance

The Digital Living Network Alliance (DLNA) [6] has developed open technical specifications to ensure interoperability among digital TVs, STBs, Blu-ray players, and other handheld devices (e.g., tablets, smartphones) so that these devices can seamlessly stream multimedia content to each other over point-to-point wireless (e.g., 802.11) and wired (e.g., Ethernet and MoCA) connections. Such an approach for sharing media content is clearly more attractive and instantaneous than transferring all media to a central controller in a home network.

7.4 ADAPTIVE VIDEO STREAMING

Many 802.11 devices start with the best modulation and coding scheme (MCS), which corresponds to the highest data rate, and adjust this rate dynamically to cope with variations in signal quality and range. Such rate variations may be permissible for data/voice applications but can be detrimental for video delivery. Techniques that involve reducing the video frame rate (which reduces the data rate since less frames are sent) are not desirable since this may result in perceptible video quality degradation, especially when streaming live events, such as sports and music concerts. Loss-free protocols that ensure ordered delivery of packets are desirable for compressed video transport because these videos are highly sensitive to information loss. Unlike UDP, which does not add much to IP other than specifying the source and destination ports of the process and a checksum to detect corrupted packets, TCP is a connection-oriented transport protocol that guarantees delivery of packets. This reliability is achieved by first establishing a session and then retransmitting any corrupted or lost packets. HTTP is largely based on TCP. HTTP allows standard caching (which can scale to a large number of connections) and eliminates any issues with firewalls since it uses the same ports as Web browsing.

Adaptive bit rate streaming is widely deployed in HTTP video streaming systems (e.g., Adobe, Microsoft, Apple, and Move) and has been demonstrated by ESPN3 to be robust enough to broadcast over 10 million simultaneous live streams at the 2010 World Cup. Adaptive streaming is a combination of scalable-quality video coding and progressive download. The encoded video is typically fragmented into 2- or 10-second chunks (or "streamlets") with varying levels of video quality, each level requiring a specific bit rate. These chunks are then switched dynamically to adapt the video quality to fluctuating link quality, channel capacity, network congestion, video scene complexity, and end-user device resources. In doing so, serious video artifacts, such as choppy playback, freeze frames, and frame breakup, are minimized, while smooth playback at the video's natural frame rate is maintained. For example, when network bandwidth decreases in a lossy network, the client can request the server to deliver a chunk with a lower quality (hence, at a lower rate). On a managed network, since packet losses are not common, bit rate variation can be minimized by appropriately changing the quality of the chunks (Figure 7.3). Longer chunks can also be used, as opposed to the 2- to 10-second chunks typically used for operations over the public Internet. Because all chunks are precoded in advance, the average rate can be precomputed. This allows the server to send the appropriate video chunk encoded at a rate suitable for the connection without relying on real-time transcoding. Thus, the bit-rate scalability of adaptive streaming may reliably transport bandwidth-intensive HD and multiview 3D-HD video content over a variety of networks in the future. It is especially well suited for VBR-coded videos. In addition, the technology is a key enabler for efficient and cost-effective multiscreen video delivery to the TV, laptop, tablet, and smartphone as TV programs and the Internet converge over a variety of managed and unmanaged IP networks (Figure 7.3).

The key to the success of adaptive streaming is the ability to switch to the transmission of a lower quality video chunk that may not necessarily result in

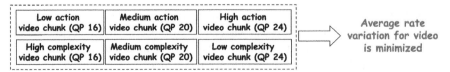

Figure 7.3   Minimizing bit rate variation using adaptive streaming.

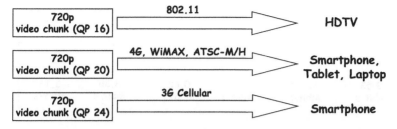

A higher quantization parameter (QP) value implies more lossy encoding

Figure 7.4   Multiscreen framework for adaptive streaming.

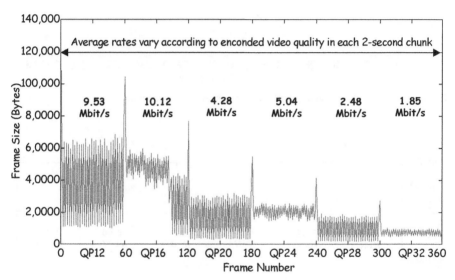

Figure 7.5   Average rate variation for six 2-second chunks. Peak frame size at the start of each chunk indicates the presence of an I-frame.

perceptible video quality degradation (Figure 7.4). In general, a QP value that ranges between 16 and 24 may not lead to perceptible differences in visual quality, although the difference in the rates can be substantial (Figure 7.5). However, a chunk encoded with QP 40 can lead to perceptible video quality degradation (Figure 7.6). In some cases, having a single quality level with one QP value may reduce the average bit rate variation (Figure 7.7) with no compromises to the video quality. It is therefore important to evaluate the performance of adaptive streaming under various network conditions and how this impacts video quality (and hence, end-user quality of experience), especially when transporting HD videos.

7.4.1   Video Quality and Chunk Efficiency

Many commercial adaptive streaming systems employ an intracoded I-frame at the start of each chunk, which may reduce the overall video compression efficiency and

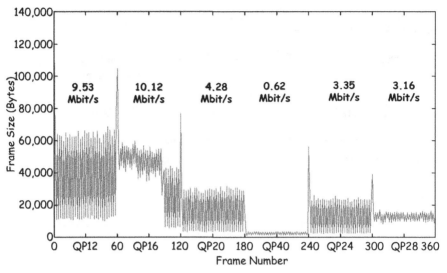

Figure 7.6   Perceptual visual quality degradation for QP 40 chunk.

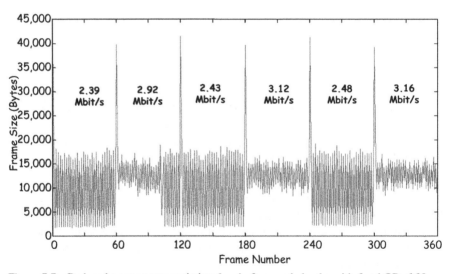

Figure 7.7   Reduced average rate variation for six 2-second chunks with fixed QP of 28.

lead to a high startup latency for each chunk. It also requires a customized video player at the receiver to estimate the bandwidth and request the appropriate chunk from the server. In this section, we analyze the efficiency and bit rates of the chunks generated by an encoder/fragmenter. Several SD and HD VBR videos were tested and they are listed in Table 7.1.

TABLE 7.1   Sample VBR Videos

Name	Resolution	Frame rate (Hz)	Duration (second)	Average bit rate (Mbit/s)	Content complexity
Water	$1280 \times 720p$	30	12	2.463	Moderate activity with no scene change. Moderate to high scene complexity with bridge and water subjects sandwiched between blurred foreground and background.
FCL	$1280 \times 720p$	24	72	0.957	Moderate activity with five scene changes. High scene complexity.
Office	$1280 \times 720p$	30	20	9.711	No motion, moderate complexity.
300	$1280 \times 720p$	24	106	4.113	Moderate activity but high number of scene changes and includes audio.

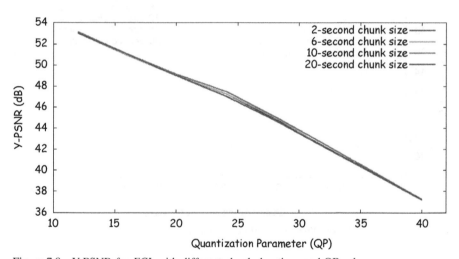

Figure 7.8   Y-PSNR for *FCL* with different chunk durations and QP values.

7.4.2   Video Quality for Different VBR Chunk Durations

In Figures 7.8–7.10, the peak-signal-to-noise ratio (PSNR) for each frame is averaged over the entire VBR video. The PSNR compares the maximum pixel value (the desired signal) with the difference in the pixel values of the original uncompressed and decompressed videos (the noise) [7]. Although there are three components in the PSNR computation (Y, U, and V), the human eye is most sensitive to the luminance (Y) component. Hence, the Y-PSNR is the more popular metric and is used

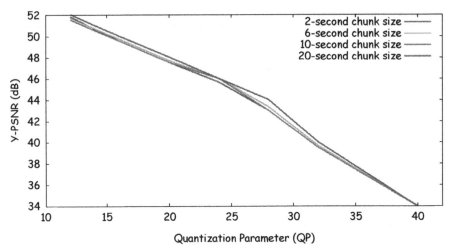

Figure 7.9   Y-PSNR for *Water* with different chunk durations and QP values.

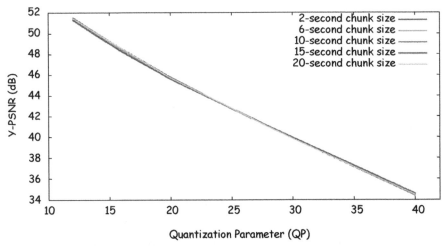

Figure 7.10   Y-PSNR for *Office* with different chunk durations and QP values.

in the figures. As can be seen, chunk duration has little impact on the video quality. However, the QP value should be carefully calibrated to obtain an acceptable Y-PSNR. The *Water* video suggests higher Y-PSNR variation for more "challenging" content.

7.4.3   Chunk Rate versus Chunk Duration

The average chunk rate in bits per second (bit/s) is the ratio of the chunk file size (in bits) and the chunk duration (in seconds). The chunk duration is the ratio of the number of video frames in the chunk and the frame rate. As can be seen from Figures

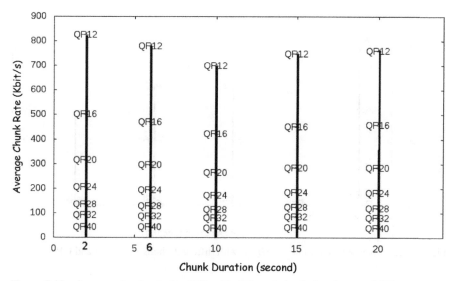

Figure 7.11   Average chunk rate for *FCL* with different chunk durations and QP.

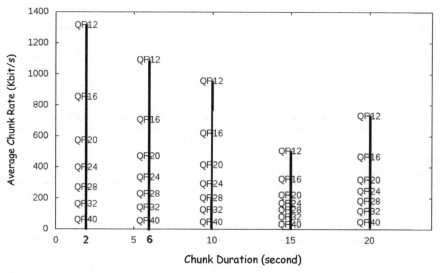

Figure 7.12   Average chunk rate for *Water* with different chunk durations and QP.

7.11–7.13, shorter chunk sizes generally lead to higher average chunk rates, and this is primarily due to the I-frame at start of each chunk. The bit rate variability becomes lower for chunks encoded with high QP values (i.e., lower quality chunks). Content clearly plays a role since any chunk may contain high activity or video complexity, resulting in a higher average chunk rate. The *Office* video contains no motion, and thus, removes the dependency on content. It proves that longer chunks are less

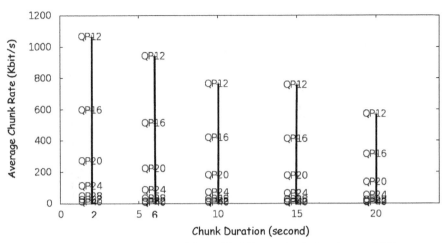

Figure 7.13 Average chunk rate for *Office* with different chunk durations and QP.

TABLE 7.2 VBR Chunk Efficiency Comparison for Different Chunk Sizes (*Water*)

Quantization parameter (QP)	Total size for all 2-second chunks (Mbyte)	Total size for all 10-second chunks (Mbyte)	Unfragmented video (Mbyte)
20	9.138	8.241	6.335
28	4.275	4.020	3.649
40	0.960	0.917	0.912

TABLE 7.3 VBR Chunk Efficiency Comparison for Different Chunk Sizes (*Office*)

Quantization parameter (QP)	Total size for all 2-second chunks (Mbyte)	Total size for all 10-second chunks (Mbyte)	Unfragmented video (Mbyte)
20	5.927	5.380	5.196
28	1.003	0.770	0.688
40	0.181	0.125	0.105

efficient than shorter chunks. In this case, the average chunk rate is almost constant for QP 28 and above.

7.4.4 Chunk Efficiency versus Chunk Duration

In this section, we compare the chunk efficiency (in terms of the chunk file size) for different chunk durations using different encoding modes. Tables 7.2 and 7.3 show the case for H.264 VBR video chunks. Clearly, longer chunks lead to higher efficiency (i.e., smaller file size), and this is independent of the content. Significant savings are achievable for videos with no motion (72%) or encoded with low

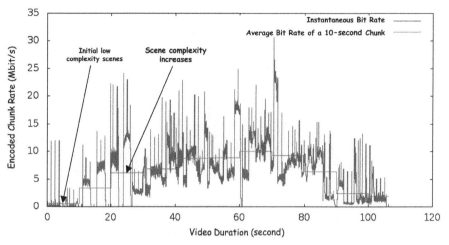

Figure 7.14   Average bit rate for 10-second chunks (*300* encoded with QP 20).

QP values (42%). These results confirm the observations obtained in the previous section.

7.4.5   Instantaneous and Average Rates for Different Chunk Durations

Fragmenting and buffering chunks may result in significant differences in the instantaneous and average rates. These rates are independent of the available network bandwidth and can be used to determine video scene complexity and estimate bandwidth requirements. The instantaneous rate is ratio of the size of the encoded frame (in bits) and the frame interval (in seconds). For a 30 Hz video, the frame interval becomes 33.33 ms. The average chunk rate is related to the minimum amount of buffering needed for adaptive streaming. As seen from Figure 7.14, the instantaneous rate can be three times higher than the average chunk rate. This difference reduces as the chunk duration reduces (Figure 7.15). Although there are typically more overheads for 2-second chunks (since more I-frames are required for 2-second than 10-second chunks), the instantaneous rates match the corresponding rates for the 10-second chunks quite closely. However, a higher bit rate variation is observed for 2-second chunks compared with 10-second chunks. Figure 7.16 illustrates the instantaneous rates for the unfragmented video, which incurs the least overheads. By comparing with Figures 7.14 and 7.15, it is quite evident that the instantaneous rates are closely matched, which implies that the fragmentation of the encoded video into smaller chunks did not introduce excessive overheads.

7.4.6   Wireless Live Streaming

The following sections describe a systematic evaluation on the performance of adaptive streaming for supporting SD and HD video transmission over 802.11 and 802.16

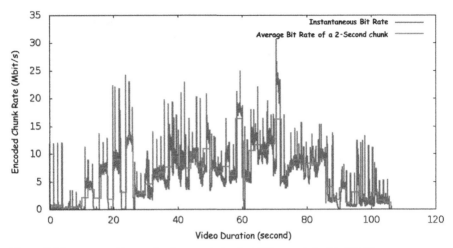

Figure 7.15 Average bit rate for 2-second chunks (*300* encoded with QP 20).

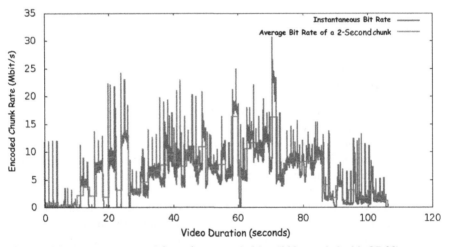

Figure 7.16 Instantaneous rate for unfragmented video (*300* encoded with QP 20).

wireless networks. The experimental setup for Apple's live streaming is shown in Figure 7.17. The iPad video playback is performed using QuickTime on HTML5. We ensure that sufficient memory resources are available on the iPad tablet to avoid unintentional switching to low quality chunks. The *300* movie trailer (720p, 24 Hz, 106 seconds) was used, and the H.264 video was preencoded with QP 20 (high quality) and QP 40 (low quality). A demo is available in Reference [8].

Because the video contains a series of dark scenes followed by some higher complexity scenes (see Figure 7.14), a switch in quality from QP 20 to QP 48 was initiated for Figure 7.18 when the network bit rate is low. This implies that content plays a role in the chunk switching: when high complexity scenes appear, lower

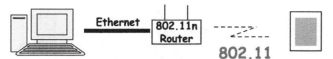

Figure 7.17   Experimental setup for adaptive streaming over 802.11.

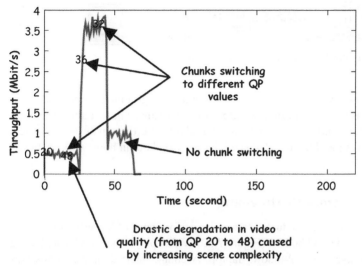

Figure 7.18   Chunk switching shown by changes in QP values.

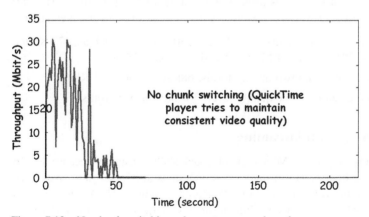

Figure 7.19   No chunk switching when rates are unshaped.

quality chunks are requested by QuickTime to reduce the chunk bit rate. The Quick-Time player bandwidth estimator works better for detecting a bit rate increase. After the switch to QP 36 and then QP 32, chunks are received at a very fast rate. Hence, there is no need to switch to a lower QP even when the bit rate reduces later. Figure 7.19 illustrates when high bit rates are available and no chunk switching takes place. In this case, as the rates vary rapidly, many chunks are already buffered and

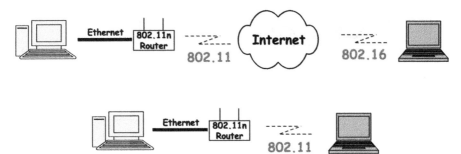

Figure 7.20 Experimental setups for smooth streaming.

switching to a lower QP is unnecessary. Although the high initial request rate may not be desirable as the network supports more video streams, it helps reduce instances of chunk switching under high bandwidth variability. Further analysis of Apple's live streaming over the iPhone and 4G cellular is carried out in homework problems 7.10-7.12.

7.4.7 Wireless Smooth Streaming

Several experiments were conducted to test the performance of Silverlight's smooth streaming over 802.11 and 802.16 (Figure 7.20). Video playback is performed using the Silverlight player on a Windows laptop. A chunk interval of 2 seconds was chosen in all the tests. At the HTTP server, rates are recorded at 1s intervals. The 802.16 service typically provides a maximum wireless rate of 3 Mbit/s. The *300* movie trailer (720p, 24 Hz, 106s) was used with varying levels of fragmentation:

- VBR video containing eight quality levels: 230 Kbit/s, 331 Kbit/s, 477 Kbit/s, 688 Kbit/s, 991 Kbit/s, 1.4 Mbit/s, 2.1 Mbit/s, and 3 Mbit/s;
- VBR video containing two quality levels: 688 Kbit/s and 331 Kbit/s;
- VBR video containing one quality level: 3 Mbit/s or 688 Kbit/s.

7.4.8 802.16 Smooth Streaming

When the video is capped at 3 Mbit/s, the playback stalls constantly because 802.16 cannot handle the high video bit rate. Streaming stops at 375 seconds, over three times greater than the video duration of 106 seconds (Figure 7.21). When the video is capped at 688 Kbit/s, playback stalling occurs less frequently because 802.16 can now handle the video bit rate, although streaming still stops at 122 seconds, 16 seconds greater than the video duration of 106 seconds (Figure 7.22). When the number of fragmentation levels is increased from 1 to 2, playback stalling occurs less frequently. Streaming stops at 120 seconds, 14 seconds greater than the video duration of 106 seconds (Figure 7.23). As the number of levels increases to 8 (Figure 7.24), playback becomes smoother, but the video quality degrades frequently. In this case, frequent chunk switching among the eight levels presents higher bit rate variability and higher rates compared to one and two levels. The rates are higher because

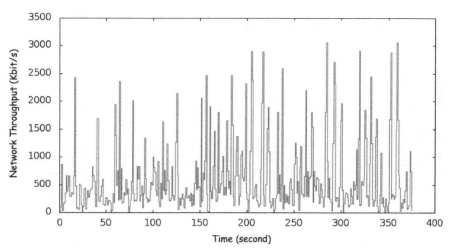

Figure 7.21   802.16 smooth streaming (1-level *300* video capped to 3 Mbit/s).

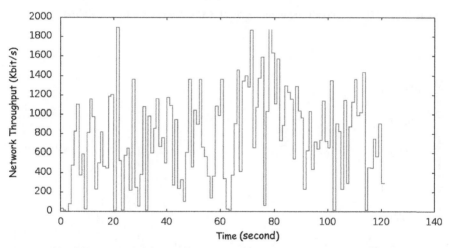

Figure 7.22   802.16 smooth streaming (1-level *300* video capped to 688 Kbit/s).

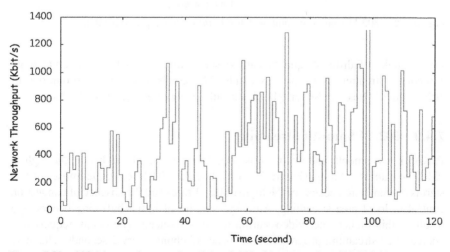

Figure 7.23   802.16 smooth streaming (2-level *300* video capped to 688 Kbit/s).

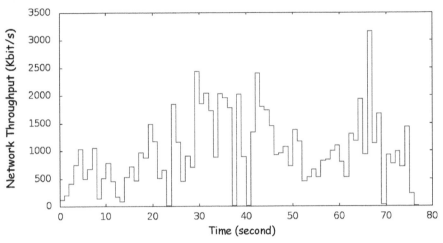

Figure 7.24   802.16 smooth streaming (8-level *300* video capped to 688 Kbit/s).

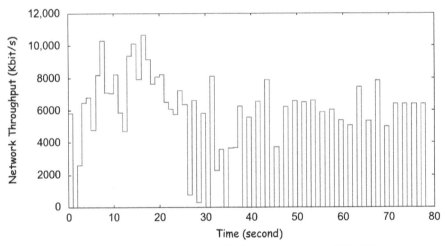

Figure 7.25   802.11 smooth streaming (1-level *300* video capped to 3 Mbit/s).

more chunks of different quality can be sent within a fixed time interval (e.g., 1 second). Streaming stops at 77 seconds, 29 seconds less than video duration of 106 seconds. Thus, the Silverlight player buffers about 29 seconds of video.

7.4.9   802.11 Smooth Streaming

Playback was smooth and quality was good because 802.11 can handle the highest video bit rate of 3 Mbit/s (Figure 7.25). Unlike the 802.16 case, the impact of fast start is evident here because of the higher 802.11 rates and the absence of congestion control in local area streaming. Although significant initial buffering occurs as a result of this higher rate, video startup delay is reduced. The chunk request rate decreases as streaming progresses: one 2-second chunk every 2 seconds. This may

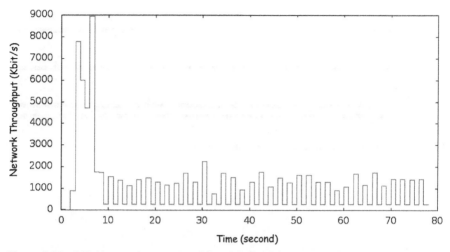

Figure 7.26   802.11 smooth streaming (2-level *300* video capped to 688 Kbit/s).

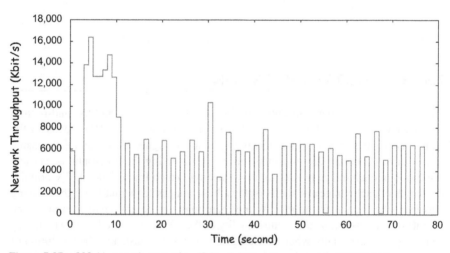

Figure 7.27   802.11 smooth streaming (8-level *300* video capped to 688 Kbit/s).

be important for scaling the HTTP streams to serve many concurrent connections. For two-level fragmentation capped at 688 Kbit/s (Figure 7.26), playback was also smooth and quality was good. Streaming stops at 78 seconds, 28 seconds less than the video duration of 106 seconds. For eight-level video chunks (Figure 7.27), playback was smooth and quality was good. In this case, 802.11 can handle the maximum video bit rate at the highest quality level. Higher bit rate variability and higher rates are observed compared with 1 and 2 levels. Streaming stops at 77 seconds, 29 seconds less than the video duration of 106 seconds. Once again, significant initial buffering occurs, and a lower chunk request rate is observed as streaming progresses. Figures 7.28 and 7.29 show the higher chunk request rates at the start and the subsequent slowdown.

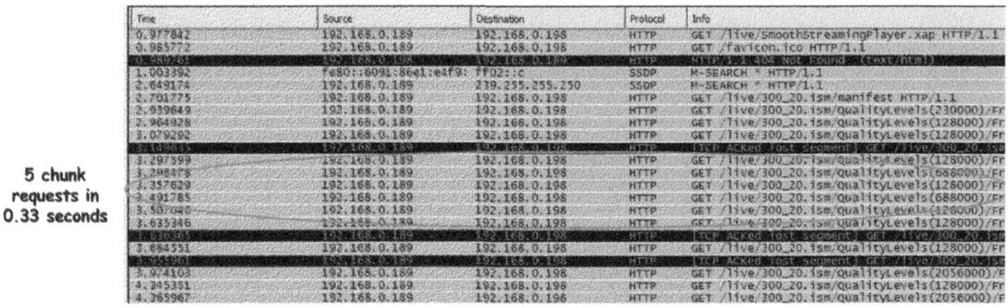

5 chunk requests in 0.33 seconds

Figure 7.28   Higher chunk request rate at the start of streaming.

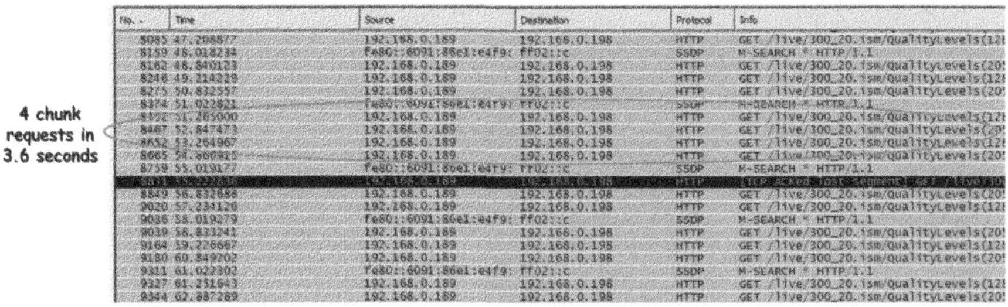

4 chunk requests in 3.6 seconds

Figure 7.29   Lower chunk request rate at the later part of streaming.

7.5   3D VIDEO TRANSMISSION

Pay TV service providers currently employ side-by-side (SbS) and top-and-bottom (TaB) view packing technologies to deliver stereoscopic 3D (S3D) videos to the home. These frame-compatible technologies are designed to consume the same bandwidth as standard 2D transmissions by subsampling the horizontal or vertical resolution of each image by half to fit into a single video frame. The 3D TV doubles the length of each image and then displays those images sequentially for the viewing glasses at the regular frame rate. While there may not be any visible artifacts, the overall spatial resolution and picture quality of the S3D video is reduced, which may be perceptible, especially when servicing HD videos. For instance, the resolution of a 1080p video (about 2 million pixels) is immediately reduced to a level similar to 720p (about 1 million pixels) with view packing. Conversely, the ability to increase the number of views without increasing bandwidth or compromising resolution enables new 3D services, as illustrated in Figure 7.30.

7.5.1   View Multiplexing

In general, temporal or spatial multiplexing of the views can be used to support 3D services without increasing bandwidth. In temporal multiplexing, the full spatial resolution of each view is preserved at the expense of reduced temporal resolution. On the other hand, spatial (frame-compatible) multiplexing preserves the temporal resolution but reduces the spatial resolution (Figures 7.30 and 7.31). View packing within a video frame is a necessary step to support spatial multiplexing. The packing

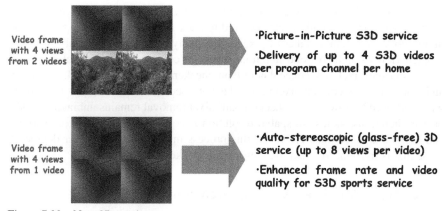

Video frame with 4 views from 2 videos

•Picture-in-Picture S3D service

•Delivery of up to 4 S3D videos per program channel per home

Video frame with 4 views from 1 video

•Auto-stereoscopic (glass-free) 3D service (up to 8 views per video)

•Enhanced frame rate and video quality for S3D sports service

Figure 7.30   New 3D services.

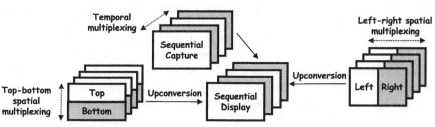

Figure 7.31   Spatial and temporal view multiplexing.

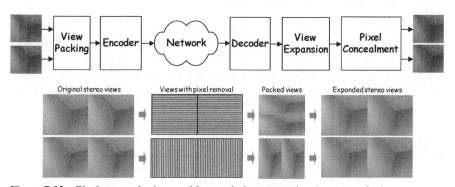

Figure 7.32   Pixel removal, view packing, and view expansion (upconversion).

method must be structured so that the views can be unpacked (expanded) easily without unnecessary bit overheads (Figures 7.32).

In SbS multiplexing, alternate columns of pixels in each view are removed and the remaining pixels packed together. In TaB multiplexing, alternate rows of pixels in each view are removed and the remaining pixels packed together. As an alternative, alternate pixels from each row of pixels can be removed, giving rise to a checkerboard (quincunx) pattern, which can be row or column packed. A single iteration of pixel removal is usually applied to pack two stereo views in a single frame. Two or more iterations are required to pack four or more views. Each iteration removes the same number of pixels, effectively reducing the spatial resolution of each view by half. An odd iteration of row or column pixel removal results in fewer adjacent pixels for concealment than checkerboard pixel removal. However, an iteration of row–column pixel removal is unbiased (i.e., removing pixels column-wise, followed by row-wise). Checkerboard pixel removal remains unbiased for odd or even iterations, as loss in spatial resolution is equally distributed in the vertical and horizontal directions. However, this process may result in a higher degree of noise that is introduced when the views are unpacked.

7.5.2 H.264 Multiview Coding Extension

A multiview coding (MVC) bitstream comprises a base view and one or more nonbase (enhancement) views. To improve coding efficiency, a nonbase view may utilize other views for inter-view prediction. Currently, the JM reference software [9] is the only open source code that implements the H.264 MVC extension. The software supports two or more views for S3D and autostereoscopic 3D displays. Although full-resolution views are more common, packed views can also be used. Views are input as separate videos into the encoder, which produces a single file with temporally multiplexed MVC-encoded views. The views are output as separate videos at the decoder. Several VBR HD videos encoded in H.264 are tested and they are listed in Tables 7.4 and 7.5.

7.5.3 MVC Inter-View Prediction

The basic concept of inter-view prediction is to exploit temporal motion tools, such as disparity vectors. This is enabled by the *MVCEnableInterView* option in the JM software. Figure 7.33a shows the simulcast H.264 coding of the left and right views of a S3D video using the IPB group of pictures (GOP) structure. Figure 7.33b shows the MVC coding of the left and right views of the same video. For simulcast H.264

TABLE 7.4 Sample VBR H.264 Videos

Name	HD resolution	Frame rate (Hz)	Duration (s)	Average bit rate (Mbit/s)
Magicforest	1920×1080p	25	636	7.717
Magicforest Left View	1440×1080p	25	636	6.467
Magicforest Right View	1440×1080p	25	636	5.042
Avatar	1280×720p	30	208	1.257
Forest	1280×720p	25	253	3.020
Coke	1280×720p	24	121	2.376
Flowers	960×720p	30	384	0.759
Fountain	1280×720p	30	67	6.750

TABLE 7.5   Sample VBR H.264 Videos

Name	HD resolution	Aspect ratio	2D or S3D	Content complexity
Magicforest	1920 × 1080p	16:9	SbS	Moderate activity. High scene complexity.
Magicforest Left View	1440 × 1080p	4:3	2D	Moderate activity. High scene complexity. Original base view of *Magicforest*.
Magicforest Right View	1440 × 1080p	4:3	2D	Moderate activity. High scene complexity. Original nonbase view of *Magicforest*.
Avatar	1280 × 720p	16:9	SbS	High activity. Moderate scene complexity.
Forest	1280 × 720p	16:9	SbS	Moderate activity. High scene complexity.
Coke	1280 × 720p	16:9	SbS	High activity. Moderate scene complexity.
Flowers	960 × 720p	4:3	TaB	Low activity. Low scene complexity.
Fountain	1280 × 720p	16:9	2D	Low activity. High scene complexity.

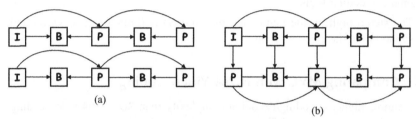

Figure 7.33   (a) Simulcast H.264 coding; (b) MVC coding.

coding, temporal redundancy (between frames) and intraview spatial redundancy (within a frame) are exploited to achieve coding gain. However, for MVC coding, in addition to temporal redundancy and intraview spatial redundancy, inter-view temporal redundancy between the views within the same frame is exploited to achieve higher coding efficiency. In order to reduce complexity, only the nonbase view is inter-view predicted using the base view. The base view is encoded in a similar way to single view H.264 encoding. This also helps existing H.264 decoders (not supporting MVC) to decode single views from an MVC-encoded 3D video, thereby maintaining backward compatibility.

Since the prediction is adaptive, the best predictor among the temporal and inter-view references can be selected on a MB basis. MVC makes use of the flexible reference picture management capabilities that are already available in H.264 by making decoded pictures from other views available as reference pictures for inter-view prediction. Specifically, reference picture lists are maintained for each picture to be decoded in a given view. Each list is initialized as usual for the single-view video, which may include temporal reference pictures used to predict the current picture. Additionally, inter-view reference pictures are included in the list and made available for prediction of the current picture. MVC does not allow prediction of a picture in one view at a given time using a picture from another view at a different time.

7.5.4 MVC Inter-View Reordering

Just as it is possible for an encoder to reorder the positions of the reference pictures in a reference picture list (including temporal reference pictures), MVC can also place inter-view reference pictures at any desired position in the list. The core MB level and lower-level decoding modules of the MVC decoder are the same, regardless of whether a reference picture is a temporal reference or an inter-view reference.

7.5.5 MVC Profiles

Currently, there are two profiles defined by MVC: the multiview high profile and the stereo high profile. Both profiles are based on the high profile of H.264 with the following differences:

- The multiview high profile supports multiple views and does not support interlace coding tools.
- The stereo high profile (supported by Blu-ray) is limited to two views and supports interlace coding tools.

7.5.6 Comparing MVC with H.264 Video Coding

MVC requires higher encoding/decoding complexity than 2D H.264 video coding because multiple views from different videos are encoded/decoded simultaneously in order to exploit spatial redundancy. However, as shown in Table 7.6, the improvement in coding efficiency for MVC is marginal when compared with H.264 (less than 1% for two-view S3D). The video quality is hardly affected. In addition to the higher bit overheads associated with MVC coding (due to the need for inter-view prediction), the key reasons for this phenomenon are discussed in the next section. Note that the combined encoded file sizes for the left and right views are similar for

TABLE 7.6 MVC versus H.264 for *Magicforest Left View* and *Magicforest Right View* (100 Frames, QP = 28, IPB GOP)

	MVC high profile[a]		H.264 high profile	
	Left view	Right view	Left view	Right view
Average Y-PSNR	35.72 dB	37.10 dB	35.72 dB	37.06 dB
Average U-PSNR	39.94 dB	40.30 dB	39.95 dB	40.27 dB
Average V-PSNR	42.87 dB	43.52 dB	42.87 dB	43.53 dB
I-coded bits	1,324,344	0	1,324,280	1,008,848
P-coded bits	10,640,320	7,327,864	10,649,728	6,532,576
B-coded bits	1,797,944	889,616	1,800,760	878,384
Total bits (1 view)	13,762,800	8,217,656	13,774,960	8,420,000
Total bits (2 views)	21,980,456		22,194,960	

[a]Multiview high profile and stereo high profile give the same results.

TABLE 7.7   2-Packed[a] MVC versus 2-Packed[b] and 4-Packed[c] H.264 for *Coke* and *Forest* (120 Frames, QP = 28, IPB GOP)

	MVC high profile		H.264 high profile		
	2-packed left views	2-packed right views	2-packed left views	2-packed right views	4-packed left and right views
Average Y-PSNR	37.25 dB	36.84 dB	38.98 dB	38.87 dB	37.20 dB
Average U-PSNR	44.09 dB	43.94 dB	44.89 dB	44.80 dB	44.05 dB
Average V-PSNR	46.32 dB	46.15 dB	46.48 dB	46.45 dB	46.25 dB
I-coded bits	384,296	0	561,072	596,792	795,208
P-coded bits	7,836,872	8,648,240	11,406,552	12,161,496	16,195,624
B-coded bits	1,033,632	567,216	4,538,224	4,857,096	2,141,432
Total bits (2 views)	9,254,976	9,215,624	16,506,024	17,615,560	
Total bits (4 views)	18,470,600				19,132,440

[a]Comprises 2 views per video frame, 640 × 720 overall resolution per frame.

[b]Comprises 2 views per video frame, 1280 × 720 overall resolution per frame.

[c]Comprises 4 views per video frame, 1280 × 720 overall resolution per frame.

both MVC and H.264. Table 7.7 shows that the gains in coding efficiency improve as the number of views per frame increases, or equivalently, as the video resolution for each view decreases (3.5% for four views). In addition, the H.264 file size for four-packed video is roughly the same as a two-packed video if the video resolution is the same.

7.5.7   Correlation between Left and Right Views in S3D Videos

Figure 7.34a shows that the correlation between two SbS views for *Avatar* and *Magicforest* (first 200 frames) is weak with less than 20% giving a Y-PSNR of 40 dB or more. The right (nonbase) view is then horizontally shifted one column of pixels at a time relative to a static left (base) view. The Y-PSNR of the right view is computed using the left view as a reference. This is repeated for up to 10 columns to the left and right of the static view. The maximum Y-PSNR improvement relative to the case when there is no shift (i.e., Figure 7.34a) is selected among the 20 Y-PSNR values. The results are shown in Figure 7.34b. Less than 10% of the frames for *Avatar* show an improvement of over 3 dB. For the higher resolution *Magicforest Left View* and *Magicforest Right View*, the correlation is the weakest (Figure 7.34c,d). These results suggest that inter-view pixel interpolation methods may lead to degradation in video quality, especially for higher resolution videos. In addition, these methods create dependency between the views and require more processing time

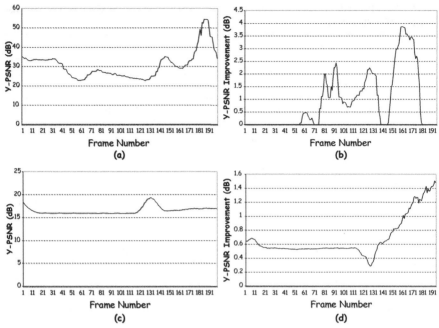

Figure 7.34 (a) Y-PSNR of right view when compared with left view for *Avatar*. (b) Maximum Y-PSNR improvement when shifting right-view of *Avatar*. (c) Y-PSNR of *Magicforest Right View* when compared with *Magicforest Left View*. (d) Maximum Y-PSNR improvement when shifting *Magicforest Right View*.

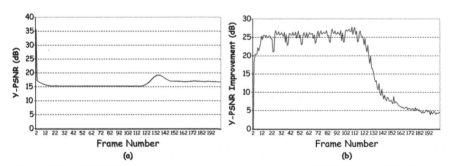

Figure 7.35 (a) Y-PSNR between current and previous frame of *Magicforest Left View*. (b) Maximum Y-PSNR improvement when shifting previous frame of *Magicforest Left View*.

compared with intraview methods. Thus, MVC may not provide significant coding efficiency when compared with H.264. The process of view shifting is repeated in Figure 7.35 except that this is conducted for successive frames, as opposed to left and right views. As can be seen, there is more correlation with the current frame when the previous frame is shifted. The drop in Y-PSNR in Figure 7.35b is due to a gradual scene change that is applied in a vertical fashion.

7.5.8   View Expansion via Pixel Interpolation

Pixel interpolation is a concealment technique that is employed by many HDTV vendors to improve the visual quality and frame rate of fast action movies and sports, even when the original video is captured at a lower frame rate. Many 3D HDTVs are able to convert 1080p 24 Hz SbS Blu-ray output to 240 Hz (120 Hz for each view) using pixel interpolation. Unlike frame repetition or pixel copy, using pixel interpolation to generate intermediate views or frames smoothes motion detail and enhances image clarity and sharpness. This can be done spatially (i.e., within a frame) or temporally (i.e., between frames). To illustrate spatial interpolation, we employ a MB of 8×8 pixels as shown in Figure 7.36. Spatial interpolation after one iteration of pixel removal is straightforward and can be directly applied to SbS and TaB S3D videos—simply use the nearest available pixels for interpolation. Spatial interpolation after two iterations of pixel removal is somewhat challenging. If the same checkerboard pattern is successively applied to the original and packed views, a pattern identical to row–column removal (Figure 7.37b) results. However, if complementary checkerboard is used, a biased pattern results. Figure 7.37a shows a new checkerboard pixel removal method where alternate columns of pixels are shifted vertically by one pixel. Figure 7.38 shows the results of concealing the pixels. In Figure 7.38a, the black cell uses the weighted average of the top and bottom neighbors for interpolating its value. The gray cell uses left, right-top, and right-bottom neighbors, whereas the white cell uses left-top, left-bottom, and right neighbors. Likewise in Figure 7.38b, the black cell uses left and right neighbors for interpolation. The gray cell uses top and bottom neighbors, whereas the white cell uses left-top, left-bottom, right-top, and right-bottom neighbors.

1	2	3	4	5	6	7	8
9	10	11	12	13	14	15	16
17	18	19	20	21	22	23	24
25	26	27	28	29	30	31	32
33	34	35	36	37	38	39	40
41	42	43	44	45	46	47	48
49	50	51	52	53	54	55	56
57	58	59	60	61	62	63	64

Figure 7.36   Original macroblock of 8×8 pixels.

(a)

	2				6		
			12				16
	18				22		
			28				32
	34				38		
			44				48
	50				54		
			60				64

(b)

1		3		5		7	
17		19		21		23	
33		35		37		39	
49		51		53		55	

Figure 7.37   Remaining pixels after two iterations of pixel removal (4× compression). (a) Checkerboard removal. (b) Row–column removal.

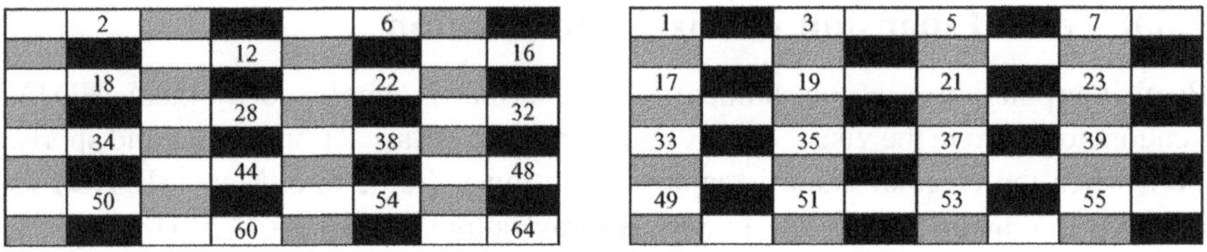

	2				6		
		12				16	
18				22			
		28				32	
34				38			
		44				48	
50				54			
		60				64	

(a)

1		3		5		7	
17		19		21		23	
33		35		37		39	
49		51		53		55	

(b)

Figure 7.38 Concealing pixels after two iterations of pixel removal. (a) Checkerboard removal. (b) Row–column removal.

TABLE 7.8 Checkerboard versus Row–Column Interpolation for *Flowers, Coke, Forest,* and *Avatar* (120 Frames Each)

Average Y-PSNR	Original left view versus interpolated left view (two iterations of checkerboard, 4× compression)	Original left view versus interpolated left view (row–column removal, 4× compression)
Flowers		
Y	51.82 dB (46.64 dB)	53.81 dB (46.64 dB)
U	56.92 dB (54.41 dB)	59.42 dB (54.41 dB)
V	58.24 dB (57.71 dB)	60.30 dB (57.71 dB)
Coke		
Y	41.72 dB (35.89 dB)	45.82 dB (35.89 dB)
U	54.51 dB (53.25 dB)	58.95 dB (53.25 dB)
V	55.32 dB (55.32 dB)	59.68 dB (55.32 dB)
Forest		
Y	37.36 dB (32.13 dB)	40.90 dB (32.13 dB)
U	50.84 dB (46.43 dB)	54.33 dB (46.43 dB)
V	52.95 dB (51.34 dB)	56.61 dB (51.34 dB)
Avatar		
Y	43.25 dB (41.70 dB)	43.97 dB (41.70 dB)
U	55.08 dB (51.75 dB)	56.15 dB (51.75 dB)
V	55.21 dB (52.17 dB)	56.42 dB (52.17 dB)

7.5.9 Pixel Interpolation Results

We compare the visual quality after two iterations of pixel removal in Table 7.8. As can be seen, row–column pixel removal consistently achieves better visual quality than checkerboard pixel removal (possibly due to more structured pixel removal and interpolation), although the overall average Y-PSNR for both methods remains high. Checkerboard pixel removal also results in less efficient H.264 coding. This can be confirmed by evaluating the H.264 file sizes for *Magicforest Left View* after one iteration of row, column, checkerboard (column-packed), and

TABLE 7.9   Interpolation Performance for *Avatar* (120 Frames)

Average Y-PSNR	Original left view versus interpolated left view (row removal, 2× compression) (dB)	Original left view versus interpolated left view (row–column removal, 4× compression) (dB)	Original left view versus interpolated left view (row–column removal followed by column removal, 8× compression) (dB)
Checkerboard			
Y	50.78	44.27	37.29
U	58.28	52.63	48.75
V	59.66	54.32	50.73
Row–Column			
Y	53.53	46.41	44.30
U	59.91	56.54	52.90
V	61.37	58.18	54.61

checkerboard (row packed) removal. The JM-encoded file sizes are 8,087,176 bits, 8,223,248 bits, 9,467,664 bits, and 9,165,680 bits, respectively. The experiment was repeated for *Fountain* using the H.264 encoder, and the sizes are 15,467,192 bits, 15,572,912 bits, 17,098,240 bits, and 17,085,912, respectively. Note that the row-packed video is most efficiently encoded, possibly because fewer rows are removed compared with the column-packed case. The Y-PSNR values in parentheses are obtained using column or row copy methods (i.e., no interpolation). These methods are currently employed when views are unpacked and expanded, and are clearly less superior than interpolation schemes. In this case, the Y-PSNR values are the same regardless of whether checkerboard or row–column pixel removal is performed during packing. Table 7.9 illustrates the performance of checkerboard and row–column interpolation under multiple iterations of pixel removal. As can be seen, up to eight times compression efficiency is achievable while maintaining a high average Y-PSNR. Again, row–column interpolation achieves better quality than checkerboard interpolation.

7.5.10   Inter-View versus Intraview Pixel Concealment

Inter-view pixel concealment methods may perform poorly when compared with intraview methods. To confirm this, we evaluate the performance of intraview pixel interpolation and inter-view pixel copy for *Magicforest Left View* and *Magicforest Right View*. We subject the base and nonbase views to one iteration of complementary information removal (e.g., even row pixel removal on left view and odd row pixel removal on right view). For intraview interpolation, two vertical neighboring pixels are selected. For inter-view pixel copy, the missing pixel from the base view

TABLE 7.10 Intraview versus Inter-View Interpolation (100 Frames)

Average Y-PSNR	Intraview interpolation (dB)	Inter-view interpolation (dB)
Magicforest Left View		
Y	31.00	19.12
U	44.87	31.93
V	49.79	39.98
Avatar base view		
Y	57.09	44.99
U	61.00	54.00
V	62.76	55.12

is obtained from the corresponding pixel in the nonbase view. A similar procedure is repeated for *Avatar* with one iteration of row removal. In this case, inter-view interpolation uses two vertical neighboring pixels from the base view, two horizontal neighboring pixels from the nonbase, as well as the pixel in the nonbase view corresponding to the position of the missing pixel in the base view. As seen in Table 7.10, the visual quality using inter-view pixel interpolation is severely degraded, again demonstrating the lack of direct correlation between the views.

7.5.11 Interframe versus Intraview Pixel Interpolation

Figure 7.35b suggests that interframe pixel interpolation may achieve better performance. Figure 7.39 shows the results obtained for the column-packed *Magicforest Left View* (one iteration). In Figure 7.39a, intraview interpolation is performed using the left and right neighboring pixels. In Figure 7.39b, the corresponding pixel from the previous frame is used to replace the missing pixel. As can be seen, the video quality improves over intraview interpolation until a gradual scene change occurs. The correlation with the previous frame weakens as a result of this scene change, reducing the effectiveness of interframe pixel copy. Figure 7.39c shows an integrated solution where the left and right pixels, and the corresponding pixel from the previous frame are averaged to predict the missing pixel's value. In this case, the video quality becomes more consistent.

7.5.12 Impact of Quantization on Interpolated S3D Videos

We perform one iteration of column removal for *Magicforest Left View* (resulting in an SbS 3D vdeo), encode and decode the video using H.264, and then apply spatial pixel interpolation to the decoded video. As shown in Figure 7.40, employing less aggressive quantization (e.g., using a low QP value) will not improve the video quality of S3D views but increase the video file size instead. This is because the video quality is dominated by pixel removal in view packing. Losing half the pixels in a view with each iteration of pixel removal can have a detrimental impact on video quality. Table 7.11 shows that for a similar Y-PSNR, a higher QP value of 34

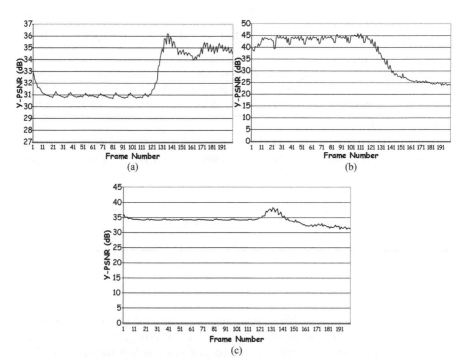

Figure 7.39 Concealing pixels after one iteration of column removal. (a) Intraview pixel interpolation. (b) Interframe pixel copy. (c) Interframe pixel interpolation.

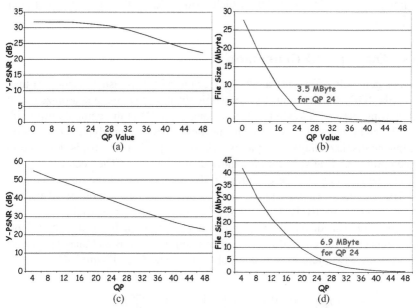

Figure 7.40 Assessing video quality and H.264 encoded file size for *Magicforest Left View* (10 frames). (a) Y-PSNR versus QP (interpolated video). (b) Encoded file size versus QP (packed video). (c) Y-PSNR versus QP (original video). (d) Encoded file size versus QP (original video).

TABLE 7.11 Impact of Quantization on Original and Column-Packed *Magicforest Left View*

Video	Resolution	Quantization parameter (QP)	Encoded file size (bits)	Decoded Y-PSNR (dB)	Decoded U-PSNR (dB)	Decoded V-PSNR (dB)
10 frames						
Original	1440 × 1080p	34	1,577,256	31.70	40.40	43.79
Column removed	720 × 1080p	24	3,403,104	31.19[a]	40.49[a]	43.81[a]
100 frames						
Original	1440 × 1080p	34	4,121,064	31.20	37.62	41.08
Column removed	720 × 1080p	24	16,912,128	30.47[a]	40.08[a]	43.10[a]

[a]Includes spatial pixel interpolation.

can be used for the original video, giving rise to an encoded file size that is two to four times smaller than the packed video. This is effectively four to eight times smaller since the number of pixels in the packed video is half the original. Thus, spatial multiplexing is significantly less bandwidth efficient than temporal multiplexing of S3D views. This is because the spatial multiplexing is more disruptive to the encoder than temporal multiplexing. This can be compared with the use of rate capping in some video encoders, which result in less efficient encoding or poor video quality.

7.5.13 Anaglyph 3D Generation

Anaglyphs generally offer wide-angle 3D viewing with less visual discomfort than S3D. Unlike S3D, which requires stereoscopic displays that are synchronized with active viewing glasses, a key advantage of anaglyph 3D is the compatibility with legacy TVs and regular displays. This means one can watch anaglyph 3D videos on a smartphone display using cheap passive glasses. Equation 7.1 [10] is designed to generate optimized anaglyph from the individual S3D views. The vector $[r_a, g_a, b_a]$ represents the red, blue and green color (RGB) components of the optimized anaglyph. The vectors $[r_1, g_1, b_1]$ and $[r_2, g_2, b_2]$ represent color components of the left and right views, respectively. For optimized anaglyph, there is partial color reproduction (not of red shades) and almost no retinal rivalry (which is caused by brightness differences of the colored objects). The interpolated TaB and anaglyph views for *Flowers* after two iterations of interpolation are shown in Figure 7.41.

$$\begin{bmatrix} r_a \\ g_a \\ b_a \end{bmatrix} = \begin{bmatrix} 0 & 0.7 & 0.3 \\ 0 & 0 & 0 \\ 0 & 0 & 0 \end{bmatrix} \begin{bmatrix} r_1 \\ g_1 \\ b_1 \end{bmatrix} + \begin{bmatrix} 0 & 0 & 0 \\ 0 & 1 & 0 \\ 0 & 0 & 1 \end{bmatrix} \begin{bmatrix} r_2 \\ g_2 \\ b_2 \end{bmatrix}. \qquad (7.1)$$

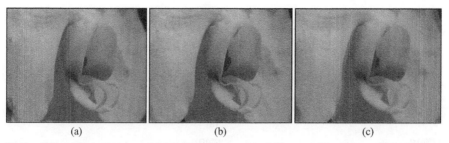

| (a) | (b) | (c) |

Figure 7.41   Interpolated snapshots for column-packed *Flowers* (4× compression). (a) Top view. (b) Bottom view. (c) Anaglyph.

TABLE 7.12   Optimized Anaglyph H.264 Coding Efficiency

	Left view (Bytes)	Right view (Bytes)	Optimized anaglyph (Bytes)	Overhead (%)
Forest (120 frames)	2,094,977	2,244,695	2,652,667	22.25
Coke (120 frames)	844,876	879,426	1,042,503	20.92
Flowers (120 frames)	199,663	193,719	247,114	25.64
Magicforest Left/Right Views (350 frames)	13,718,082	10,032,586	11,025,961	−7.15

7.5.14   H.264 Coding Efficiency for Anaglyph Videos

The H.264 coding efficiency for anaglyph videos is shown in Table 7.12. We employ an encoder with a QP of 28. The average size is computed as (left view size + right view size)/2. The relative overhead is then computed as (anaglyph size-average size)/ (average size) × 100%. Interpolation is first performed on the raw (YUV) video files of *Forest*, *Coke*, and *Flowers* to convert the packed views to the original resolution. The YUV file is then converted to RGB format for anaglyph generation, which is converted back to YUV for H.264 coding. As can be seen, there is a 20–25% increase in overheads if the interpolated views are used. However, the overheads are negative if the original views of *Magicforest* are used. Note that in this case, the H.264 encoded anaglyph file size can be predicted since it is roughly equal to the sum of 1/3 left view file size and 2/3 right view file size, consistent with Equation 7.1.

7.5.15   Delta Analysis

The use of delta values to represent the difference between the original pixel value and value obtained using interpolation can enhance the visual quality of the re-constructed video. In theory, the delta values can range from −255 to +255 depending on the value of the original pixel and value obtained using interpolation. In order to represent the range of values from −255 to +255, 9 bits are required, which implies more overheads than sending the actual pixel value! In practice, fewer bits are

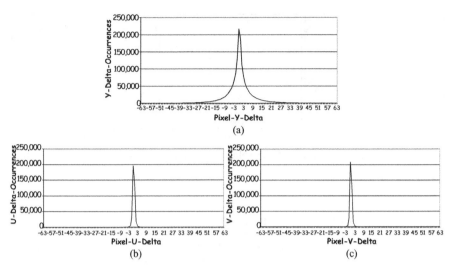

Figure 7.42   Y-, U-, and V-delta occurrences for *Magicforest* (100 frames). (a) Y-delta occurrences. (b) U-delta occurrences. (c) V-delta occurrences.

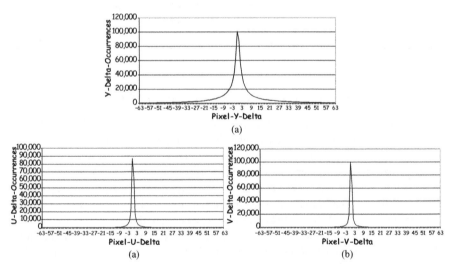

Figure 7.43   Y-, U-, and V-delta occurrences for *Fountain*. (a) Y-delta occurrences. (b) U-delta occurrences. (c) V-delta occurrences.

required due to correlation between adjoining pixels. For example, Figure 7.42 shows the delta occurrence results for the Y, U, and V components of the base view of *Magicforest*. 8 bits are required to represent the exact pixel values of the Y component (delta values range from −66 to +74), but only 5 bits are required for the U and V components (delta values range from −11 to +14 and from −9 to +10 respectively). Figure 7.43 shows the results for *Fountain*. It can be observed from these two videos that the probability distribution for the Y values is broader than the variation for U and V values due to more aggressive subsampling for the U and V components (we employ the $4:2:0$ color format). Following this observation, the set-based delta approach is applied to the Y component to improve the Y-PSNR. Four bits are allocated to represent the delta to reduce the overheads using the following approximations.

- **Set 0 Delta:** Use 4 bits to represent the least significant 4 bits (bits 0–3) of the delta value. Only delta values in the range −8 to +7 can be represented using this approach. Delta values less than −8 and greater than +7 are upperbounded to −8 and +7, respectively.

- **Set 1 Delta:** Use 4 bits to represent bits 4 through 1 of the delta value. Even numbers in the range −16 to +15 can be represented using this approach. If the original delta value is an odd number, the value it finally takes is equal to the quotient obtained by performing an integer division of delta with 2.

- **Set 2 Delta:** Use 4 bits to represent bits 5 through 2 of delta. Multiples of 4 in the range −32 to +31 can be represented using this approach. If the original delta value is not a multiple of 4, the value it finally takes is equal to the quotient obtained by performing an integer division of delta with 4.

The results are shown in Table 7.13. For *Magicforest*, the average Y-PSNR monotonically decreases from Set 0 to Set 2. However, for *Fountain*, Set 2 yields the maximum Y-PSNR. This phenomenon can be explained as follows. For *Magicforest*, pixels with delta values ranging from −16 to +15 represent 93% of the total pixels. For *Fountain*, pixels with delta values ranging from −16 to +15 represent only 80% of the total pixels.

7.5.16 Disparity Vector Generation

Disparity estimation is useful for recovering the depth map of a pair of stereo views and can potentially improve the video quality of S3D videos. Quality metrics may include depth range, vertical misalignment, and temporal consistency. The distinction between motion and disparity vectors is shown in Figure 7.44. We employ the disparity vector to recreate the nonbase view from the base view. This may be used to further reduce the bandwidth requirements or improve the video quality of the

TABLE 7.13   Delta Performance after One Iteration of Checkerboard Pixel Removal

	Average Y-PSNR for *Magicforest* (dB)	Average Y-PSNR for *Fountain* (dB)
Original versus interpolated	35.55	25.24
Original versus interpolated with Set 0 Delta	40.36	26.94
Original versus interpolated with Set 1 Delta	42.73	28.30
Original versus interpolated with Set 2 Delta	44.06	31.06
Original versus interpolated with Set 3 Delta	41.55	35.21
Original versus interpolated with Set 4 Delta	38.44	34.83

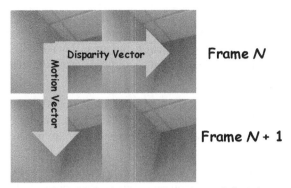

Figure 7.44   Distinction between motion and disparity vectors.

TABLE 7.14   Average Y-PSNR Using Disparity Vectors (15 Frames)

	N	Average Y-PSNR for *Avatar* (dB)	Average Y-PSNR for *Magicforest* (dB)
Left View versus Right View (no disparity vectors)	NA	30.06	16.94
Right View versus Right View generated	4	30.40	27.49
Right View versus Right View generated	2	33.48	32.47
Right View versus Right View generated	1	36.69	37.42

base view. Two views of an S3D video are passed to a disparity vector generation algorithm that partitions each frame of the nonbase view into pixel blocks, each of size $N \times N$. For each pixel block in the nonbase view, the algorithm attempts to find the best match based on the sum of absolute differences (SAD), with blocks of the base view in the same row (i.e., vertical parallax between the two views is assumed to be zero). The block under investigation in the nonbase view is scanned against a total of 256 blocks from the base view (ranging from a left shift of 128 pixels to a right shift of 127 pixels). The pixel shift value (ranging from 0 to 255) corresponding to the best block match is recorded as the disparity vector. For a block size of $N \times N$, (Width/N) $\times$ (Height/N), disparity vectors are computed for each frame. This translates to $1/N^2$ reduction in the overheads required to represent disparity vectors when compared with the data required to represent the raw pixels. The disparity data generated as described above can be efficiently coded using platelet-based coding methods [11]. MVC coding methods are not recommended for coding disparity data, as they may cause coding errors leading to visible distortions in the synthesized view. The generation of the right (nonbase) view from the left (base) view using disparity vectors is straightforward. For each block in the right view, the corresponding disparity vector is used to obtain the block from left view that minimizes the difference in pixel values. As seen from Table 7.14, the improvement in Y-PSNR can be significant for smaller block sizes and this is correlated with subjective evaluation.

7.6 MEDIA-ACTIVATED WIRELESS COMMUNICATIONS

The last digital island, the TV, is finally joining the PC, tablet, and smartphone as an Internet-connected device. The broadband TV brings all of the Internet to the TV, including the full range of video Web sites. These developments will have a profound impact on the distribution and consumption of digital media. Media-activated wireless communications is a new field that will gain importance in the future. Media-activation can help improve the navigation, selection, and playback of videos, photos, and music. As an example, Google Talk personalizes the Web search engine with speech input from the user. Google also provides a voice-recognition application that converts speech commands to actions via devices with speech input such as a smartphone. Media-activated communications can also be nonverbal. For example, popular immersive games employ video sensors on game consoles to recognize human body movement, while hand gestures can also be analyzed to navigate TV or video thumbnails. Microsoft's Kinect Sensor is one of the biggest commercial successes, bringing games, entertainment, and TV navigation to life in extraordinary new ways by capturing hand gestures and human body motion via video sensors. In this case, video communications is wireless but more limited in range and more tolerant to transmission errors. There is no notion of transmission bandwidth, but resolution and clarity in the video capture plays an important role in relaying critical information for processing. Such communications are intuitive, hands-free, and laid-back. These advantages can be far more powerful than those provided by any RF wireless technology. Media-activated communications are ideal for the home or personal office environments. In addition to point-to-point communications, point-to-multipoint communications involving multiple voice or video sensors is also possible. The pervasive use of media-activated communications may eventually remove the need for a human–machine interface, such as a remote.

7.6.1 Leanback TV Navigation Using Hand Gestures

Interactive TV, unlike PC interaction or immersive games, should not involve active or lean-forward user participation. Hand gesture recognition is a key technology that will enable a relaxed TV experience than the touchscreen of a smartphone/tablet or a wireless keyboard, both requiring the user to lean forward. Hand gestures can be used to navigate, search, and control media thumbnails. These gestures can be captured using a regular webcam and allow the user to access the TV anywhere in the living room. "Swipe-and-search" can aid media discovery, while autocorrection may reduce "retyping." Simple DVR functions, such as zoom, pause, and deactivation, can be executed quickly without the need to hunt for a button on a remote. In addition, the technology can enable Web search on any browser on smart TVs.

7.6.2 Multiuser and Multiscreen Media Sharing Using 802.11

Mobile devices can be used for multiuser sharing of photos and other media on social interactive TV. Figure 7.45 shows an interoperable social interactive TV

Figure 7.45   802.11-based smartphone user-TV remote.

system using any 802.11-enabled Android smartphone or tablet. A demo is available in Reference [12]. Multiple images can be transferred and displayed instantly using the smartphone remote. The capabilities of the remote can be divided into individual and shared usage. In individual usage, the smartphone's touchscreen can be used to transfer photos from the smartphone to the TV wirelessly. This is achieved by using a finger to select and "flick" the photo from the touchscreen in the direction of the TV. The photos can be stored on the smartphone or downloaded from a remote website (e.g., Facebook and Flickr) or a local networked server. As an alternative, the user can also exploit the accelerometer feature in the smartphone to "throw" a group of images at the TV simultaneously. Other features include the ability to manipulate the photos on the TV with vertical/horizontal scroll, rotate, expand, or zoom out using the smartphone.

In shared usage, two or more users can employ all the capabilities in individual usage with the added benefit of being able to quickly share photos on the TV. Mobile devices from different vendors can be used as long as the device runs on the Android operating system. In addition, once the photo is transferred and displayed on the TV, any user can navigate to any selected photo, even if the photo originated from another user using a different smartphone or tablet. Such a system can also be extended to multiplayer games.

The technology offers a unique and fun social TV experience that allows the user to share local and Facebook photos and other media with family members and friends anywhere in the living room on a shared screen. Besides enabling a multiuser social media environment, it also allows seamless transfer of media in a multiscreen environment: small screen (e.g., smartphone), medium screen (e.g., tablet), and big screen (e.g., HDTV).

The ability to share media from a full range of social media websites (e.g., Twitter, Facebook, and Linkedin) on the big screen will have a profound impact on the distribution and consumption of digital media in the home and office. This will encourage ubiquitous sharing of social media that fosters bonding of family members and friends, similar to the impact made by Skype video calling on the broadband TV. It will also be useful for sharing media and documents in office meetings and conferences, eliminating time-consuming installations of pull-down screens and expensive projectors. More importantly, this approach expands the capabilities of emerging smart TVs, which cater to one user at a time.

REFERENCES

[1] YouTube Video Downloads, 2011, http://youtube-global.blogspot.com/2011/05/thanks-youtube-community-for-two-big.html.
[2] HTC Autostereoscopic Smartphone, 2011, http://www.htc.com/www/product/evo3d/specification.html.
[3] Nintendo Autostereoscopic 3D Game Console, 2011, http://www.nintendo.com/3ds/hardware.
[4] G. Venkatesen, "Multimedia Streaming over 802.11 Links," *IEEE Wireless Communications Industry Perspectives*, April 2010.
[5] IEEE 802.1Qat Working Group, http://www.ieee802.org/1/pages/802.1at.html.
[6] Digital Living Network Alliance, http://www.dlna.org/about_us/roadmap/DLNA_Whitepaper.pdf.
[7] B. Bing, *3D and HD Broadband Video Networking*, Artech House, August 2010.
[8] Adaptive Streaming Demo, 2011, http://www.youtube.com/watch?v=9ye107EZz30IIS.
[9] JM H.264/MVC Reference Software, http://iphome.hhi.de/suehring/tml.
[10] Optimized Anaglyph Generation, http://3dtv.at/Knowhow/AnaglyphComparison_en.aspx.
[11] Y. Morvana, P. de With, and D. Farina, "Platelet-based Coding of Depth Maps for the Transmission of Multiview Images," *Proceedings of SPIE, Stereoscopic Displays and Applications*, 2006.
[12] Smartphone Remote Demo, 2011, http://www.youtube.com/watch?v=fyoVjbKjDQc.

HOMEWORK PROBLEMS

7.1. The setup for a 2D HDTV shows two complementary image snapshots of several videos, placed side by side to allow the user to select which image view is more desirable. Explain whether a similar selection method is useful for 3D videos.

7.2. Some 802.11 vendors allow the possibility of disabling the ACK that is sent by the receiver after each data frame is successfully received. This may lead to higher throughput and lower delay at layer 2, potentially benefiting real-time traffic, such as video. Explain how such a feature may impact adaptive video streaming.

7.3. Sending uncompressed videos do have some benefits. For example, it removes the need for a video codec and the latency due to coding operations. With multigigabit rates available in some emerging wireless standards (e.g., 802.11ac and ad), uncompressed video streaming can be supported. Explain whether it is possible to improve the video quality of these videos for final viewing by the user.

7.4. The following figures show the instantaneous and average bit rates for the *300* H.264 video when encoded with a QP of 40. What can you conclude about these rates when compared with Figures 7.14 and 7.15 (same video encoded with a lower QP of 20).

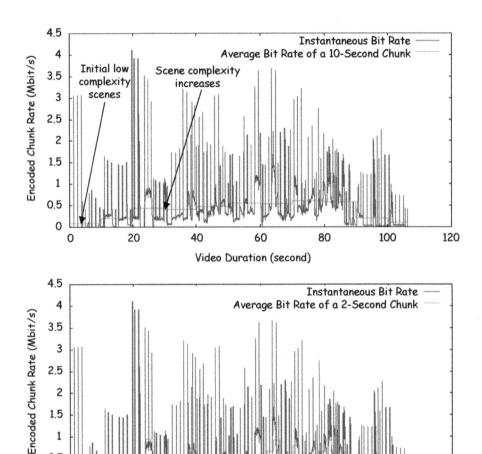

7.5. Full-resolution temporal multiplexing is shown to be more bandwidth efficient than spatial multiplexing by several orders of magnitude despite the doubling of frame rate. Temporal pixel interpolation based on motion vectors can be used to reduce the coded frame rate by half or less, further improving bandwidth utilization. Other important advantages of temporal multiplexing include minimum postprocessing (e.g., no upconversion) for playback and simplified preprocessing (e.g., no view packing). Current 3D HDTVs and HDMI interfaces support temporal multiplexing with two full-resolution views occupying each frame. Explain how 802.11aa can be used to support temporal multiplexed 3D videos efficiently.

7.6. Two-packed views may lead to higher inefficiency in H.264 coding when compared with single-packed views. From Table 7.8, can we conclude that four-packed views are encoded more efficiently in H.264 than two-packed views? Note that the Y-PSNR for both cases are quite similar. By comparing the MVC efficiencies for Tables 7.8 and 7.9, can we conclude that MVC is more efficient as more views are packed in a frame?

7.7. Unlike 2D videos, the key subjects in 3D videos can be emphasized, with depth information and background objects need not be blurred. Explain whether this implies that the overall video quality of 3D videos is better than 2D.

7.8. Some people claim that half-resolution SbS or TaB formatting is an interim solution for transporting 3D videos with an existing network infrastructure. Explain whether the existing infrastructure can support full-resolution temporal multiplexed 3D videos.

7.9 When packing S3D views using the row, column, checkerboard methods, the same number of pixels is removed in each iteration. Explain why row packed views lead to the best Y-PSNR when the views are expanded.

7.10 The figure below shows an experimental setup to study the performance of Apple's live streaming. A 720p HD video is fragmented into three quality levels (350, 800, 1500 Kbit/s) and streamed from an HTTP server via a wide-area high speed Ethernet connection to a 4G cellular network (path 1). The same video may also be streamed locally via an 802.11n connection (path 2). A smooth transition with no interruption in video playback on the iPhone is observed when the connection switches from 802.11n to 4G but the quality level changes from 1500 to 350 Kbit/s.

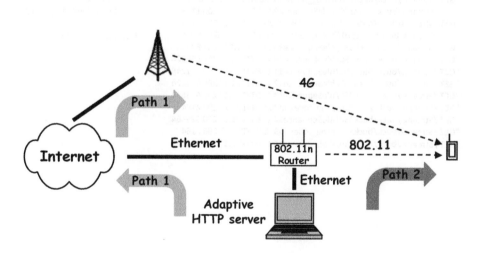

The following HTTP GET requests (sent by the iPhone to the server) were obtained when the connection switched from 802.11n to 4G. Explain why duplicate chunks (i.e., chunks 8 and 9) are requested by the iPhone when a change in video quality is requested. How will this impact bandwidth consumption and video playback? Why is the playlist (indicating the sequence of video chunks) requested several times even when there is no change in quality?

```
"GET /Streams/buck/all.m3u8 HTTP/1.1" 304 -
"GET /Streams/buck/all.m3u8 HTTP/1.1" 206 323
"GET /Streams/buck/all.m3u8 HTTP/1.1" 200 323
"GET /Streams/buck/buck1500/prog_index.m3u8 HTTP/1.1" 200 2164
"GET /Streams/buck/buck1500/fileSequence0.ts HTTP/1.1" 200 2329508
"GET /Streams/buck/buck1500/fileSequence1.ts HTTP/1.1" 200 2401512
"GET /Streams/buck/buck1500/fileSequence2.ts HTTP/1.1" 200 3278344
"GET /Streams/buck/buck1500/fileSequence3.ts HTTP/1.1" 200 2819812
"GET /Streams/buck/buck1500/fileSequence4.ts HTTP/1.1" 200 2093004
"GET /Streams/buck/buck1500/fileSequence5.ts HTTP/1.1" 200 1688992
"GET /Streams/buck/buck1500/prog_index.m3u8 HTTP/1.1" 200 2164
"GET /Streams/buck/buck1500/fileSequence6.ts HTTP/1.1" 200 1695384
"GET /Streams/buck/buck1500/prog_index.m3u8 HTTP/1.1  200 2164
"GET /Streams/buck/buck1500/fileSequence7.ts HTTP/1.1" 200 2595904
"GET /Streams/buck/buck1500/prog_index.m3u8 HTTP/1.1" 200 2164
"GET /Streams/buck/buck1500/fileSequence8.ts HTTP/1.1" 200 1329348
"GET /Streams/buck/buck1500/prog_index.m3u8 HTTP/1.1" 200 2164
"GET /Streams/buck/buck1500/fileSequence9.ts HTTP/1.1" 200 1158644
"GET /Streams/buck/buck350/prog_index.m3u8 HTTP/1.1" 200 2164
"GET/Streams/buck/buck350/fileSequence8.ts HTTP/1.1" 200 513428
"GET /Streams/buck/buck350/fileSequence9.ts HTTP/1.1" 200 483912
"GET /Streams/buck/buck350/fileSequence10.ts HTTP/1.1" 200 558924
"GET /Streams/buck/buck350/fileSequence11.ts HTTP/1.1" 200 751248
"GET /Streams/buck/buck350/fileSequence12.ts HTTP/1.1" 200 634124
"GET /Streams/buck/buck350/fileSequence13.ts HTTP/1.1" 200 578852
"GET /Streams/buck/buck350/prog_index.m3u8 HTTP/1.1" 200 2164
"GET /Streams/buck/buck350/fileSequence14.ts HTTP/1.1" 200 625664
"GET /Streams/buck/buck350/prog_index.m3u8 HTTP/1.1" 200 2164
"GET /Streams/buck/buck350/fileSequence15.ts HTTP/1.1" 200 565128
```

Starts at
highest quality

802.11 → 4G
Jumps to lowest quality

Duplicate chunks

The following GET requests were obtained when streaming the video over a local 802.11n connection (no 4G connection). The maximum bit rate of the 802.11n connection is 144 Mbit/s. The video quality switches quite regularly. How effective is the bandwidth estimation algorithm in the iPhone? Explain why more duplicate chunks are requested when switching to higher quality.

```
"GET /Streams/buck/ HTTP/1.1" 200 178
"GET /Streams/buck/all.m3u8 HTTP/1.1" 206 2
"GET /Streams/buck/all.m3u8 HTTP/1.1" 206 323
"GET /Streams/buck/all.m3u8 HTTP/1.1" 304 -
"GET /Streams/buck/all.m3u8 HTTP/1.1" 206 323
"GET /Streams/buck/all.m3u8 HTTP/1.1" 200 323
"GET /Streams/buck/buck1500/prog_index.m3u8 HTTP/1.1" 200 2164
"GET /Streams/buck/buck1500/fileSequence0.ts HTTP/1.1" 200 2329508
"GET /Streams/buck/buck1500/fileSequence1.ts HTTP/1.1" 200 2401512
"GET /Streams/buck/buck350/prog_index.m3u8 HTTP/1.1" 200 2164
"GET /Streams/buck/buck350/fileSequence0.ts HTTP/1.1" 200 672664
"GET /Streams/buck/buck350/fileSequence1.ts HTTP/1.1" 200 739592

...
"GET /Streams/buck/buck350/fileSequence5.ts HTTP/1.1" 200 587876
"GET /Streams/buck/buck350/fileSequence6.ts HTTP/1.1" 200 696916
"GET /Streams/buck/buck800/prog_index.m3u8 HTTP/1.1" 200 2164
"GET /Streams/buck/buck800/fileSequence4.ts HTTP/1.1" 200 1160148
"GET /Streams/buck/buck800/fileSequence5.ts HTTP/1.1" 200 983804
"GET /Streams/buck/buck800/fileSequence6.ts HTTP/1.1" 200 1083256
"GET /Streams/buck/buck800/fileSequence7.ts HTTP/1.1" 200 1558520
"GET /Streams/buck/buck800/prog_index.m3u8 HTTP/1.1" 200 2164
"GET /Streams/buck/buck800/fileSequence8.ts HTTP/1.1" 200 801444
"GET /Streams/buck/buck800/prog_index.m3u8 HTTP/1.1" 200 2164
"GET /Streams/buck/buck800/fileSequence9.ts HTTP/1.1" 200 729252
"GET /Streams/buck/buck800/prog_index.m3u8 HTTP/1.1" 200 2164
"GET /Streams/buck/buck800/fileSequence10.ts HTTP/1.1" 200 853332
"GET /Streams/buck/buck1500/prog_index.m3u8 HTTP/1.1" 200 2164
"GET /Streams/buck/buck1500/fileSequence4.ts HTTP/1.1" 200 2093004
"GET /Streams/buck/buck1500/fileSequence5.ts HTTP/1.1" 200 1688992
"GET /Streams/buck/buck1500/fileSequence6.ts HTTP/1.1" 200 1695384
"GET /Streams/buck/buck1500/fileSequence7.ts HTTP/1.1" 200 2595904
"GET /Streams/buck/buck1500/fileSequence8.ts HTTP/1.1" 200 1329348
"GET /Streams/buck/buck1500/fileSequence9.ts HTTP/1.1" 200 1158644
"GET /Streams/buck/buck1500/fileSequence10.ts HTTP/1.1" 200 1365068
```

Starts at
highest quality

Jumps to
lowest quality

Medium quality

Back to
highest quality

Explain the variation in the rates when one-level, 10-second video chunks and the unfragmented video (all encoded at the same bit rate) are streamed using HTTP over 4G cellular (no 802.11n connection). Video playback is smooth for both cases. Recall that one-level fragmentation does not introduce significant overheads when compared to the unfragmented video.

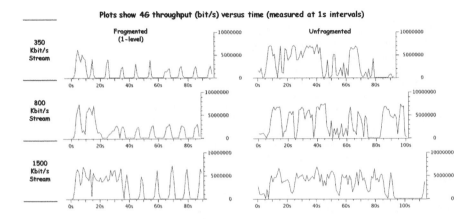

7.11 The following plots were obtained when Apple's live streaming was used to send a 596-second 720p HD video to the iPhone over separate 4G and 802.11n connections. The video was fragmented into four-level chunks (478, 992, 1756, 4884 Kbit/s) of 10-second duration. Explain the variation of the rates. Calculate the amount of buffering on the iPhone required to play back the entire video. Which network requires more buffering?

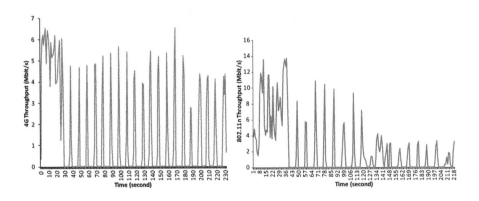

7.12 The following plot shows the case when four-level, 120-second video chunks with audio
(478, 992, 1756, 4884 Kbit/s quality levels) are received on an iPhone using Apple's
live streaming over 4G cellular. The TCP maximum segment length is 1448 bytes, and
the congestion window size is set to 128 bytes. The video bit rate starts at 1756 Kbit/s,
dips to 992 Kbit/s for about 6 seconds (with a corresponding reduction in video quality),
and then returns to 1756 Kbit/s. Why are the video chunks not received at fixed intervals
when the chunk duration is fixed? Estimate the chunk duration. What is the impact of
a change in video quality on the number of video chunks received?

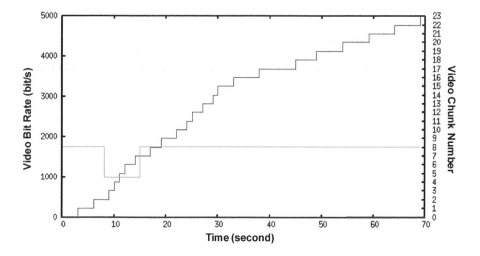

The following plot shows the case when the duplicate chunks are suppressed. What is
the impact of a change in video quality on the number of video chunks received? How
will this impact video playback? Bonus: Suggest a method to suppress the duplicate
chunks.

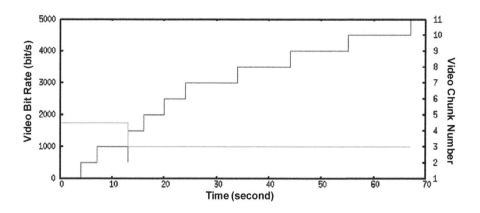

When the video chunk duration is varied and the quality level is fixed at 1756 Kbit/s, the following table is obtained. The server output includes the transmitted video plus all network overheads such as Ethernet, TCP/IP headers, and TCP ACKs. The GET overhead includes the TCP/IP overhead. As can be seen, the output overhead is fairly consistent, roughly 5% (regardless of chunk duration). Why does the overall video file size and the playlist file size increase as the chunk duration is reduced? Why are the lowest number of playlist requests and the highest percentage of GET overhead associated with the shortest chunk duration of 2 seconds? Explain why the 2-second and 10-second chunks lead to a lower amount of sent/received data (see Server Input+Output column) when compared to the 4-second and 5-second chunks. Hint: Think about wireless retransmissions.

If the chunk duration is fixed at 5 seconds, the playlist file size becomes fixed at 920 bytes. As shown in the second table below, while the output overhead stays at about 5% (regardless of video bit rate), the lowest number of playlist requests and the lowest percentage of GET overhead are associated with the highest video bit rate of 4884 Kbit/s. Explain this phenomenon. ■

Live Streaming Performance for 1756 Kbit/s Video Bit Rate and Varying Chunk Durations

Chunk Duration (s)	Number of Chunks	Overall Video File Size (Bytes)	Playlist File Size (Bytes)	Number of Playlist Requests	Total Playlist Overhead (Bytes)	Server Output (Bytes)	Output Overhead (Bytes)	% Output Overhead	Server Input (Bytes)	Total GET Overhead (Bytes)	% GET Overhead (Bytes)	Server Input/ Server Output	Server Input+ Output (Bytes)
2	60	26950552	2157	3	6471	28294622	1337599	4.73%	378216	28326	7.49%	0.0134	28672838
4	30	26933820	1136	12	13632	28375166	1427714	5.03%	662073	18459	2.79%	0.0233	29037239
5	24	26930812	920	13	11960	28271395	1328623	4.70%	403905	16413	4.06%	0.0143	28675300
10	12	26923480	515	10	5150	28180166	1251536	4.44%	292272	11040	3.78%	0.0104	28472438

Live Streaming Performance for 5-second Chunk Duration and Varying Video Bit Rates

Video Bit Rate (Kbit/s)	Overall Video File Size (Bytes)	Number of Playlist Requests	Total Playlist Overhead (Bytes)	Server Output (Bytes)	Output Overhead (Bytes)	% Output Overhead	Server Input (Bytes)	Total GET Overhead (Bytes)	% GET Overhead (Bytes)	Server Input/ Server Output	Server Input+ Output (Bytes)
478	8510670	14	12880	8960297	436747	4.87%	121517	16679	13.7%	0.0136	9081814
992	15356780	15	13800	16141078	770498	4.77%	369390	17450	4.72%	0.0229	16510468
1756	26930812	13	11960	28271395	1328623	4.70%	403905	16413	4.06%	0.0143	28675300
4884	74612876	3	2760	78233638	3618002	4.62%	491110	13346	2.72%	0.0063	78724748

CHAPTER 8

GREEN COMMUNICATIONS IN WIRELESS HOME AREA NETWORKS

Contributed by Bob Heile, PhD, ZigBee Alliance

The ZigBee Alliance has taken the lead in delivering open energy management solutions for the home. ZigBee was launched in 2002 to develop open standards for wireless sensor networks. The idea is simple: make it easy and cost-effective to put sensors and controllers in everything and give them a way to connect and exchange information for a variety of purposes. By doing this, we achieve an Internet of Things. By connecting a variety of applications, from heating and air conditioning, lighting systems, home security systems, to things like remote controls, consumers gain greater control of their home and see a variety of convenience, safety and comfort benefits, as well as an opportunity save money. Given that our own homes are a key factor in any smart grid equation, the balance of this chapter is devoted to some of the important things targeted at homes in terms of applications and standards development. Consider the following:

- Over 50% of electricity is consumed in the home.
- The home will be the location of the largest number of distribution generation devices.
- The home will be the home base for electric vehicles (EVs).

Impacting consumer behavior in the home will be essential in reducing the growth in electric demand and, because of the sheer number of residences and consumers, workable demand response, and load control requires a high level of information, communication, automation, and participation.

8.1 ZIGBEE OVERVIEW

ZigBee is an organization supporting the development and promotion of a family of license-free standards based on wireless technology that targets inexpensive, low power applications utilizing wireless sensor and control networks or what is now

Broadband Wireless Multimedia Networks, First Edition. Benny Bing.
© 2013 John Wiley & Sons, Inc. Published 2013 by John Wiley & Sons, Inc.

being commonly referred to as the Internet of Things. These networks have several very important attributes. They must be:

- Self-organizing and self-healing delivering sustained long-term operation without operator intervention
- Scalable up to many thousands of devices
- Energy efficient delivering exceptional battery life (up to self-life) and be able to take advantage of harvested energy sources
- Reliable and extremely tolerant to interference
- Low cost, low complexity
- Able to support multiple layers of security from open to 128-bit AES.

Today, there are fully integrated ZigBee chipsets delivering these attributes and capable of running on battery power for multiple years and available at very low cost. In addition, ZigBee devices can be networked easily. As many as 65,536 devices can be connected in a single mesh network. It can take as little as 30 ms to join a network, less than 15 ms to go from sleeping to active, and less than 15 ms to access a channel. The ZigBee standards family, which utilizes the IEEE 802.15.4 standard for its base communications needs, performs very well in a variety of environments. Communication range varies based on a variety of factors, including antennae and power, but is typically from 10 to 40 m within a building and up to hundreds of meters in an open field. The key ZigBee attributes are shown in Table 8.1.

ZigBee mesh networks form autonomously between a set of ZigBee network nodes. There are three categories defined in the ZigBee stack that can be used stand-alone or made part of a product with a sensing or controlling function like a programmable display thermostat. These are:

- **ZigBee router:** Component that associates with other ZigBee routers and participates in multihop routing of messages to form a mesh network.
- **ZigBee coordinator:** Only one is required for each ZigBee network; initiates network formation and acts as the network coordinator; any ZigBee router can act as the coordinator and can return to acting as a router once network is formed.
- **ZigBee end device:** Optional network component that delivers base communications and network participation for a simple low-cost device like a

TABLE 8.1 ZigBee Frequency Bands and Data Rates

Frequency band	Geographic region	Data rate (Kbit/s)	Channel number
780 MHz	China	250	0
868.3 MHz	Europe	20	0
902 MHz to 928 MHz	Americas	40	1 to 10
950 MHz	Japan	40	0
2.405 to 2.480 MHz	Worldwide	250	11 to 26

window or door monitor but does not participate in packet routing in the mesh network.

The ZigBee network specification is designed to use minimal hardware resources and implement core functions on even 8-bit microcontrollers. Memory requirements vary based on features used in a particular device. Coordinators/routers require more RAM than end devices, for example. ZigBee technology is used in everything from industrial light switches to light fixtures, appliances, home electronics, and even light bulbs. The ZigBee Alliance currently supports 10 application standards:

- ZigBee Building Automation (efficient commercial spaces)
- ZigBee Remote Control (advanced remote controls)
- ZigBee Smart Energy (home energy savings)
- ZigBee Health Care (health and fitness monitoring)
- ZigBee Home Automation (smart homes)
- ZigBee Input Device (easy-to-use touchpads, mice, keyboards, wands)
- ZigBee Light Link (LED lighting control)
- ZigBee Retail Services (smarter shopping)
- ZigBee Telecom Services (value-added services)
- ZigBee Network Devices (assist and expand ZigBee networks).

The ZigBee Certified Product program verifies compliance to these standards and ensures device interoperability. The full suite of standards for both the network and applications are available for free download at ZigBee.org.

8.2 SMART GRID CHALLENGES

The power grid of tomorrow must deliver both energy and information end-to-end and do so bidirectionally (Figure 8.1). Energy will be generated at traditional

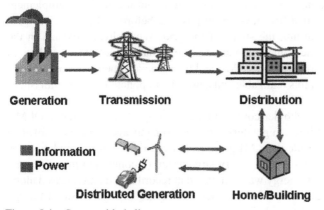

Figure 8.1   Smart grid challenges.

generation facilities and in local or distributed generation facilities. Information covering every aspect of electricity from generation to consumption will also be conveyed throughout the system. The motivation for this monumental change is broad and is driven by a desire to ensure the future power needs of a region by improving power reliability and quality, generation and transmission efficiency, expand the of use of renewable energy, and implement new load shifting and energy efficiency programs at the point of use that reduce consumption on a per customer basis, while boosting consumer awareness and choice, to name but a few. Additionally, this new smart grid is key to more effectively managing and meeting our carbon footprint goals and building a workable EV infrastructure.

In the United States, there are a number of groups and agencies engaged in developing solutions. Perhaps the most visible has been the work of the National Institute of Standards and Technology (NIST) and the Department of Energy (DoE) over the last year in developing the Smart Grid Roadmap. That work now continues as the Smart Grid Interoperability Panel and is joined by activities from other groups like IEEE, Utilities Communications Architecture (UCA) Open Smart Grid (OpenSG), and various other industry alliances and associations who are all working on developing and promoting open standards. IEEE has a large number of activities and specific standards, both completed and in development, relating to specific pieces of the grid and to the communications needed to support its operations. In May 2009, IEEE launched Project 2030, a project exclusively devoted to the Smart Grid, to develop a standard called a Smart Grid Guide. It covers the Smart Grid architecture, information technology requirements and communication needs. The result will be a useful compendium and compliment to the NIST activity on what is needed, what already exists to meet needs, the identification of the gaps, and the addition of needed new projects.

8.3 HOME AREA NETWORKS

In 2006, several major U.S. utilities made the connection that if they could become a part of the home area network (HAN), they would gain the ability to implement new customer programs that had never been possible before and they would gain detailed information regarding the status of the grid. What already existed in the HAN environment was the home automation piece. What did not exist was how to do advanced metering infrastructure (AMI) and the associated network communications in a way to make all this realizable and a useful component in the overall Smart Grid architecture (Figure 8.2).

In 2007, a large stakeholder community, consisting of chip suppliers, OEMs, metering companies, utilities, regulators, and government agencies, assembled in the ZigBee Alliance to tackle the AMI piece and developed what is now known as ZigBee Smart Energy. Even though a very large group of competing interests were at the table, ZigBee Smart Energy was completed at record speed and released along with an initial set of ZigBee Certified products in May 2008. ZigBee Smart Energy delivers a number of benefits including:

Figure 8.2   ZigBee home area network.

- Metering (real-time measurements, historical info, etc.)
- Support for distributed generation (solar, wind, etc.)
- Demand response (DR) and load control
- Pricing (multiple units and currencies, price tiers, etc.)
- Customer messaging
- Device support for programmable communicating thermostats (PCTs), load controllers, energy management systems, in-home displays (IHDs), smart appliances, and more
- Security level options for data which is consumer only, utility only, or shared
- Support for water and gas metering applications.

The network communications piece is handled in a few ways. In all situations, you need a gateway device supporting two communications streams joining the utility AMI central database to devices in the HAN. The gateway also acts as the trust center and firewall in the ZigBee Smart Energy implementation to protect assets on the grid side of the network and to allow for different security levels depending on traffic types. Backhaul communications from the gateway can be over the Internet or over a utility-owned network, although most utilities in the United States favor using their own network from a security and maintenance point of view since most plan to include load disconnect switches as part of the installation and need to serve all customers, not just those with broadband connections.

The implementation of these utility networks is quite varied, although most use some form of wireless communications to a concentration point and then an available backhaul technology like LTE, GPRS, digital subscriber line (DSL), power line communications (PLC), WiMAX, and so on back to the central database. Having a single standard for the backhaul is less urgent since this portion of the network connects fixed infrastructure. Nonetheless, there is active work ongoing to develop such a standard in IEEE 802.15.4g.

The communications between the gateway and the HAN is over a ZigBee Smart Energy wireless sensor network, which meets several important requirements established by the utilities. HAN products need to be based on open standards in a competitive market with no single vendor locked in. They need to be low cost, support high levels of security and ideally support a global reach. Plus, there needs to be a large pool of people who understand the technology. For many of the utilities in North America, the meter with integrated load disconnect is the logical gateway. There are currently over 40 million ZigBee Smart Energy meters deployed or in the process of being deployed.

Some obvious use cases are clear. For ZigBee Smart Energy-equipped devices (HVAC, appliances, smart plugs, etc.), consumers can request a meter read and historical data by using an IHD, PCT, set-top box, computer, or smart phone. For legacy devices lacking ZigBee Smart Energy, consumers have options to retrofit with ZigBee-based smart plugs that can meter and report electricity usage. The end result is the consumer can now begin to quickly associate what it costs to operate various devices in the house (Figure 8.3).

8.3.1 Time of Use

Another scenario focuses on time of use pricing. During peak high price periods, consumers can manually power devices off and/or change temperature settings. They can also program devices equipped with ZigBee Smart Energy to automatically respond to these utility price signals or they can voluntarily allow the utility to have

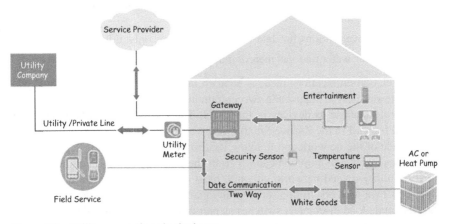

Figure 8.3 Utility connections in the home.

direct access and control to devices while maintaining the ability to override utility control. When there is enough advance notice of a peak event, heating and refrigeration equipment can be set to preheat/cool before the event and reduce usage during high rates. The result is utilities gain the ability to manage peak events in a more efficient and effective way, and consumers see little loss in convenience or comfort—two important elements to instilling lasting changes in consumer behavior.

8.3.2   Electric Vehicles

If EVs become a significant part of our future transportation infrastructure, then the same tools allowing the consumer to become a more intelligent user of energy is essential for maintaining grid performance with EVs. These tools will ensure Smart Grid integrity, as well as set the stage for energy storage and recapture via EVs. Because vehicles are obviously mobile, standards become critical to achieve anything useful and ubiquitous.

EVs are not just another consumer appliance; they are significantly more challenging. Most people normally arrive at their homes in a fairly narrow window of time every evening. This time period is already associated with peak energy use. Each vehicle can represent as much as a 6 kW load, drawing significantly more energy than the average single-family home. If everyone parks their EV and plugs it in during this peak period, a local feeder or transformer will likely fail. Adding to that, what if you get home, are low on energy and know you have somewhere to go soon? What if you want the lowest price? What if you plug in somewhere else? ZigBee Smart Energy helps both the consumer and the utility intelligently manage these scenarios.

EVs offer an interesting Smart Grid opportunity beyond being eco-friendly transportation. They can be used to store energy. EVs can charge up during low use periods, and then can supply power to the grid during periods of high demand. With the owners' permission, utilities can selectively target vehicles with sufficient spare power to cover a peak event. This new capability results in better management of resources and is significantly more environmentally friendly. Consumers can also benefit independently of the utility by using cheaper night time electricity from their cars to power their homes during peak daytime periods.

The relationship between the Smart Grid, homes and consumers is an important and tightly coupled part of the overall system. The energy stakeholders have been enhancing the ZigBee Smart Energy standard to serve the growing set of needs required of the relationship, including the ability to run over a variety of mainstream communications protocols, wired and wireless, like IEEE 1901, 802.3, 802.11, and 802.16 in addition to the original 802.15.4.

In the coming months, more progress needs to be made on a number of areas relating to this part of the Smart Grid architecture and design. A few of the major ones are:

- EV infrastructure
- Implementing the programs enabled by smart meters
- Getting consumer buy-in on the value of the programs

- Resolving data ownership and privacy issues
- Completing the next phase of the HAN system definition.

8.4   FUTURE CHALLENGES

Harmonization of the many different groups and interests in the Smart Grid community on these issues will be a major challenge. To that end, the ZigBee Alliance is using its open consensus process to gather input from experts around the world. Every week, new organizations are joining and participating in this development. The standards community is being engaged through a growing number of collaboration agreements signed over the last year. In fact, the ZigBee Alliance recently took the extraordinary step of making drafts publicly available and seeking comment from those who have not had the opportunity to be a part of the drafting process (see download button on ZigBee home page http://www.zigbee.org).

When it comes to defining and building the Smart Grid, the hope is that through efforts like this, it is possible to not only achieve consensus solutions on specific requirements, but a harmonized and coordinated systems approach as well. As with anything, time will tell. But, with a lot of work and a bit of luck, we can achieve an extensible core architecture capable of growing and evolving as effectively as the Internet does today.

HOMEWORK PROBLEM

8.1. ZigBee is based on the 802.15.4 wireless personal area network (wireless PAN) standard and is intended for fixed applications, including industrial controls, telemetry, and in-building/home automation. It offers a range of 30 m (100 ft), very low power consumption, and low data rates. The 2.4 GHz band is by far the most popular band for Zigbee. The complete protocol stack occupies less than 4 bytes for the user device and less than 32 bytes for the coordinator (which can directly communicate with up to 255 user devices in a mesh network). ZigBee radios are therefore cheap to develop (less than $1 per chipset) and can be embedded into a light bulb with a battery life measured in years. Because ZigBee transmissions occur so infrequently, they cause minimum interference to other wireless devices using the same frequency band. A key characteristic of ZigBee networks is the ability to scale to many sensor devices. Hence, ZigBee is not limited to cable replacement applications. Explain how unnecessary features can be removed from 802.11 to achieve comparable cost and capabilities as ZigBee. ∎

GLOSSARY

3D	Three dimensional
3G	Third generation
3GPP	3G partnership project
4G	Fourth generation
AAS	Adaptive antenna system
A-BFT	Association beamforming training
ABR	Adaptive bit rate
ACK	Acknowledgment
ACMA	Australian Communications and Media Authority
AES	Advanced encryption standard
AGC	Automatic gain control
AIFS	Arbitration IFS
AM	Acknowledged mode
AMC	Adaptive modulation and coding
AMI	Advanced metering infrastructure
AMPE	Authenticated mesh peering exchange
AODV	Ad hoc on-demand distance vector
AP	Access point
APECC	Antenna pattern envelope correlation coefficient
ARQ	Automatic repeat request
AS	Authentication server
ASEL	Antenna selection
AT	Announcement time
ATSC	Advanced Television Systems Committee
ATSC-M/H	ATSC-mobile/handheld
AVB	Audio video bridging
AVC	Advanced Video Coding
AWGN	Additive white Gaussian noise
AWS	Advanced wireless services
BA	Block ACK
BAR	BA request
BCC	Binary convolutional code

Broadband Wireless Multimedia Networks, First Edition. Benny Bing.
© 2013 John Wiley & Sons, Inc. Published 2013 by John Wiley & Sons, Inc.

BCCH	Broadcast control channel
BCH	Bose Chaudhuri Hocquenghem
BER	Bit error rate
BLAST	Bell laboratories layered space–time
BLER	Block error rate
BPSK	Binary phase-shift keying
BRP	Beam refinement protocol
BS	Base station
BSA	Basic service area
BSS	Basic service set
BSSID	BSS identity
CA	Carrier aggregation
CAZAC	Constant amplitude zero autocorrelation
CBC	Cipher block chaining
CBP	Coexistence beacon protocol (802.22) or contention-based period (802.11)
CBR	Constant bit rate
CC	Carrier component
CCA	Clear channel assessment
CCCH	Common control channel
CCI	Cochannel interference
CCK	Complementary code keying
CCMP	Counter-mode CBC message authentication code protocol
CCR	Congestion control request
CD	Cyclic delay
CDD	Cycle delay diversity
CDMA	Code division multiple access
CE	Consumer electronics
CEPT	Committee of Post and Telecommunications Administrations
CF	Contention free
CID	Connection ID
CIF	Common intermediate format
CINR	Carrier to interference plus noise ratio
CoMP	Coordinated multipoint transmission and reception
CP	Cyclic prefix
CPE	Customer premise equipment
CQI	Channel quality information
CRC	Cyclic redundancy check
CS	Convergence sublayer
CSD	Cyclic shift diversity
CSI	Channel state information
CSMA	Carrier sense multiple access
CSMA/CA	CSMA with collision avoidance
CSMA/CD	CSMA with collision detection

CTC	Convolutional turbo coding
CTS	Clear-to-send
DCCH	Dedicated control channel
DCD	DL channel descriptor
DCF	Distributed coordination function
DCI	DL control information
DCT	Discrete cosine transform
DES	Data encryption standard
DFE	Decision feedback equalizer
DFH	Dynamic frequency hopping
DFIR	Diffuse infrared
DFS	Dynamic frequency selection
DFT	Discrete Fourier transform
DHCP	Dynamic host configuration protocol
DIFS	DCF IFS
DL	Downlink
DLFP	DL frame prefix
DLNA	Digital Living Network Alliance
DMB	Digital multimedia broadcasting
DMZ	Diversity MAP zone
DN	Destination network
DOCSIS	Data over cable service interface specification
DoE	Department of Energy
DoS	Denial-of-service
DR	Demand response
DRM	Digital rights management
DRS	Demodulation reference signal
DRX	Discontinuous reception
DS	Downstream or distribution system (when referring to 802.11)
DSA	Dynamic service addition
DSL	Digital subscriber line
DSM	Distribution system medium
DSRC	Dedicated short-range communications
DSS	Distribution system service
DSSS	Direct-sequence spread spectrum
DTCH	Dedicated traffic channel
DTIM	Traffic indication map
DTV	Digital TV
DTX	Discontinuous transmission
DVB	Digital video broadcasting
DVB-H	DVB-Handheld
DVB-T	Digital Video Broadcast-Terrestrial
DwPTS	DL pilot time slot

EAP	Extensible Authentication Protocol
EAS	Emergency alert system
EBS	Educational broadband service
ECC	Elliptic curve cryptographic
ED	Energy detection
EDCA	Enhanced distributed channel access
EDGE	Enhanced Data rates for GSM Evolution
EIV	Extended IV
EKS	Encrypted key sequence
E-MBMS	Enhanced multimedia broadcast multicast service
E-MBS	Enhanced multicast broadcast service
eNB	Evolved NodeB
EPC	Evolved packet core
EQP	Extended quiet period
ESG	Electronic service guide
ESS	Extended service set
ESSID	ESS identity
ETSI	European Telecommunications Standards Institute
E-UTRA	Evolved Universal Terrestrial Radio Access
EV	Electric vehicle
EVDO	Evolution-data optimized
FAP	Femto access point
FCC	Federal Communications Commission
FCFS	First come first served
FCH	Frame control header
FCS	Frame check sequence
FDD	Frequency division duplex
FDMA	Frequency division multiple access
FEC	Forward error correction
FFR	Fractional frequency reuse
FFT	Fast Fourier transform
FHDC	Frequency hopped diversity coding
FHSS	Frequency-hopping spread spectrum
FIPS	Federal Information Processing Standard
FM	Frequency modulation
FSO	Free-space optics
FST	Fast session transfer
FTTH	Fiber-to-the-Home
FUSC	Full usage of subchannels
GANN	Gate announcement
GAS	Generic advertisement service
GCM	Galois counter mode
GI	Guard interval

GOP	Group of pictures
GPRS	General packet radio service
GPS	Global positioning system
GSM	Global system for mobile communications
GTK	Group temporal key
HAN	Home area network
HARQ	Hybrid ARQ
HCCA	HCF controlled channel access
HCF	Hybrid coordination function
HD	High definition
HDCP	High-bandwidth Digital Content Protection
HDMI	High-definition multimedia interface
HetNet	Heterogeneous network
HEVC	High efficiency video coding
HiperMAN	High performance metropolitan area network
HNB	Home node B
HQ	High quality
HSPA	High-speed packet access
HSS	Home subscriber server
HT	High throughput
HTTP	Hypertext transfer protocol
HWMP	Hybrid Wireless Mesh Protocol
IBSS	Independent BSS
ICI	Intercarrier interference
ICIC	Intercell interference coordination
ICV	Integrity check vector
IEC	International Electrotechnical Commission
IEEE	Institute of Electrical and Electronic Engineers
IETF	Internet Engineering Task Force
IFS	Interframe space
IGMP	Internet group management protocol
IHD	In-home display
IMS	IP multimedia subsystem
IMT-A	IMT-Advanced
IP	Internet Protocol
IPSec	IP security
ISI	Intersymbol interference
ISM	Industrial, scientific and medical
ISO	International Standardization Organization
ISP	Internet service provider
ITS	Intelligent transportation systems
ITU	International Telecommunication Union
ITU-R	ITU Radio Standardization Sector

ITU-T	ITU Telecommunication Standardization Sector
IV	Initialization vector
JM	Joint model
JVT	Joint Video Team
LAN	Local area network
LBS	Location-based services
LBT	Listen before talk
LDPC	Low-density parity-check
LED	Light-emitting diode
LOS	Line-of-sight
LTE	Long term evolution
LTE-A	LTE-Advanced
LTF	Long training field
LTS	Long training sequence
MA	Mesh authenticator
MAC	Medium access control
MAI	Multiaccess interference
MAP	Bandwidth allocation message
MB	Macroblock
MBCA	Mesh beacon collision avoidance
MB-SFN	Multicast/Broadcast Single Frequency Network
MBSS	Mesh basic service set
MCCA	MCF-controlled channel access
MCCAOP	MCCA opportunity
MCCH	Multicast control channel
MCF	Mesh coordination function
MCS	Modulation and coding scheme
MCW	Multiple codeword
MDA	Mesh deterministic access
MDAOP	MDA opportunity
MIB	Management information base
MIC	Message integrity code
MIMO	Multiple input multiple output
MITM	Man-in-the-middle
MK	Master key
MKD	Mesh key distributor
MKK	Radio Equipment Inspection and Certification Institute
MLME	MAC layer management entity
MLSE	Maximum likelihood sequence estimator
MME	Mobility management entity
MOCA	Multimedia over Coax Alliance
MPDU	MAC protocol data unit
MPEG	Moving Picture Experts Group

MPM	Mesh peering management
MR-BS	Multihop relay BS
MS	Mobile station
MSDU	MAC service data unit
MTCH	Multicast traffic channel
MTU	Maximum transmission unit
MUD	Multiuser detection
MU-MIMO	Multiuser MIMO
MV	Motion vector
MVC	Multiview video coding
NACK	Negative ACK
NAT	Network address translator
NAV	Net allocation vector
NCTA	National Cable and Telecommunications Association
NIST	National Institute of Standards and Technology
NRT	Nonreal time
NRTPS	NRT polling service
NTSC	National Television System Committee
OEM	Original equipment manufacturer
OFDM	Orthogonal frequency division multiplexing
OFDMA	Orthogonal frequency division multiple access
OLSR	Optimized link state routing
OMVC	Open Mobile Video Coalition
OpenSG	Open Smart Grid
OSI	Open systems interconnection
OSPF	Open shortest path first
PA	Power amplifier
PAE	Port access entity
PAL	Protocol adaptation layer
PAPR	Peak-to-average power ratio
PBCC	Packet binary convolutional coding
PBSS	Personal BSS
PCCH	Paging control channel
PCF	Point coordination function
PCFICH	Physical control format indicator channel
PCP	PBSS central point
PCRF	Policy and charging rules function
PCT	Programmable communicating thermostat
PD	Protecting device
PDCCH	Physical DL control channel
PDCP	Packet data convergence protocol
PDSCH	Physical DL shared channel
PDU	Protocol data unit

PERR	Path error
P-GW	Packet-network gateway
PHICH	Physical HARQ indicator channel
PHS	Payload header suppression
PHY	Physical layer
PIC	Parallel interference cancellation
PIFS	PCF IFS
PKM	Privacy key management
PLC	Power line communications
PLCP	PHY convergence procedure
PLMN	Public land mobile network
PLP	Physical layer pipe
PLR	Packet loss rate
PMD	Physical medium dependent
PMI	Precoding matrix indicator
PMK	Pairwise master key
PN	Pseudonoise
PoE	Power over Ethernet
POP	Point of presence
PPD	Primary protecting device
PPDU	PLCP (PHY) protocol data unit
PRACH	Physical random access channel
PREP	Path (Set-up) Reply
PREQ	Path (Set-up) Request
PS	Physical slot
PSDU	PLCP service data unit
PSK	Preshared key
PSMP	Power save multipoll
PSNR	Peak signal to noise ratio
PTK	Pairwise transient key
P2P	Peer to peer
PUCCH	Physical UL control channel
PUSC	Partial usage of subchannels
PUSCH	Physical UL shared channel
QAM	Quadrature amplitude modulation
QoS	Quality of service
QP	Quantization parameter
QPSK	Quadrature phase-shift keying
RADIUS	Remote access dial-in user service
RAN	Radio access network
RANN	Root announcement
RB	Resource block
RBG	RB group

RC4	Rivest Cipher 4
RCPI	Received channel power indicator
RD	Reverse direction
RE	Resource element
RERR	Route error
RF	Radio frequency
RFID	RF identification
RGC	Request-grant cycle
RI	Rank indicator
RIFS	Reduced IFS
RIV	Resource indication value
RLC	Radio link control
RMS	Root mean square
RNTP	Relative narrowband transmit power
ROHC	Robust header compression
RRC	Radio resource control
RRE	Remote radio equipment
RREP	Route Reply
RREQ	Route Request
RRM	Radio resource measurement
RS	Relay station
RSA	Rivest, Shamir, and Adleman
RSNA	Robust Security Network Association
RSS	Receiver sector sweep
RSSI	Received signal strength indicator
RSVP	Resource reservation protocol
RT	Real time
RTP	RT transfer protocol
RTPS	RT polling service
RTS	Request-to-send
SAD	Sum of absolute differences
SAE	System architecture evolution
SAP	Service access point
SBS	Side-by-side
SBTC	Shortened block turbo code
SC	Single carrier
SC-FDMA	SC frequency division multiple access
SCH	Superframe control header
SCTP	Stream control transmission protocol
SCW	Self-coexistence window
SD	Standard definition
SDR	Software defined radio
SDU	Service data unit

SFBC	Space–time frequency block coding
SFCMM	Scalable full channel mobile mode
SFID	Service flow ID
S-GW	Serving gateway
SI	Service interval
SIC	Successive interference cancellation
SIFS	Short IFS
SIG	Signal field
SINR	Signal-to-interference-noise ratio
SIPTO	Selected IP traffic offload
SISO	Single input single output
SLS	Sector level sweep
SME	Station management entity
SNR	Signal-to-noise ratio
S-OFDMA	Scalable OFDMA
SP	Service period
SPD	Secondary PD
SR	Scheduling request
SRP	Stream reservation protocol
SRS	Sounding reference signal
SS	Subscriber station
SSF	Spectrum sensing function
SSID	Service set identity
SSPN	Subscription service provider network
STA	Station
STB	Set-top box
STBC	Space–time block coding
STC	Space–time coding
STF	Short training field
STS	Short training sequence
SU-MIMO	Single-user MIMO
TAB	Top-and-bottom
TBTT	Target beacon transmission time
TCP	Transmission control protocol
TDD	Time division duplex
TDM	Time division multiplexing
TDMA	Time division multiple access
TD-SCDMA	Time-duplex synchronous CDMA
TEK	Traffic encryption key
TG	Task group
TID	Traffic identifier
TIM	Traffic indication map
TKIP	Temporal key integrity protocol

TLS	Transport layer security
TM	Transparent mode
TPC	Transmit power control
TPM	Trusted platform module
TS	Transport stream
TSF	Timing synchronization function
TSPEC	Traffic specification
TTL	Time-to-live
TUSC	Tile usage of subchannels
TXOP	Transmission opportunity
UCA	Utilities Communications Architecture
UCD	UL channel descriptor
UCI	User control information
UCP	Uncoordinated coexistence protocol
UCS	Urgent coexistence situation
UDP	User datagram protocol
UE	User equipment
UGS	Unsolicited grant service
UHF	Ultra high frequency
UL	Uplink
UM	Unacknowledged mode
UMA	Unlicensed mobile access
UMTS	Universal mobile telecommunications system
U-NII	Unlicensed national information infrastructure
UpPTS	UL pilot time slot
UPS	Universal power supply
US	Upstream
USB	Universal serial bus
UTP	Unshielded twisted pair
UWB	Ultra-wideband
VBR	Variable bit rate
VCEG	Video coding experts group
VHF	Very high frequency
VHT	Very high throughput
VLAN	Virtual LAN
VoIP	Voice over IP
VPN	Virtual private network
VSB	Vestigial sideband
WAVE	Wireless access in vehicular environments
WBE	WiGig bus extension
WCDMA	Wideband CDMA
WDE	WiGig display extension
WEP	Wired equivalent privacy

WFA	Wi-Fi Alliance
WG	Working group
Wi-Fi	Wireless Fidelity
WiGig	Wireless Gigabit Alliance
WiMAX	Worldwide Interoperability for Microwave Access
WLAN	Wireless LAN
WMAN	Wireless metropolitan area network
WPA	Wi-Fi protected access
WPAN	Wireless personal area network
WRAN	Wireless regional area network
WSA	Whitespace Alliance
WSE	WiGig serial extension
Y-PSNR	Luminance PSNR

INDEX

Broadband Wireless Multimedia Networks, First Edition. Benny Bing.
© 2013 John Wiley & Sons, Inc. Published 2013 by John Wiley & Sons, Inc.

Printed and bound by CPI Group (UK) Ltd, Croydon, CR0 4YY

16/04/2025

14658364-0004